AF581690

Fundamental Knowledge on Forgotten Species: An Exploration of Data from Rarely Studied Captive Animals

Fundamental Knowledge on Forgotten Species: An Exploration of Data from Rarely Studied Captive Animals

Editors

Kris Descovich
Caralyn Kemp
Jessica Rendle

MDPI • Basel • Beijing • Wuhan • Barcelona • Belgrade • Manchester • Tokyo • Cluj • Tianjin

Editors
Kris Descovich
School of Veterinary Science
University of Queensland
Gatton
Australia

Caralyn Kemp
School of Environmental and
Animal Sciences
Unitec Institute of Technology
Auckland
New Zealand

Jessica Rendle
College of Science, Health,
Engineering and Education
Murdoch University
Murdoch
Australia

Editorial Office
MDPI
St. Alban-Anlage 66
4052 Basel, Switzerland

This is a reprint of articles from the Special Issue published online in the open access journal *Journal of Zoological and Botanical Gardens* (ISSN 2673-5636) (available at: www.mdpi.com/journal/jzbg/special_issues/forgotten_species).

For citation purposes, cite each article independently as indicated on the article page online and as indicated below:

LastName, A.A.; LastName, B.B.; LastName, C.C. Article Title. *Journal Name* **Year**, *Volume Number*, Page Range.

ISBN 978-3-0365-7223-9 (Hbk)
ISBN 978-3-0365-7222-2 (PDF)

Contents

About the Editors

Kris Descovich

Dr. Descovich is an animal scientist working at the interface between welfare, behaviour, management and zoology. She is currently a Research Fellow in UQ's School of Veterinary Science working within the Animal Welfare Standards Project, a large international collaboration between Australia and China that aims to improve livestock management and welfare. Kris has a broad range of professional industry experience, having worked within zoos, veterinary clinics, and animal shelters. Dr Descovich's research is diverse and encompasses a range of topics in animal welfare science including captive wildlife management, pain identification in animals, novel welfare methods, and applied animal ethology.

Caralyn Kemp

Dr. Kemp is an applied ethologist and anthrozoologist, currently employed as a lecturer at Unitec Te Pukengā, Aotearoa New Zealand. Caralyn's research interests are diverse, investigating topics such as responses of primates to sensory cues, animal training techniques, olfaction, human-animal interactions in zoos, enrichment, social behaviour of dogs in dog parks, cognitive bias, primate vocalisations, use of horse grimace scales, and donations given by zoo visitors. She has been fortunate to work across a variety of animal sectors and in different countries. Caralyn is particularly interested in improving the connections between research and management in animal facilities to allow for evidence-based decision making for the improvement of animal welfare.

Jessica Rendle

Dr. Rendle is an animal welfare scientist with a career spanning across species, sectors and countries. Jess's research has largely focused on zoo and wild animal health and welfare, specifically focussing on the management of disease risk in wild and captive macropod populations. With a broad interest across animal species, health and disease, Jess has forged a career path which has moved between domestic, production, captive and wild animals - with animal welfare science always being at the core of her work.

Preface to "Fundamental Knowledge on Forgotten Species: An Exploration of Data from Rarely Studied Captive Animals"

Dear Colleagues,

Zoological institutions contribute a large amount of fundamental and applied knowledge on a diverse array of animal species. Despite this significant contribution, research conducted within zoos or other captive wildlife facilities has historically been skewed toward charismatic mammals, which comprise only a small proportion of the species that are held in captive collections. Modern zoos play an important role in animal welfare, conservation, and environmental education; therefore, this shortfall in knowledge may have large, unseen, and negative impacts on these "forgotten species". Hypothesis-driven, experimental research plays a key role in filling these knowledge gaps; however, other avenues of data collection exist which may be equally important. These include observational data (collected without experimental interventions), operational data (data collected within the general management activities of a facility), and incidental data (data collected for one purpose which may reveal further important information when explored in more detail). These unpublished datasets may provide fundamental information on species for which comparatively little is known.

The aim of this Special Issue is to encourage the reporting and publication of data on rarely studied species within captive facilities including zoos, aquaria, and wildlife rescue centres. Manuscripts may focus on fundamental or applied animal data including, but not limited to, information on biology, development, health, behaviour, anatomy, enrichment, and reproduction. Manuscripts describing data on non-mammalian species including birds, non-avian reptiles, amphibians, fish, and insects are particularly encouraged.

Kris Descovich, Caralyn Kemp, and Jessica Rendle
Editors

Journal of
Zoological and Botanical Gardens

MDPI

Editorial

Fundamental Knowledge on Forgotten Species: An Exploration of Data from Rarely Studied Captive Animals

Kris Descovich [1,*], Caralyn Kemp [2] and Jessica Rendle [3,4]

[1] School of Veterinary Science, The University of Queensland, Gatton, QLD 4343, Australia
[2] School of Environmental and Animal Sciences, Unitec Institute of Technology, Te Pūkenga, Auckland 1025, New Zealand
[3] School of Veterinary Medicine and Science, University of Nottingham, Sutton Bonington Campus, Loughborough LE12 5RD, UK
[4] School of Veterinary Medicine, College of Science, Health, Engineering and Education, Murdoch University, Perth, WA 6150, Australia
* Correspondence: k.descovich1@uq.edu.au

Zoological institutions contribute a large amount of fundamental and applied knowledge on a diverse array of animal species. Despite this significant contribution, published research conducted within zoos or other captive wildlife facilities has historically been skewed toward charismatic mammals [1], which comprise only a small proportion of the species that are maintained in zoological collections, and are not reflective of taxonomic group sizing. Modern zoos play an important role in developing effective animal welfare, conservation, and environmental education; therefore, this shortfall in knowledge on "forgotten species" may have large, unseen, and negative impacts. The aim of this special issue was to encourage the reporting and publication of data on rarely studied species within captive facilities. This collection of 14 papers brings to light new information on a diverse range of taxonomic groups, from reptiles and birds, to amphibians and sharks.

1. Non-Avian Reptiles

Reptiles are a broad taxonomic group that are well-represented in zoo collections but for which there is limited experimental evidence for conditions that support good welfare. Enrichment is considered an essential component of appropriate captive husbandry for mammals, yet research on this aspect of welfare has been largely overlooked for reptiles, including the monitor lizards (Varanidae). To provide a base of knowledge for informing enrichment design, Howard and Freeman [2] undertook a scoping review of the physiological, cognitive, and behavioral abilities of Varanidae to suggest enrichment methods that may be appropriate and effective. They stressed the need for greater publishing and sharing of findings to promote positive quality of life for these species in captivity. Additionally, also with a focus on Varanidae, Waterman et al. [3] monitored the effect of food- and scent-based enrichment on three monitor lizard species, including Komodo dragons (*Varanus komodoensis*), reporting an increase in exploratory behaviour, with scent-based enrichment being as effective for encouraging natural behaviours as food. The effect of enrichment and environmental change was also explored by Turner et al. [4], who monitored the behaviour of three tortoise species after an enclosure size increase, the addition of floor substrate, or handling protocol adjustments. These changes primarily altered social interactions, but larger, more positive, environmental changes are proposed to improve behavioural diversity. Reptile social behaviour was also studied by Walsh et al. [5], who compared differences in sociality and congregation behaviour between captive and wild American alligators (*Alligator mississippiensis*). Social behaviours were much more frequent and diverse in the wild population, while captive activity budgets were dominated by a small number of non-social behaviours. The results of these studies show that there is more work zoos can do to improve the welfare of reptiles in their collection and promote full behavioural repertoires, as is encouraged in mammals.

Citation: Descovich, K.; Kemp, C.; Rendle, J. Fundamental Knowledge on Forgotten Species: An Exploration of Data from Rarely Studied Captive Animals. *J. Zool. Bot. Gard.* **2023**, *4*, 50–52. https://doi.org/10.3390/jzbg4010005

Received: 19 December 2022
Accepted: 19 December 2022
Published: 9 January 2023

2. Birds

Birds such as the southern ground hornbill (*Bucorvus leadbeateri*) are intelligent and long-lived, which can present challenges for maintaining welfare in captivity. Brereton et al. [6] examined the effect of enrichment on the behaviour of two captive hornbills. Carcass provision resulted in long periods of food manipulation and plastic mirrors encouraged stalking and mirror pecking, similar to behaviours observed in wild hornbills, suggesting a positive effect of these enrichment types. In the paper by Bryant et al. [7], enclosure use by two blue-throated macaws (*Ara glaucogularis*) was explored, specifically the effect of UVA- and UVB-rich lighting on indoor area use. Macaws significantly increased the time they spent near the enriched lighting, suggesting indoor areas can be enhanced through lighting choice. Lastly, Thomas et al. [8] detailed the veterinary treatment provided to a zoo-housed Verreaux's eagle owl (*Bubo lacteus*) after toe constriction caused by plastic litter. While positive health outcomes were achieved, this case study highlights the dangers of macroplastic pollution to wildlife, even to those housed in a captive setting.

3. Amphibians

Two amphibian papers were represented in this Special Issue, both focused on behavioural indicators of stress and welfare. Dias et al. [9] developed the first ethogram for *Xenopus longipes* frogs through observation of a group of 24 individuals from this critically endangered species. This ethogram was then used to measure activity budgets and behavioural response to restraint during a routine health check. Many behaviours were significantly impacted in the period post restraint, suggesting health assessments should be non-invasive whenever possible. Similarly, Carter et al. [10] used food intake as a measure of stress in the terrestrial amphibian, *Herpele squalostoma*, after environmental disturbance imposed by floor substrate change for routine cleaning. Regardless of the food prey offered, substrate disturbance had a significant suppressive effect on feeding, and this behavioural indicator may be useful for future studies on the welfare of this caecilian species.

4. Sharks

The final paper in this special issue examined activity levels and three-dimensional space use in five captive sharks, all of different species. Hart et al. [11] found that area usage in the 'xy plane' was fairly consistent; however, time spent at different depths was uneven. Although space use and activity largely reflected the natural behavioural biology of each species, the behaviour of the smooth dogfish (*Mustelus canis*) was found to be abnormal, demonstrating the importance of monitoring behavioural patterns in captive sharks.

5. Mammals

Although mammals are a popular research focus of captive collections, there has been a bias towards primates, ungulates, and large carnivores. Several understudied mammalian species were represented by papers within this special issue. Free et al. [12] assessed the welfare of common cusimanse (*Crossarchus obscurus*) with an adapted 'Animal Welfare Assessment Grid'. Using resource- and animal-based measures, 21 factors were identified, and the final template was validated by retrospectively scoring the welfare of four zoo-housed individuals. With a focus on behaviour, Spiezio et al. [13] monitored two pairs of zoo-housed red pandas (*Ailurus fulgens*) using the 'Behavioural Variety Index'. Observed individuals performed approximately three quarters of all behaviours reported previously for this species and no abnormal behaviour was found. Behavioural activity, as well as space use, was also examined by Finch and Humphreys [14] for two Goodfellow's tree kangaroos (*Dendrolagus goodfellowi*), an endangered, arboreal macropod. High arboreal spaces were found to be of key importance, with more time spent at the top height by the tree kangaroos than at any other height. Lastly, the work of Truax et al. [15] focused on cognition in African crested porcupines (*Hystrix cristata*). This study used the 'loose-string task' to determine if porcupines, a cooperative breeder, can work with their partner to receive a reward. Although the porcupines were successful in the task, they did not clearly demonstrate understanding of their partner's role in task success.

The collection of research in this special issue opens the door to future studies on these species, as well as the multitude of others in need of systematic observation and empirical assessment. We thank the authors for their contributions to this issue and for their commitment to the management of their respective study species. We hope their work encourages future and on-going programs of research that shed light on optimal management of these "forgotten species".

Author Contributions: Conceptualization, K.D., C.K. and J.R.; writing—original draft preparation, K.D., C.K. and J.R.; writing—review and editing: K.D., C.K. and J.R. All authors have read and agreed to the published version of the manuscript.

Conflicts of Interest: The authors declare no conflict of interest.

References

1. Rose, P.E.; Brereton, J.E.; Rowden, L.J.; de Figueiredo, R.L.; Riley, L.M. What's new from the zoo? An analysis of ten years of zoo-themed research output. *Palgrave. Commun.* **2019**, *5*, 128. [CrossRef]
2. Howard, D.; Freeman, M.S. Overlooked and Under-Studied: A Review of Evidence-Based Enrichment in Varanidae. *J. Zool. Bot. Gard.* **2022**, *3*, 32–43. [CrossRef]
3. Waterman, J.O.; McNally, R.; Harrold, D.; Cook, M.; Garcia, G.; Fidgett, A.L.; Holmes, L. Evaluating Environmental Enrichment Methods in Three Zoo-Housed Varanidae Lizard Species. *J. Zool. Bot. Gard.* **2021**, *2*, 716–727. [CrossRef]
4. Turner, J.T.; Whittaker, A.L.; McLelland, D. Behavioural Impact of Captive Management Changes in Three Species of Testudinidae. *J. Zool. Bot. Gard.* **2022**, *3*, 555–572. [CrossRef]
5. Walsh, Z.C.; Olson, H.; Clendening, M.; Rycyk, A. Social Behavior Deficiencies in Captive American Alligators (*Alligator mississippiensis*). *J. Zool. Bot. Gard.* **2022**, *3*, 131–146. [CrossRef]
6. Brereton, J.E.; Myhill, M.N.G.; Shora, J.A. Investigating the Effect of Enrichment on the Behavior of Zoo-Housed Southern Ground Hornbills. *J. Zool. Bot. Gard.* **2021**, *2*, 600–609. [CrossRef]
7. Bryant, Z.; Konczol, E.; Michaels, C.J. Impact of Broad-Spectrum Lighting on Recall Behaviour in a Pair of Captive Blue-Throated Macaws (*Ara glaucogularis*). *J. Zool. Bot. Gard.* **2022**, *3*, 177–183. [CrossRef]
8. Thomas, L.; Dobbs, P.; Ashfield, S. Digit Entrapment Due to Plastic Waste in a Verreaux's Eagle Owl (*Bubo lacteus*). *J. Zool. Bot. Gard.* **2022**, *3*, 442–447. [CrossRef]
9. Dias, J.E.; Ellis, C.; Smith, T.E.; Hosie, C.A.; Tapley, B.; Michaels, C.J. Baseline Behavioral Data and Behavioral Correlates of Disturbance for the Lake Oku Clawed Frog (*Xenopus longipes*). *J. Zool. Bot. Gard.* **2022**, *3*, 184–197. [CrossRef]
10. Carter, K.C.; Fieschi-Méric, L.; Servini, F.; Wilkinson, M.; Gower, D.J.; Tapley, B.; Michaels, C.J. Investigating the Effect of Disturbance on Prey Consumption in Captive Congo Caecilians Herpele squalostoma. *J. Zool. Bot. Gard.* **2021**, *2*, 705–715. [CrossRef]
11. Hart, A.M.; Reynolds, Z.; Troxell-Smith, S.M. Location, Location, Location! Evaluating Space Use of Captive Aquatic Species—A Case Study with Elasmobranchs. *J. Zool. Bot. Gard.* **2022**, *3*, 246–255. [CrossRef]
12. Free, D.; Justice, W.S.M.; Smith, S.J.; Howard, V.; Wolfensohn, S. An Approach to Assessing Zoo Animal Welfare in a Rarely Studied Species, the Common Cusimanse Crossarchus obscurus. *J. Zool. Bot. Gard.* **2022**, *3*, 420–441. [CrossRef]
13. Spiezio, C.; Altamura, M.; Weerman, J.; Regaiolli, B. Behaviour of Zoo-Housed Red Pandas (Ailurus fulgens): A Case-Study Testing the Behavioural Variety Index. *J. Zool. Bot. Gard.* **2022**, *3*, 223–237. [CrossRef]
14. Finch, K.; Humphreys, A. Day Time Activity Budgets, Height Utilization and Husbandry of Two Zoo-Housed Goodfellow's Tree Kangaroos (Dendrolagus goodfellowi buergersi). *J. Zool. Bot. Gard.* **2022**, *3*, 102–112. [CrossRef]
15. Truax, J.; Vonk, J.; Vincent, J.L.; Bell, Z.K. Teamwork Makes the String Work: A Pilot Test of the Loose String Task with African Crested Porcupines (*Hystrix cristata*). *J. Zool. Bot. Gard.* **2022**, *3*, 448–462. [CrossRef]

Journal of
Zoological and Botanical Gardens

MDPI

Article

Behavioural Impact of Captive Management Changes in Three Species of Testudinidae

Jessica T. Turner [1,*], Alexandra L. Whittaker [1] and David McLelland [1,2]

[1] Roseworthy Campus, School of Animal and Veterinary Sciences, The University of Adelaide, Roseworthy, SA 5371, Australia
[2] Zoos South Australia (Royal Zoological Society of South Australia), Adelaide, SA 5000, Australia
* Correspondence: jessica.t.turner@adelaide.edu.au

Abstract: Reptile behaviour and welfare are understudied in comparison with mammals. In this study, behavioural data on three species (*Astrochelys radiata, Stigmochelys pardalis, Aldabrachelys gigantea*) of tortoises were recorded before and after an environmental change which was anticipated to be positive in nature. The environmental changes differed for each population, but included a substantial increase in enclosure size, the addition of substrate material, and a change in handling procedure. A tortoise-specific ethogram was created to standardise data collection. Focal behaviour sampling was used to collect behavioural data. Changes in the duration of performance of co-occupant interaction and object interaction in the leopard (*Stigmochelys pardalis*) and Aldabra (*Aldabrachelys gigantea*) tortoises were observed following the environmental changes. The Shannon–Weiner diversity index did not yield a significant increase after the changes but had a numerical increase which was relatively greater for the leopard tortoise group, which had experienced the greatest environmental change. The leopard tortoises also demonstrated changes in a greater number of behaviours compared to the other species, and this was sustained over the study period. However, this included a behaviour indicative of negative affect: aggression. Whilst we are unable to conclude that welfare was improved by the management changes, there are suggestions that behavioural diversity increased, and some promotion of positive social behaviours occurred.

Keywords: reptile; tortoise; welfare; diversity index; behaviour; enrichment; ethogram

Citation: Turner, J.T.; Whittaker, A.L.; McLelland, D. Behavioural Impact of Captive Management Changes in Three Species of Testudinidae. *J. Zool. Bot. Gard.* **2022**, *3*, 555–572. https://doi.org/10.3390/jzbg3040041

Academic Editors: Kris Descovich, Caralyn Kemp and Jessica Rendle

Received: 30 June 2022
Accepted: 2 November 2022
Published: 7 November 2022

1. Introduction

In recent times, zoos have shown a strong commitment to optimising the welfare of the animals in their care [1–4], and utilising accreditation scheme membership to showcase this commitment. Many zoos conduct their own welfare research and strive to implement the findings within their premises [5]. Likewise, most zoo accreditation programs, e.g., The Australian Zoo and Aquarium Association, require members to regularly conduct animal welfare assessments [5,6]. Simultaneously there has been a shift in public attitudes towards animals, with increasing expectations of high welfare standards [1–3,5].

The welfare of an animal, as described by Webster, can be considered to be 'its capacity to avoid suffering and sustain fitness' [7]. An affective state is defined as a feeling, emotion, or mood such as fear, that motivates an animal to avoid a particular environmental stimulus that is potentially detrimental to its fitness. Affective states can be positive (excitement/joy) or negative (fear/sadness) in valence. Furthermore, these states also vary in motivational intensity or arousal based on the urge to move towards or away from the eliciting stimulus [8]. Determination of welfare state includes considering the number of positive versus negative affective experiences, where 'good' welfare is determined by having more positive experiences, 'poor' welfare determined by having a greater number of negative experiences, and 'neutral' welfare assigned when there are an equal number of positive and negative experiences [9,10]. In practice, welfare is commonly assessed by looking at animal-based or resource-based measures (water, shelter, etc.) [7,11]. However, the former is likely superior, being a direct measure of welfare resulting (partially) from the resource inputs provided [11,12]. There are a number of ways that welfare can be assessed, including physiological, immunological, or behaviour-based techniques [9,13,14]. It is common to use multiple modalities in welfare assessment [9,13,14]. In a zoo environment there is a need for methods to be non-invasive, simple, and undemanding on resources [9,13]. As a result, behaviour-based methods are likely to be the most practical and have received the most research focus [13].

There are various models or frameworks which have been used as the basis of welfare assessment tools, including the Five Domains Model [15], the Five Freedoms [16], and the Welfare Quality® protocol [17]. A number of these have been trialled at zoos. However, the Five Domains Model is perhaps the most employed as part of zoological accreditation programs [5,18]. These welfare assessment protocols utilise behavioural observations to infer affective state, but this requires a good understanding of which species-specific behaviours are indicative of differently-valenced affective states [11].

To date, there has been a research taxa-bias towards mammals in studies of biology and welfare of animals in zoos [5,11,12,19,20]. Reptiles have been comparatively understudied. A reduced research focus on reptiles may be due to a combination of factors, including difficulties observing wild reptile behaviour, challenges intuitively recognising and interpreting reptile behaviours such as signs of distress, or that reptiles are perceived as less important or less intelligent [1,11,20,21]. Furthermore, a misconception that reptiles are highly tolerant of, and easily adaptable to, suboptimal captive conditions (which is not supported by the literature) may result in the provision of only the most basic husbandry requirements for captive management being considered by some [19,20].

Recently, there has been increased research focus on identifying behaviours that may be indicative of welfare state in reptiles [21–30]. However, there remains a dearth of primary studies exploring reptile behaviours, their relation to affective state, and how husbandry practices may modify expression of these behaviours. A key challenge is in identifying indicators that infer positive, as opposed to negative, affective states [13]. Given the lack of validated methods to assess reptile welfare, it is important that potential tools are explored.

There has been recent interest in using behavioural diversity measures, calculated from behavioural data, to provide an objective insight into the welfare of both individual and groups of animals by determining how much variation is shown in their behavioural repertoire [31–35]. Greater behavioural diversity is generally accepted as a positive indicator of welfare [31,34,35]. This is based on the assumption that animals displaying varied behavioural repertories are having their behavioural needs met. Alternately, when diversity is low an animal may show reduced overall behaviours due to lethargy or the performance of stereotypies [31,35].

This study opportunistically investigated changes in the activity budget and behavioural diversity of land tortoises following a change to their captive environment. Testudines were selected as 56% of species in this order are threatened, making the study of captive conditions of high importance to conservation efforts and breeding programs [23]. The environmental changes were different for each species, but included a substantial increase in enclosure size, added substrates, and a change in handling procedure. It is suggested that an animal's motivation to interact with environmental enrichment, is positively correlated with welfare [1]. More complex environments allow animals to choose how to interact with their environment, allowing greater control and agency, thus improving welfare [11,19]. Given this, it was hypothesised that the environmental changes would result in a change in behavioural expression, indicative of improved welfare.

2. Materials and Methods

2.1. Ethics Statement

Ethics approval was granted for this research by the Animal Ethics Committee of The University of Adelaide (protocol code S-2021-036), and the research was conducted in accordance with the Australian Code for the Care and Use of Animals for Scientific Purposes [36].

2.2. Population

This study investigated three established groups of tortoises housed at two locations in South Australia: Adelaide Zoo and Monarto Safari Park. Radiated tortoises (*Astrochelys radiata*; $n = 5$, adult 3 males, 2 females) housed and displayed at Adelaide Zoo, leopard tortoises (*Stigmochelys pardalis*; $n = 4$, adult, all male) not on display and housed at Monarto Safari Park, and sub-adult Aldabra tortoises (*Aldabrachelys gigantea*; $n = 5$, unknown sex), housed and displayed at Adelaide Zoo. Land tortoises were selected as this provided the greatest number of individuals within the same species. The radiated tortoises arrived in the collection in 2018 and had been in their current enclosure since 2020. It is presumed that these individuals were wild caught as they were part of a group of confiscated tortoises. The leopard tortoises arrived in 2009, and had been in their first enclosure, as described in this study, since 2018. These individuals were captive-bred and transported from Auckland Zoo, New Zealand. The Aldabra tortoises arrived in 2017 and had always been housed in the same enclosure. These individuals were captive bred and transported from La Vanille Nature Park, Mauritius. The total number of animals ($n = 14$) and species were determined by the availability within the zoo collection.

2.3. Husbandry

Diet for all groups across the study consisted of ad libitum grass hay and defined portions, fed twice a day, of hard vegetables (e.g., pumpkin, sweet potato, carrot, broccoli, cauliflower), leafy greens (lettuces and endives), Wombaroo herbivorous kangaroo pellets (Wombaroo Food Products, Glen Osmund, South Australia), and a calcium-vitamin D supplement. A minimum of two feeding zones were provided in each enclosure to accommodate all individuals and reduce competition.

All enclosures (Figure 1) were temperature-controlled and were fitted with basking lights and UV lamps. Temperature and humidity were monitored and kept at species-appropriate levels by zookeepers (unbranded generic thermometer/hygrometer, product code: IC7312: accuracy of temperature ± 1 °C, accuracy of humidity $\pm 3\%$). Enclosures were cleaned and misted daily by keepers.

Prior to the management changes, the radiated tortoise enclosure had substrates of sand and straw. The leopard tortoise enclosure had straw substrate over concrete. The original Aldabra tortoise enclosure had a dirt substrate and a mock rock pool. For diagrams of each enclosure before and after the environmental change see Appendix A.

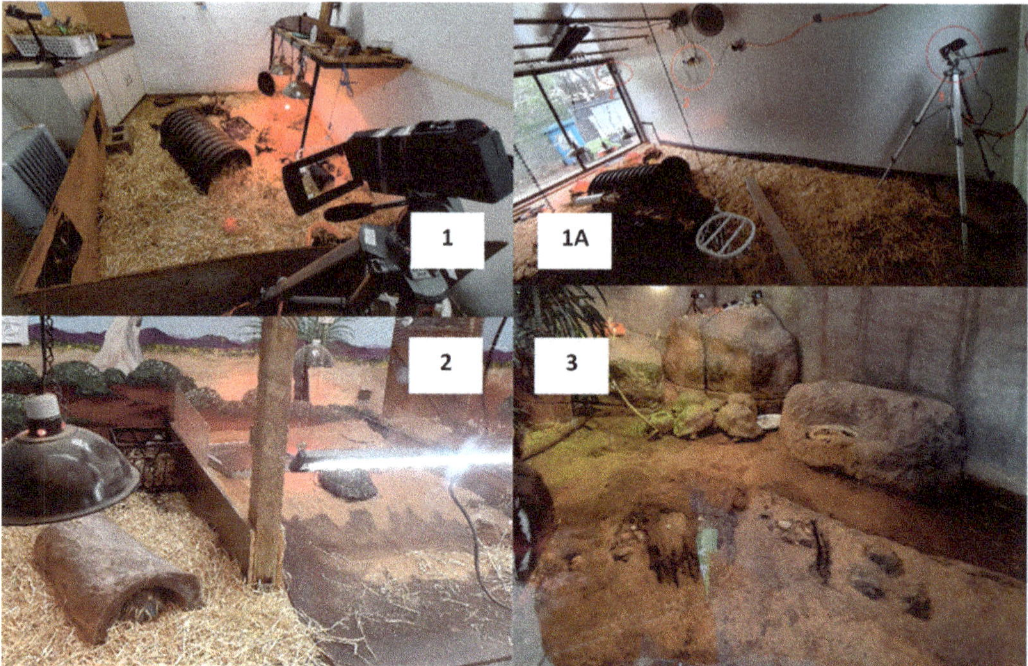

Figure 1. Photographs of enclosures **1** (the original leopard tortoises enclosure), **1A** the indoor component of the new leopard tortoise enclosure, **2** radiated tortoise enclosure, and **3** Aldabra tortoise enclosure.

2.4. Study Design

This study utilised the opportunity of planned changes to the captive management of the three groups by animal management teams at each zoo (Adelaide Zoo and Monarto Safari Park) during the period of this study. These changes were uninfluenced by the research team. The leopard tortoises were moved to a new enclosure that provided a substantial increase in enclosure size and diversity (original 7 m^2), comprising a climate controlled indoor area (13 m^2), a roofed open-air area (30 m^2), and a large uncovered naturally vegetated outdoor area (230 m^2). Depending on weather conditions, the tortoises had access to all areas.

The radiated tortoises received dried leaves as an additional substrate and a marked reduction in the frequency of manual handling. The leaves were plane tree and various species of *ficus*, selected due to their non-toxicity to tortoises and general unpalatability. It had been common for the keepers to pick up the tortoises and move them to the feeding locations when food was offered; this practice was ceased, allowing tortoises to move around the enclosure with greater choice and control.

The environmental changes for the Aldabra tortoises were the addition of an enrichment crate filled with straw, and sand added as a substrate to two areas of the enclosure.

2.5. Behavioural Data Collection

Data collection was split into three time points (Figure 2), pre-environmental change (1), post-change (between 10–21 days after change) (2), and approximately seven months (230–250 days) after the environmental change (3). After the environmental change, no behavioural observations were made for at least a week to allow tortoises to habituate to the new environment and reduce the likelihood of confounding results due to an acute stress response, or reaction to novelty. Data for the third time point were only collected for the leopard and Aldabra tortoises; it was considered that additional management changes subsequent to the second time point for radiated tortoises would have confounded interpretation.

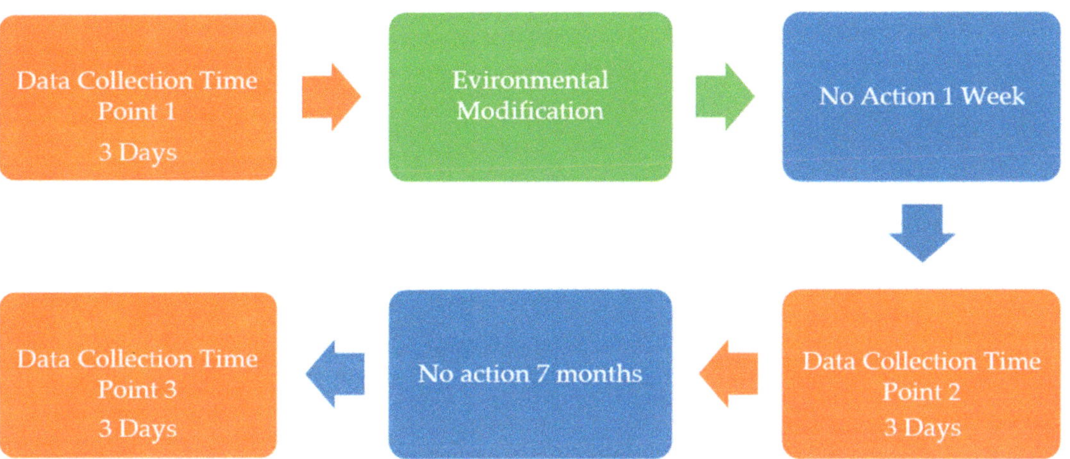

Figure 2. Timeline of data collection for the evaluation of tortoise behaviour in response to a change in environment.

Video recordings were taken of animals in all housing locations using camcorders (HC-V180, Panasonic Corporation, Kadoma, Osaka, Japan). In the larger outdoor leopard tortoise enclosure, there were occasions when manual data recording was required since the camera's field of view did not capture the full area.

Focal behavioural sampling was performed on every animal in each group using the video footage or direct in-person observation. Each tortoise was viewed for two minutes, and behaviour was continuously sampled. This occurred once every hour between 8 a.m. and 5 p.m. (9 data collection points), over a consecutive three-day period (54 min of observation per animal, per time point). For data analysis, sampling time points were grouped by time of day: morning (8–10 a.m.), midday (10 a.m.–2 p.m.), and afternoon between 2 p.m. and 5 p.m. Behaviour was catalogued using the Zoo Monitor App [37]. The frequency and total duration of each behaviour were recorded for every individual. Only one behaviour could be selected at a time for each individual. The data were compiled by one observer to exclude potential inter-observer variations, following an ethogram with set behavioural descriptions, outlined in Table 1, which was a modified version of an ethogram that has been used in previous studies [8,38,39]. Inter-rater reliability between the observer in the current study and another observer was conducted in an unpublished parallel study on the same tortoise groups. The observers reached 80% consensus on identification of the behaviours on a subset of the data.

2.6. Data Analysis

In order to explore the effects of the covariates of temperature, species, time of day, and timepoint, a multivariate General Linear Model (GLM) in the program SPSS was used with the duration of each behaviour of interest being taken as the dependant variable [40]. A Bonferroni correction was applied to account for multiple comparisons. $p < 0.05$ was taken as the significance level.

Table 1. Ethogram used for behavioural analysis modified from [8,38,39].

Behaviour	Description
Resting	Includes basking, sleeping, or resting on ground with no weight on limbs, for 3 s or more. No other activities being performed. All instances of resting are included whether awake or asleep.
Walking	Two or more steps in one direction. One foot removed from ground at a time.
Digging	One or more limbs (front or hind legs) moving substrate. Motion must be repeated twice or more to be counted as digging.
Standing still	Tortoise must be bearing its weight on 1 or more legs for 2 s or more
Bathing	Includes submerging whole or part of body for more than 3 s
Enrichment and object interaction	Touching or playing with any object provided for play and or enrichment. This includes interaction with food dispensers (ball or similar, NOT a stationary food bowl and NOT the act of eating), using or touching tunnels/hides/ramps/logs/etc. If tortoise is inside a tunnel when set observation period starts and they cannot be reasonably seen, this is counted as individual not observed. However, if the tortoise moves into tunnel during the observation period and remains in there, this is counted as use of tunnel.
Eating/Drinking	Any eating or drinking activity where food/water is consumed, or food is chewed.
Vocalisations	Any audible noises made by the tortoises by nose or throat. This excludes defaecation and/or digestion noises.
Co-occupant interaction	A positive or neutral interaction with another tortoise. This includes climbing, leaning, touching, non-aggressive approach, head bobbing, etc.
Co-occupant aggression	Any negative aggressive action, or attempted action, towards another occupant including shell ramming, charging, displacement, hooking, aggressive social posturing, scratching, biting.
Stereotypies/Abnormal behaviour	Includes pacing and other repetitive behaviour and interaction with transparent boundaries. Behaviour must be completed three times consecutively.
Not Observable	For use only when focal animal is not visible.

Data were then split by species to investigate the applied management change. Given the small sample size for each species, and that the behavioural data were often non-normally distributed data, non-parametric tests were applied (Mann–Whitney or Kruskal–Wallis tests in SPSS).

Shannon–Wiener's Diversity Index was used to calculate behavioural diversity before and after the environmental change. The formula for Shannon–Weiner's diversity index is [32]:

$$H = -\sum(p_i \times \ln(p_i)) \quad (1)$$

$H_{Duration}$ and H_{Rate} [33] were calculated using Excel (Microsoft Corporation, 2021), where p_i is the duration or frequency, respectively, of ith behaviour. A higher H value represents greater behavioural diversity [41]. Instructions from Snapshot Wisconsin's tutorial [42] were followed to create the spreadsheet. A Kruskal–Wallis test was conducted to determine any statistical significance. $p < 0.05$ was taken as the significance level. Due to the limited data points for the radiated species, a Wilcoxon test was conducted instead. $p < 0.05$ was taken as the significance level.

3. Results

3.1. Activity Budget Analysis Results

The model shows multiple significant behaviours (Table 2).

Table 2. GLM significant main effects.

Factor	Behaviour	Df (df1, df2)	F-Value	Significance
Species	Aggression	1, 107	8.300	0.01
	Object Interaction	1, 107	11.704	0.01
	Eating/Drinking	1, 107	5.649	0.02
	Standing	1, 107	11.136	0.01
	Co-occupant interaction	1, 107	18.758	<0.001
	Abnormal Behaviours	1, 107	5.024	0.03
Time Point	Aggression	2, 107	4.703	0.01
	Object Interaction	2, 107	4.868	0.01
	Resting	2, 107	4.486	0.01
	Standing	2, 107	4.655	0.01
	Co-occupant interaction	2, 107	20.696	<0.001
	Walking	2, 107	6.169	0.003
Temperature	Resting	1, 75	8.223	0.005
	Co-occupant Interaction	1, 75	21.700	<0.001
Time of Day	Aggression	2, 75	31.000	<0.001
	Eating/Drinking	2, 75	6.410	0.003
	Resting	2, 75	6.808	0.002
	Co-occupant Interaction	2, 75	6.522	0.002
	Walking	2, 75	4.390	0.02

3.1.1. Temperature and Time of Day Interactions

During the study the median temperature for the radiated tortoises was 25.75 °C, range of 25–26 °C. For the leopard tortoises the median temperature was 23.05 °C, range of 20.1–27.15 °C. The median temperature for the Aldabra tortoises was 28 °C, with range of 25–29 °C. When the group data were combined, temperature had a significant interaction with co-occupant interaction and resting with the former increasing as temperature increased, and vice versa for resting behaviour. The behaviours aggression, co-occupant interaction, eating and drinking, resting, and walking were influenced by time of day. Specific differences for the combined groups are illustrated in Figure 3.

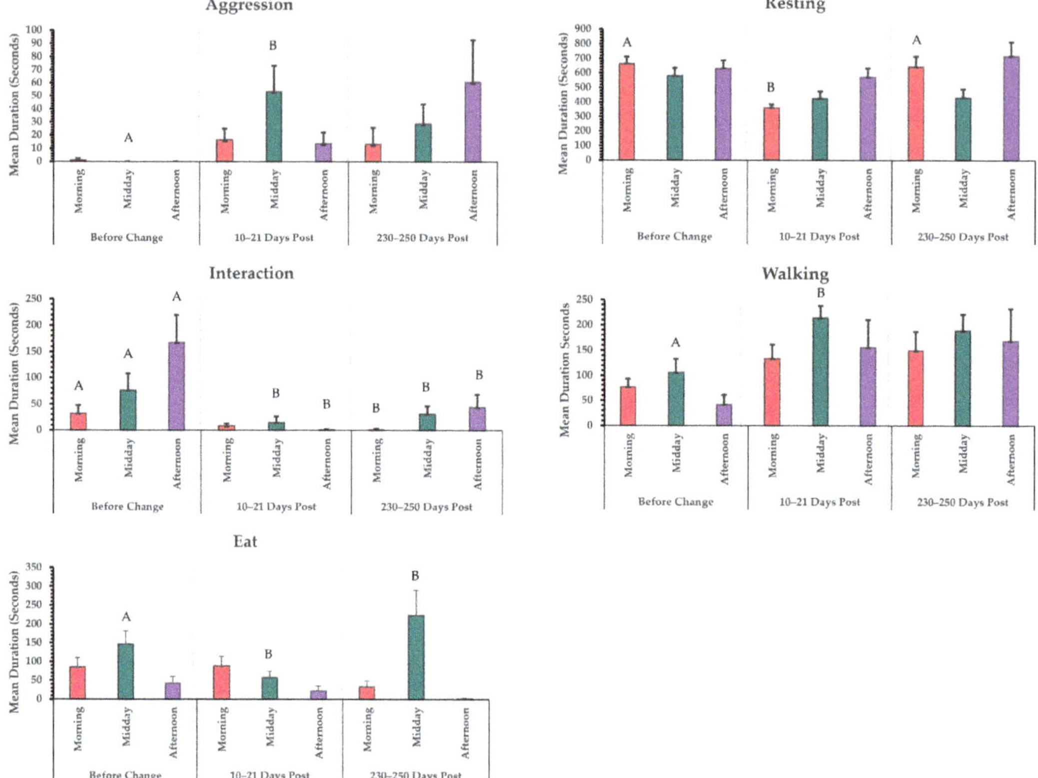

Figure 3. Bar Chart of all significant behaviours separated by time of day at three time points. Pairwise comparisons are indicated with a letter, i.e., the letter 'A' over two bars indicates no difference between those time points. Letters 'A' and 'B' over two bars indicates that these are different from each other within the same time of day comparison, i.e., morning data compared. "Interaction" refers to co-occupant interaction. Displaying mean and standard error.

3.1.2. Species and Time Point Interactions

In the radiated tortoises there were no changes in duration of the observed behaviours following the applied management change. There were differences in aggression, co-occupant interaction, eating and drinking, object interaction, resting and walking in the leopard tortoises. The Aldabra tortoises showed decreased co-occupant interaction and increased object interaction following the change. See Figure 4 for details of direction of effect and pairwise comparisons calculated using the Mann–Whitney/Kruskal–Wallis test.

3.2. Diversity Index Results

Combined data for the species H values are detailed in Figure 5. The environmental modifications did not elicit a change in $H_{Duration}$ (Wilcoxon signed rank: $Z = -1.2136$, $p = 0.55$) or H_{Rate} (Wilcoxon signed rank: $Z = -0.6742$, $p = 0.5476$) in the radiated tortoise population. Similarly, there were no differences in $H_{Duration}$ or H_{Rate} between the time points for the leopard tortoise population (Kruskal–Wallis: $\chi2(2) = 0.7$, $p = 0.705$ and $\chi2(2) = 0.81$, $p = 0.668$, respectively) or Aldabra population (Kruskal–Wallis: $\chi2(2) = 0.51$, $p = 0.775$ and $\chi2(2) = 0.039$, $p = 0.981$) (Figure 6).

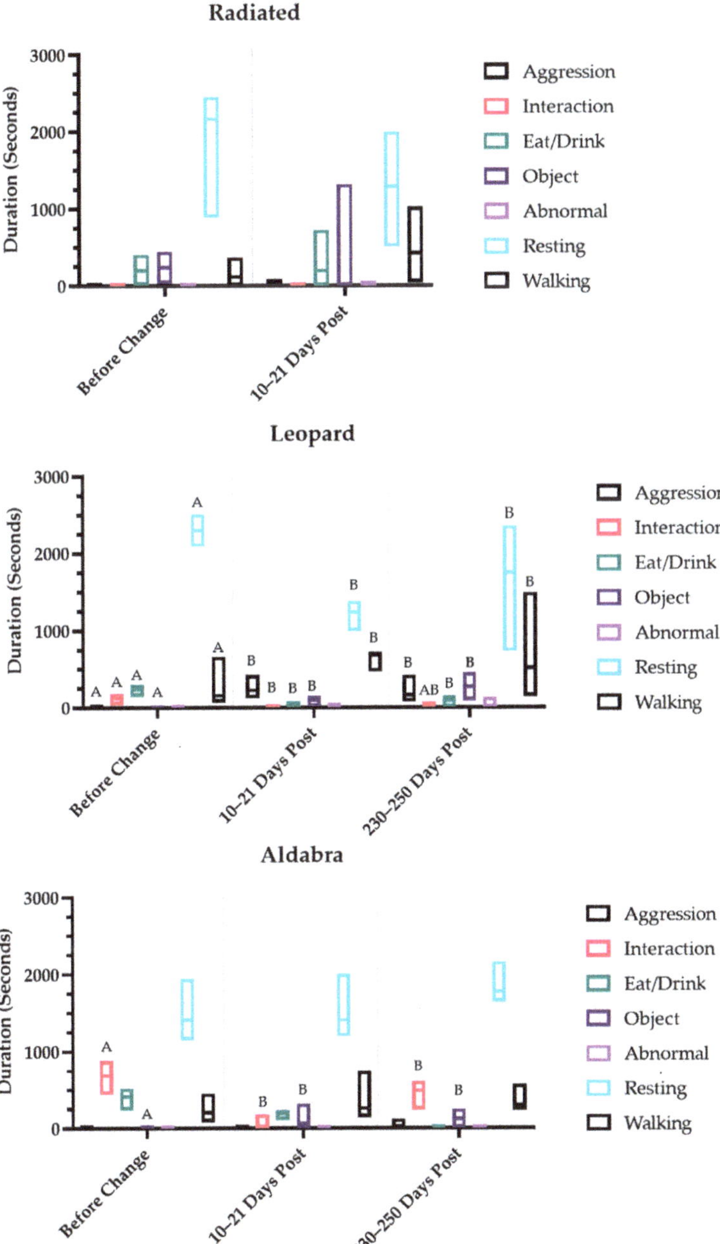

Figure 4. Box Plot of Behaviours Separated by Species at Three Time points: (1) Before Environmental Change, (2) 10–21 days Post Change, (3) 6 months Post Change. 'Interaction' refers to 'co-occupant interaction' and 'object' refers to 'object interaction'. Pairwise comparisons are indicated with a letter i.e., the letter 'A' over two timepoints indicates no difference between those time points for the same behaviour. Different letters indicate a difference between time points. Due to additional husbandry changes subsequent to the second time point, there is no third time point for radiated tortoises.

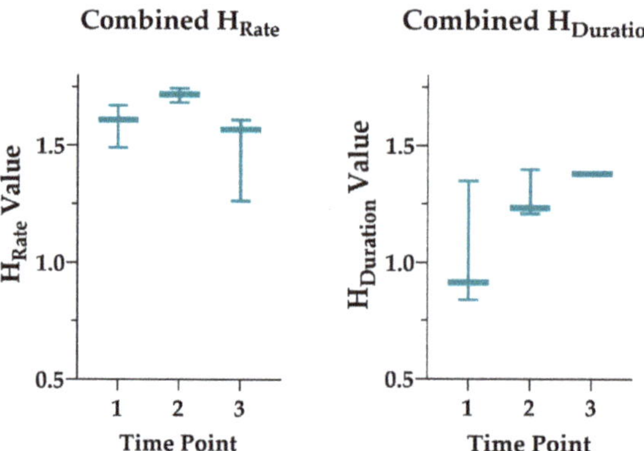

Figure 5. Box plots of diversity index H_{Rate} (**left**) and $H_{Duration}$ (**right**) for all species over three time points: (1) Before environmental change, (2) 10–21 days post-change, (3) 230–250 days post-change. Graphs present median, minimum, and maximum value.

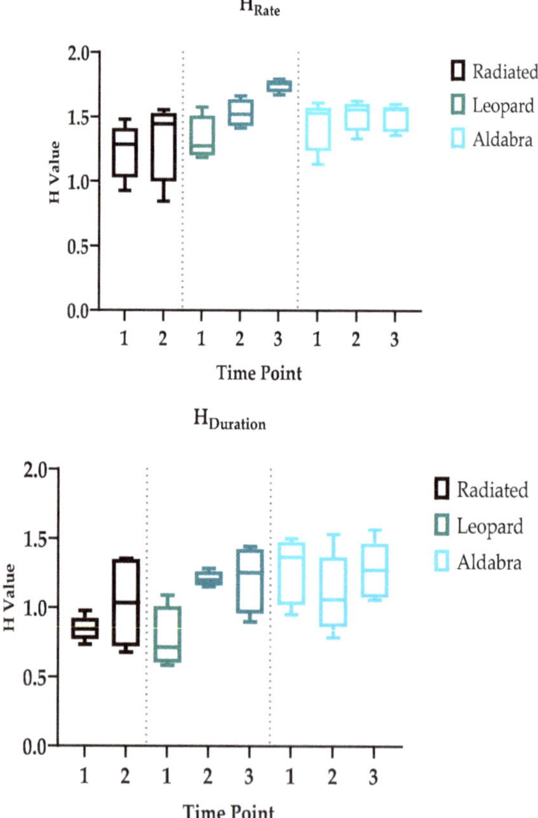

Figure 6. Diversity index box plots for H_{Rate} (**top**) and $H_{Duration}$ (**bottom**) separated by species over time (1) Before environmental change, (2) 10–21 days post-change, (3) 230–250 days post-change. Due to additional husbandry changes subsequent to the second time point, there was no third time point for radiated tortoises.

4. Discussion

In this study, we report on the impact of environmental changes in three species of testudines based on behavioural data. The management changes did not lead to an overt improvement in welfare but did elicit some changes in individual behaviours that were scored as part of the ethogram. The behaviours where differences were seen included some indicative of positive affect. However, there was also an increase in behaviours that are likely to bring about negative welfare consequences, for example aggression.

A potentially negative behaviour, co-occupant aggression, increased in the leopard group. Additionally, object interaction increased, while co-occupant interaction decreased following the management changes in the leopard and Aldabra groups, and these changes were maintained at the longer follow up time point. It could be that the resources introduced gave the tortoises something to compete over, or potentially upset the established social hierarchy. A recent study on tortoise aggression was able to identify a social hierarchical structure which was influenced by tortoise height and aggression levels [29]. There are many factors that have the potential to affect tortoise behaviour. Light and temperature affect the activity levels of tortoises. In summer they experience two daily peaks of activity, in the morning and afternoon [43–45]. In cooler winter conditions they have unimodal activity patterns and long periods of basking is required [45]. Tortoises thermoregulate behaviourally by moving through areas with different temperature gradients and will restrict movement, seek water, and/or seek shade to prevent heat stress [44,46,47]. Temperature may also affect other, non-thermoregulatory, behaviours as a trend of higher social behaviour and lower aggressive behaviour has been linked with higher body temperatures in radiated tortoises [48]. In the current study, aggression levels for the leopard species sustained an increase over time, which was not associated with enclosure temperature. This suggests another factor may be influencing this behaviour, or that enclosure temperature may not be an accurate analogue of body temperature. Co-occupant interaction and resting were the only behaviours linked to temperature in the current study, with the latter most likely being a thermoregulatory behaviour. This suggests that during data collection the enclosure temperature remained stable enough to minimally impact behaviour outside of thermoregulatory behaviours. This is unsurprising, as the temperatures were controlled at the zoological facility. The impact of the changed management strategy was most strongly seen at midday, with the most significant changes to behaviour seen in Figure 4. Whilst it could be assumed that this is due to temperature differences, this link was not supported.

Increased duration of walking with a decrease in resting in the leopard group could be interpreted as increased exploratory behaviour as a result of the increased space in the new enclosure. However, it would require spatial mapping to determine if this has arisen due to an increase in range traversed, or the making of more frequent smaller trips. Decreases in the behaviour eating and drinking in the leopard group suggest an adverse impact but should be interpreted with care. This may have arisen due to slight variability in management regimes. Whilst tortoises were fed at approximately the same time each day, this could vary depending on the schedule of the keepers. Compounding this further, the tortoises generally ate all their food within a 10 min period. Unfortunately, this period did not always align with the two-minute data collection window for that hour due to the above reasons.

To date, the most common method of assessing the behavioural impacts of enrichment items or enclosure changes in reptiles has been ethogram use, to assist in recording changes in behaviour and to create an activity budget [21–23,26,33]. Additionally, there are no studies assessing the welfare implications of reduced space for tortoises and, conversely, the welfare improvements to be gained by increasing space for tortoises. This was a novel aspect to the present study. Studies of other reptiles have shown that an increase in space increases both locomotive behaviour and space use [32,49], resulting in increased welfare. In a study on captive adult corn snakes, lower space allocations were found to negatively impact reptile welfare [50]. The snakes housed in larger enclosures were more active and spent 19% of their time fully elongated. Other than spontaneous behavioural observations, a series of evoked behavioural tests were also performed to gauge welfare including the novel environment test, novel object test, reverse emergence test, and preference test, the results of which corroborated this finding [50].

In mammals, various diversity indexes have been calculated. Generally, these have shown a reduction of behavioural diversity in less complex environments [51] and increased behavioural diversity with greater group size [31]. Time of day has also been shown to impact behavioural diversity [31].

There has been less study of diversity indexes in reptiles and consequently there is no established threshold for adequate behavioural diversity for reptiles in captivity [32,33], although a study on geckos yielded mean index values between 1 and 3 and were interpreted as adequate scores [33]. In the current study there were no differences in the diversity index across time points for any of the species. There was, however, a numerical increase in the score which was comparable with the scoring in the gecko study [33]. This may indicate a positive impact of management changes which perhaps would have attained statistical significance with a greater sample size, or a longer data collection period. Notwithstanding, as discussed by Miller et al., 2020, behavioural diversity indexes are influenced by the complexity of the ethogram used and have been argued to only be comparable when the same ethogram has been used [35]. In the current study, it is noteworthy that the absolute value for diversity index was remarkably similar across the species after the changes were made, despite the difference in environments across the three groups. Yet, the relative change in index from before to after the management change was greatest in the leopard tortoises where the most extensive management change occurred.

Whilst our results for the diversity index are inconclusive, there has been substantial recent discussion about the value of behavioural diversity as an indicator of welfare state [34,35,52,53]. Importantly, it may offer a method of gauging positive welfare states in captive reptile populations—a much sought-after goal in welfare science. Usefully it can be calculated using activity budget data as seen in the current study. A caveat attached to use of diversity indexes is that the score does not discriminate between positive and maladaptive behaviours. Hence, the monitoring and knowledge of the types of behaviours being expressed is still required, although an increase in stereotypic behaviours typically results in lower diversity [34]. For example, in the current study an increase in aggressive behaviours was seen and this would normally be regarded as a behaviour likely to cause negative affect but may have contributed to an increased index score. Recording an activity budget simultaneously would allow the behavioural type (positive or maladaptive) and diversity to be tracked. Another important consideration is the choice of method used for calculating the index. In the current study, there were differences in the index calculated for rate and duration of behaviours. This may be of little concern where a change is imposed and the outcome of interest is the difference in scores but will be more critical if the absolute number is used in decision-making around animal welfare.

Limitations of this study include the small sample size due to animal availability. This may have resulted in non-significance of some of the behaviours due to inadequate power. Further, this study cannot be generalised to all reptiles due to variation in species-specific behaviour between taxa [20]. Given the behavioural biology of these species and their relatively ambling demeanour, future studies should consider increasing the length of behavioural sampling, perhaps to 5–10 min per animal. Many of the behaviours of interest, such as digging (an exploratory behaviour), occur relatively infrequently and it is possible that they were missed during the relatively small sampling window. Modification to the definition of 'interaction with transparent boundaries' (Table 1, abnormal behaviour) behaviour is also suggested to 'interaction with boundaries', as the tortoises often repeated the same motions on glass as they did on other walls (vertical digging action interspersed with walking). Due to the widely accepted definition of ITB that was included in this study, this repetitive behaviour was not included in the abnormal behaviour category of the ethogram when observed. This means that the instances of walking could have been over-reported.

The optimal method for assessing animal welfare in zoological contexts is elusive, although it is envisaged that it will incorporate behavioural observations due to their non-invasive and non-resource-intensive nature [54]. Behaviours associated with negative welfare states, such as self-harm and abnormal behaviours, can be easily observed. However, the more recent shift towards identifying indicators of positive welfare states creates many challenges. The challenge with the most potential for harm is incorrectly interpreting behaviours used to assess welfare, since incorrect interpretations allow us to perpetuate the same husbandry and management which is detrimental to welfare [54]. If behavioural diversity is the way forward for assessing reptile welfare, then further research is required. There is a need to establish species baselines, maladaptive behaviour monitoring, and standardise the methodology. Despite these issues, the diversity index shows promise as an indicator of welfare, at least when data are available before and after a change, in conjunction with ethogram data, to provide an objective measure of management change impact.

Author Contributions: Conceptualization, A.L.W., D.M. and J.T.T.; Methodology, A.L.W., J.T.T. and D.M.; Data Curation, J.T.T.; Formal Analysis, J.T.T. and A.L.W.; Writing—Draft Preparation, J.T.T.; Writing—Review and Editing, A.L.W., D.M. and J.T.T.; Supervision, A.L.W. and D.M.; Funding Acquisition, A.L.W.; Project Administration, A.L.W., D.M. and J.T.T. All authors have read and agreed to the published version of the manuscript.

Funding: This research was funded by The AVA Animal Welfare Trust, 2019. A.L.W. was supported by a Barbara Kidman Fellowship from the University of Adelaide.

Institutional Review Board Statement: The study was conducted in accordance with the NHMRC Code for the Care and Use of Animals for Scientific Purposes and approved by the Animal Ethics Committee (AEC) of The University of Adelaide (protocol code S-2021-036).

Data Availability Statement: The datasets collected and analysed during the current study are available from the corresponding author upon request.

Acknowledgments: Firstly: I would like to thank my friends and family, especially my husband Ash Turner for their love and support at this momentous occasion—my first academic publication. I would also like to thank Zoos SA for allowing access to the animals. A big thank you to Zoos Victoria and particularly Sally Sherwen, for consultations throughout the year and for allowing access to the Zoo Monitor app. A special mention to the AWCAN research group for your support throughout the project.

Conflicts of Interest: The authors declare no conflict of interest. The funders had no role in the design of the study; in the collection, analyses, or interpretation of data; in the writing of the manuscript, or in the decision to publish the results.

Appendix A. Enclosure Diagrams

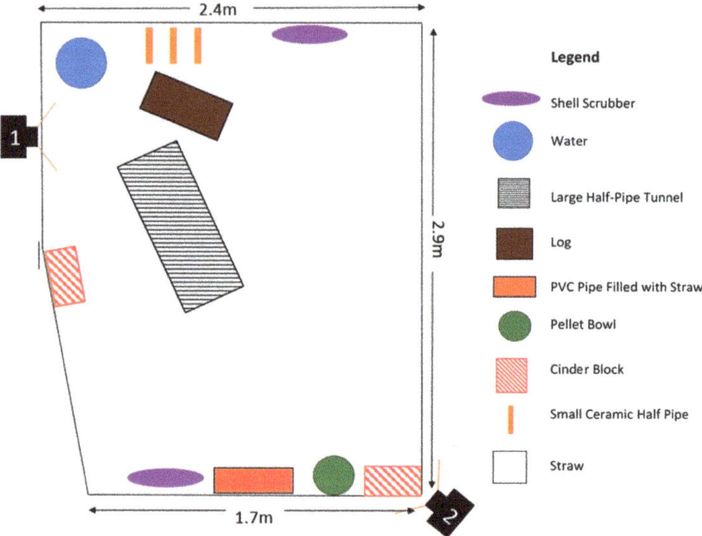

Figure A1. Original enclosure for the leopard tortoises.

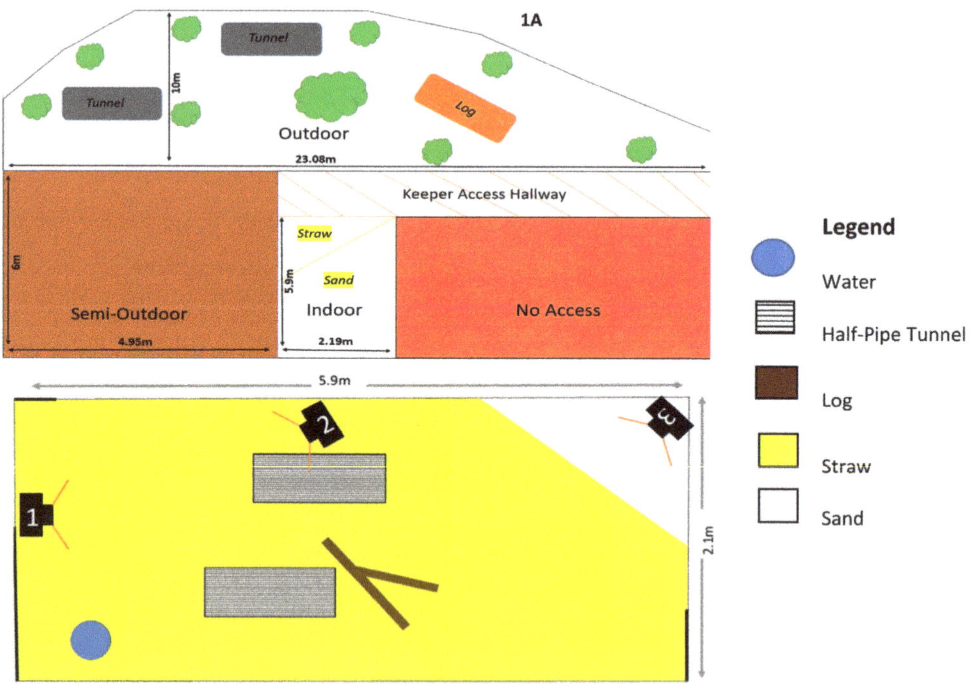

Figure A2. A, the new enclosure for the leopard tortoises post change. Top is view of full enclosure. Bottom is a detailed diagram of the indoor only section with camera placement.

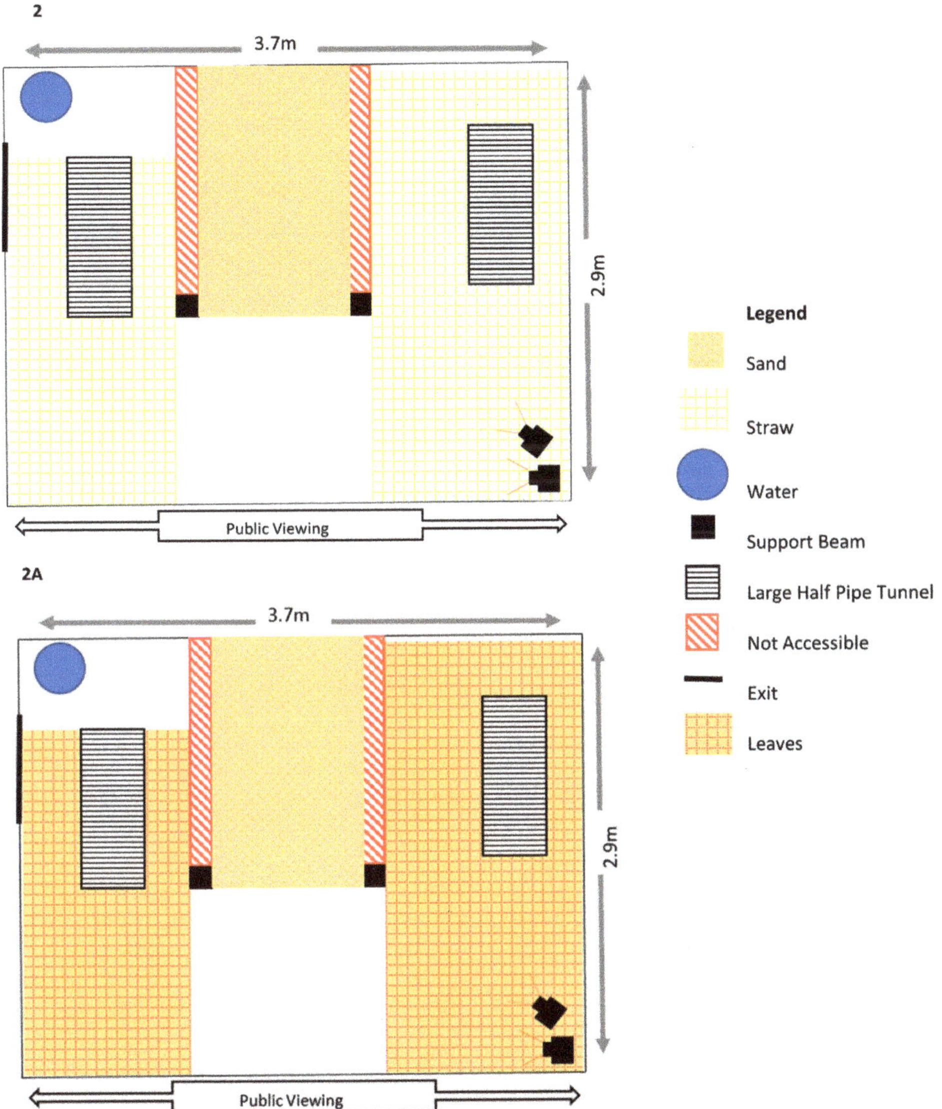

Figure A3. Diagram of Enclosure 2 & 2A. Enclosure 2 top is the unchanged enclosure for the radiated tortoises, enclosure 2A bottom is the environmentally changed enclosure for the radiated tortoises, with camera placement.

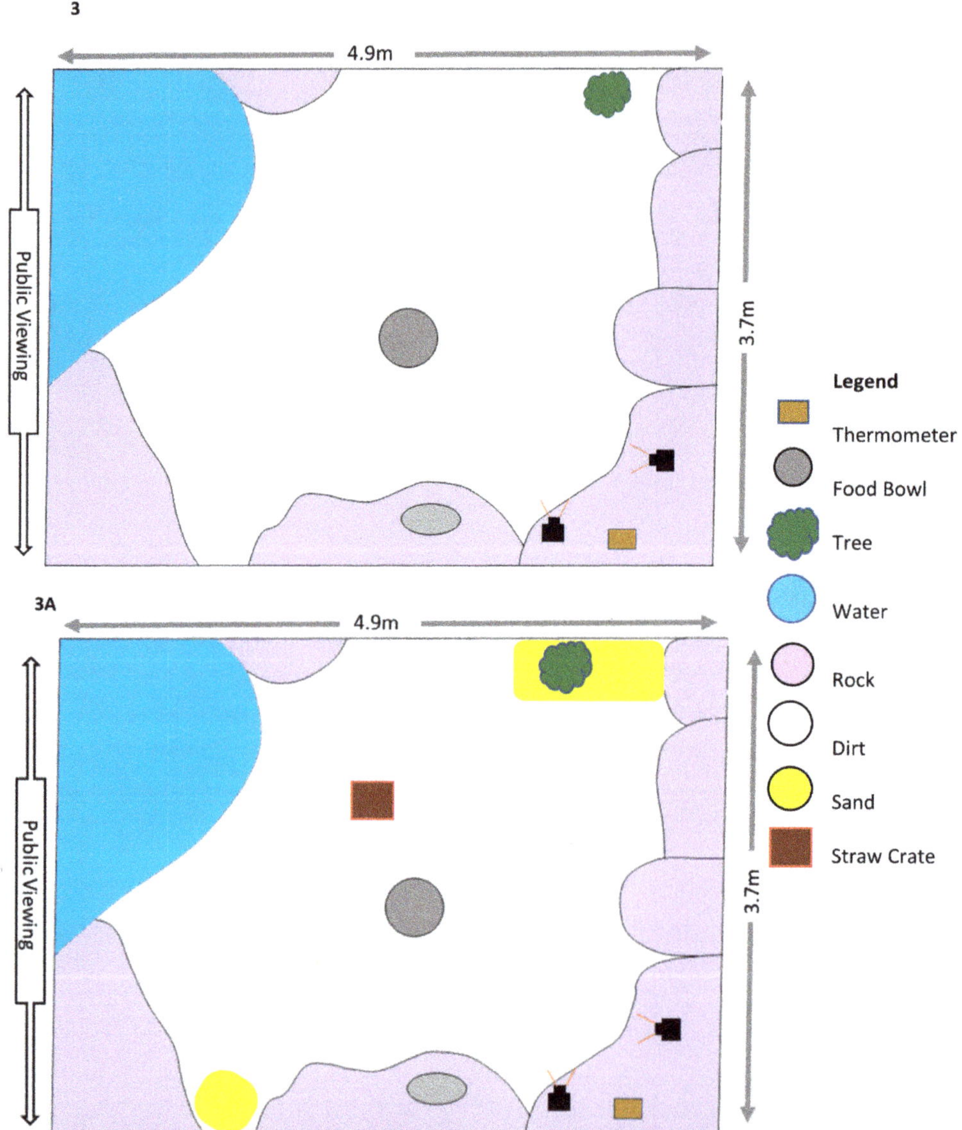

Figure A4. Diagram of Enclosure 3 & 3A. Enclosure 3 top is the unchanged enclosure for the Aldabra tortoises, enclosure 3A bottom is the changes enclosure for the Aldabra tortoises, with camera placement.

References

1. Learmonth, M.J.; Sherwen, S.; Hemsworth, P.H. Assessing preferences of two zoo-housed Aldabran giant tortoises (*Aldabrachelys gigantea*) for three stimuli using a novel preference test. *Zoo Biol.* **2020**, *40*, 98–106. [CrossRef] [PubMed]
2. Gray, J. *Zoo Ethics: The Challenges of Compassionate Conservation*; CSIRO Publishing: Clayton, Australia, 2017; p. 256.
3. Sherwen, S.; Hemsworth, L.; Beausoleil, N.; Embury, A.; Mellor, D. An Animal Welfare Risk Assessment Process for Zoos. *Animals* **2018**, *8*, 130. [CrossRef] [PubMed]
4. Zoos South Australia. Animal Welfare Charter. 2018, pp. 1–12. Available online: https://www.zoossa.com.au/wp-content/uploads/2018/08/K004-Animal-Welfare-Charter-2018-1.pdf (accessed on 5 August 2022).

5. Sayers, D.W.J. The influence of animal welfare accreditation programmes on zoo visitor perceptions of the welfare of zoo animals. *J. Zoo Aquar. Res.* **2020**, *8*, 188–193. [CrossRef]
6. Cronin, K.A. Working to Supply the Demand: Recent Advances in the Science of Zoo Animal Welfare. *J. Zool. Bot. Gard.* **2021**, *2*, 349–350. [CrossRef]
7. Webster, J. The assessment and implementation of animal welfare: Theory into practice. *Rev. Sci. Tech.* **2005**, *24*, 723–734. [CrossRef]
8. Mendl, M.; Burman, O.H.; Paul, E.S. An integrative and functional framework for the study of animal emotion and mood. *Proc. R. Soc. B Biol. Sci.* **2010**, *277*, 2895–2904. [CrossRef]
9. Whittaker, A.L.; Golder-Dewar, B.; Triggs, J.L.; Sherwen, S.L.; McLelland, D.J. Identification of Animal-Based Welfare Indicators in Captive Reptiles: A Delphi Consultation Survey. *Animals* **2021**, *11*, 2010. [CrossRef]
10. Ahloy-Dallaire, J.; Espinosa, J.; Mason, G. Play and optimal welfare: Does play indicate the presence of positive affective states? *Behav. Process.* **2018**, *156*, 3–15. [CrossRef]
11. Hill, S.P.; Broom, D.M. Measuring zoo animal welfare: Theory and practice. *Zoo Biol.* **2009**, *28*, 531–544. [CrossRef]
12. Benn, A.; McLelland, D.; Whittaker, A. A Review of Welfare Assessment Methods in Reptiles, and Preliminary Application of the Welfare Quality® Protocol to the Pygmy Blue-Tongue Skink, *Tiliqua adelaidensis*, Using Animal-Based Measures. *Animals* **2019**, *9*, 27. [CrossRef]
13. Whittaker, A.; Marsh, L. The role of behavioural assessment in determining positive affective states in animals. *CAB Rev.* **2019**, *14*, 1–13. [CrossRef]
14. Broom, D.M. Animal welfare: Concepts and measurement. *J. Anim. Sci.* **1991**, *69*, 4167–4175. [CrossRef] [PubMed]
15. Mellor, D.J.; Beausoleil, N.J.; Littlewood, K.E.; McLean, A.N.; McGreevy, P.D.; Jones, B.; Wilkins, C. The 2020 Five Domains Model: Including Human-Animal Interactions in Assessments of Animal Welfare. *Animals* **2020**, *10*, 1870. [CrossRef] [PubMed]
16. Webster, A.J.F. Farm Animal Welfare: The Five Freedoms and the Free Market. *Vet. J.* **2001**, *161*, 229–237. [CrossRef]
17. Quality, W. *Assessment Protocol for Poultry (Broilers, Laying Hens)*; Welfare Quality Consortium: Lelystad, The Netherlands, 2009.
18. Binding, S.; Farmer, H.; Krusin, L.; Cronin, K. Status of animal welfare research in zoos and aquariums: Where are we, where to next? *J. Zoo Aquar. Res.* **2020**, *8*, 166–174.
19. Burghardt, G.M. Environmental enrichment and cognitive complexity in reptiles and amphibians: Concepts, review, and implications for captive populations. *Appl. Anim. Behav. Sci.* **2013**, *147*, 286–298. [CrossRef]
20. Warwick, C. Reptilian ethology in captivity: Observations of some problems and an evaluation of their aetiology. *Appl. Anim. Behav. Sci.* **1990**, *26*, 1–13. [CrossRef]
21. Johansson, E. The Impact of Food Enrichment on the Behaviour of Turtles in Captivity. Master's Thesis, Swedish University of Agricultural Sciences, Uppsala, Sweden, 2017.
22. Therrien, C.L.; Gaster, L.; Cunningham-Smith, P.; Manire, C.A. Experimental evaluation of environmental enrichment of sea turtles. *Zoo Biol.* **2007**, *26*, 407–416. [CrossRef]
23. Bannister, C.C.; Thomson, A.J.C.; Cuculescu-Santana, M. Can colored object enrichment reduce the escape behavior of captive freshwater turtles? *Zoo Biol.* **2021**, *40*, 160–168. [CrossRef]
24. Bernheim, M.; Livne, S.; Shanas, U. Mediterranean Spur-thighed Tortoises (*Testudo graeca*) exhibit pre-copulatory behavior particularly under specific experimental setups. *J. Ethol.* **2020**, *38*, 355–364. [CrossRef]
25. Case, B.C. Environmental Enrichment for Captive Eastern Box Turtles (*Terrapene carolina carolina*). Master's Thesis, North Carolina State University, Raleigh, NC, USA, 2003.
26. Case, B.C.; Lewbart, G.A.; Doerr, P.D. The physiological and behavioural impacts of and preference for an enriched environment in the eastern box turtle (*Terrapene carolina carolina*). *Appl. Anim. Behav. Sci.* **2005**, *92*, 353–365. [CrossRef]
27. Cassola, F.M.; Henaut, Y.; Cedeño-Vázquez, J.R.; Méndez-De La Cruz, F.R.; Morales-Vela, B. Temperament and sexual behaviour in the Furrowed Wood Turtle *Rhinoclemmys areolata*. *PLoS ONE* **2020**, *15*, e0244561. [CrossRef] [PubMed]
28. Fossette, S.; Gaspar, P.; Handrich, Y.; Maho, Y.L.; Georges, J.-Y. Dive and beak movement patterns in leatherback turtles Dermochelys coriacea during internesting intervals in French Guiana. *J. Anim. Ecol.* **2008**, *77*, 236–246. [CrossRef] [PubMed]
29. Freeland, L.; Ellis, C.; Michaels, C.J. Documenting aggression, dominance and the impacts of visitor interaction on Galápagos tortoises (*Chelonoidis nigra*) in a zoo setting. *Animals* **2020**, *10*, 699. [CrossRef] [PubMed]
30. Passos, L.F.; Mello, H.E.S.; Young, R.J. Enriching Tortoises: Assessing Color Preference. *J. Appl. Anim. Welf. Sci.* **2014**, *17*, 274–281. [CrossRef]
31. Delfour, F.; Vaicekauskaite, R.; García-Párraga, D.; Pilenga, C.; Serres, A.; Brasseur, I.; Pascaud, A.; Perlado-Campos, E.; Sánchez-Contreras, G.J.; Baumgartner, K.; et al. Behavioural Diversity Study in Bottlenose Dolphin (*Tursiops truncatus*) Groups and Its Implications for Welfare Assessments. *Animals* **2021**, *11*, 1715. [CrossRef]
32. Spain, M.; Fuller, G.; Allard, S. Effects of Habitat Modifications on Behavioral Indicators of Welfare for Madagascar Giant Hognose Snakes (*Leioheterodon madagascariensis*). *Anim. Behav. Cogn.* **2020**, *7*, 70–81. [CrossRef]
33. Bashaw, M.J.; Gibson, M.D.; Schowe, D.M.; Kucher, A.S. Does enrichment improve reptile welfare? Leopard geckos (*Eublepharis macularius*) respond to five types of environmental enrichment. *Appl. Anim. Behav. Sci.* **2016**, *184*, 150–160. [CrossRef]
34. Hall, K.; Bryant, J.; Staley, M.; Whitham, J.C.; Miller, L.J. Behavioural diversity as a potential welfare indicator for professionally managed chimpanzees (*Pan troglodytes*): Exploring variations in calculating diversity using species-specific behaviours. *Anim. Welf.* **2021**, *30*, 381–392. [CrossRef]

35. Miller, L.J.; Vicino, G.A.; Sheftel, J.; Lauderdale, L.K. Behavioral Diversity as a Potential Indicator of Positive Animal Welfare. *Animals* **2020**, *10*, 1211. [CrossRef]
36. National Health and Medical Research Council (NHMRC). *Australian Code for the Care and Use of Animals for Scientific Purposes*, 8th ed.; NHMRC: Canberra, Australia, 2013.
37. *ZooMonitor*, version 4.1; Lincoln Park Zoo: Chicago, IL, USA, 2022.
38. Yon, L.; Williams, E.; Harvey, N.D.; Asher, L. Development of a behavioural welfare assessment tool for routine use with captive elephants. *PLoS ONE* **2019**, *14*, e0210783. [CrossRef]
39. Warwick, C.; Arena, P.; Lindley, S.; Jessop, M.; Steedman, C. Assessing reptile welfare using behavioural criteria. *Practice* **2013**, *35*, 123–131. [CrossRef]
40. *SPSS*, 24.0; IBM Corporation: Armok, NY, USA, 2021.
41. Henderson, P.A.; Southwood, T.R.E. *Ecological Methods*, 5th ed.; Wiley: Hoboken, NJ, USA, 2016; pp. 579–580.
42. Wisconsin, S. Measuring Biodiversity: Google Sheets Tutorial. Available online: https://dnr.wisconsin.gov (accessed on 15 September 2021).
43. Rose, F.L.; Judd, F.W. Activity and home range size of the Texas tortoise, Gopherus berlandieri, in south Texas. *Herpetologica* **1975**, *31*, 448–456.
44. Agha, M.; Augustine, B.; Lovich, J.E.; Delaney, D.; Sinervo, B.; Murphy, M.O.; Ennen, J.R.; Briggs, J.R.; Cooper, R.; Price, S.J. Using motion-sensor camera technology to infer seasonal activity and thermal niche of the desert tortoise (*Gopherus agassizii*). *J. Therm. Biol.* **2015**, *49–50*, 119–126. [CrossRef] [PubMed]
45. McMaster, M.K.; Downs, C.T. Thermoregulation in leopard tortoises in the Nama-Karoo: The importance of behaviour and core body temperatures. *J. Therm. Biol.* **2013**, *38*, 178–185. [CrossRef]
46. Judd, F.W.; Rose, F.L. Aspects of the Thermal Biology of the Texas Tortoise, Gopherus berlandieri (Reptilia, Testudines, Testudinidae). *J. Herpetol.* **1977**, *11*, 147. [CrossRef]
47. Falcón, W.; Hansen, D.M. Island rewilding with giant tortoises in an era of climate change. *Philos. Trans. R. Soc. B Biol. Sci.* **2018**, *373*, 20170442. [CrossRef] [PubMed]
48. Terespolsky, A.; Brereton, J.E. Investigating the Thermal Biology and Behaviour of Captive Radiated Tortoises. *J. Vet. Med. Anim. Sci.* **2021**, *4*, 1–6.
49. Wheler, C.L.; Fa, J.E. Enclosure utilization and activity of Round Island geckos (*Phelsuma guentheri*). *Zoo Biol.* **1995**, *14*, 361–369. [CrossRef]
50. Hoehfurtner, T.; Wilkinson, A.; Walker, M.; Burman, O.H.P. Does enclosure size influence the behaviour & welfare of captive snakes (*Pantherophis guttatus*)? *Appl. Anim. Behav. Sci.* **2021**, *243*, 105435. [CrossRef]
51. Calvert, S.; Haskell, M.; Wemelsfelder, F.; Lawrence, A.B.; Mendl, M.T. The Effect of Substrate-Enriched and Substrate-Impoverished Housing Environments On the Diversity of Behaviour in Pigs. *Behaviour* **1996**, *133*, 741–761. [CrossRef]
52. Frézard, A.; Pape, G.L. Contribution to the welfare of captive wolves (*Canis lupus lupus*): A behavioral comparison of six wolf packs. *Zoo Biol.* **2003**, *22*, 33–44. [CrossRef]
53. Miller, L.J.; Lauderdale, L.K.; Bryant, J.L.; Mellen, J.D.; Walsh, M.T.; Granger, D.A. Behavioral diversity as a potential positive indicator of animal welfare in bottlenose dolphins. *PLoS ONE* **2021**, *16*, e0253113. [CrossRef] [PubMed]
54. Watters, J.V.; Krebs, B.L.; Eschmann, C.L. Assessing Animal Welfare with Behavior: Onward with Caution. *J. Zool. Bot. Gard.* **2021**, *2*, 75–87. [CrossRef]

Journal of
Zoological and Botanical Gardens

MDPI

Article

Teamwork Makes the String Work: A Pilot Test of the Loose String Task with African Crested Porcupines (*Hystrix cristata*)

Jordyn Truax [1], Jennifer Vonk [1,*], Joy L. Vincent [1] and Zebulon Kade Bell [2]

1 Department of Psychology, Oakland University, 654 Pioneer Drive, Rochester, MI 48309, USA
2 Department of Psychology, Louisiana State University of Alexandria, 8100 Highway 71 South, Alexandria, LA 71302, USA
* Correspondence: vonk@oakland.edu; Tel.: +1-248-370-2318

Abstract: Comparative researchers have heavily focused their studies of social cognition on species that live in large social groups, while neglecting other potential predictors of social cognition. African crested porcupines (*Hystrix cristata*) are relatively rare among mammals in that they are cooperative breeders that pair for life. Little is known about their social cognition, but they are good candidates for exploring cooperative behavior due to the need to coordinate behavior to cooperatively raise young. Cooperation, as defined in this study, is the process by which two or more participants perform independent actions on an object to obtain a reward for all parties. Humans are thought to outperform all other species in the frequency and magnitude of cooperative behaviors. Yet, only by studying a variety of species can researchers fully understand the likely selection pressures for cooperation, such as cooperative breeding. Here, we pilot tested the feasibility of the popular loose-string task with a mated pair of African crested porcupines, a task that required the porcupines to cooperatively pull ropes to access an out of reach platform baited with food rewards. Other species presented with this task were able to work together to receive rewards but did not always demonstrate understanding of the role of their partner. The porcupines achieved success but did not appear to coordinate their actions or solicit behavior from their partner. Thus, similar to other species, they may achieve success in this task without taking their partner's role into account. This study demonstrates that the loose string task can be used to assess cooperation in porcupines. However, further experiments are needed to assess the porcupine's understanding of their partner's role under this paradigm.

Keywords: cooperation; loose-string task; rodents; synchronized actions; cooperative breeding; pair bonds

Citation: Truax, J.; Vonk, J.; Vincent, J.L.; Bell, Z.K. Teamwork Makes the String Work: A Pilot Test of the Loose String Task with African Crested Porcupines (*Hystrix cristata*). *J. Zool. Bot. Gard.* **2022**, *3*, 448–462. https://doi.org/10.3390/jzbg3030034

Academic Editors: Kris Descovich, Jessica Rendle and Caralyn Kemp

Received: 31 March 2022
Accepted: 12 September 2022
Published: 16 September 2022

Publisher's Note: MDPI stays neutral with regard to jurisdictional claims in published maps and institutional affiliations.

1. Introduction

Cooperation is of particular interest to comparative psychologists because individuals are expected to be self-interested; yet, cooperative behaviors have been observed in countless species in the wild (e.g., African wild dogs, *Lycaon pictus*: [1]; carrion crow, *Corvus corone*: [2]; chimpanzees, *Pan troglodytes*: [3]; Florida scrub-jays, *Aphelocoma coerulescens*: [4]; lions, *Panthera leo*: [5]). The definition of cooperation has varied (e.g., [3,6,7]), but we define it as the process by which two or more participants perform independent but coordinated actions to obtain a reward for all parties. Cooperation benefits an individual when there is a greater chance of success, in terms of short-term consequences and in lifetime reproductive success, working with another individual compared to when working alone. Cooperation is expected in species that engage in repeated interactions with the same individuals and can remember and track the outcomes of those interactions. Thus, cooperation is seen as particularly beneficial in social species as groups can be comprised of related individuals [8] or long-lasting reciprocating partners [9], and it has been extensively studied in nonhuman primates, notably chimpanzees (e.g., [10–16]. Tests of less social species and non-primates can help to establish the evolutionary timeline for the emergence of precursors to cooperation, and to identify factors that may predict the presence of these capabilities [12,17,18].

When assessing the capacity for cooperation in other species, it is critical that partners can learn to coordinate their actions [19]. Here, we present a pilot test of the capacity to cooperate in the previously unstudied African crested porcupine (*Hystrix cristata*).

Despite the increasing breadth of species studied by comparative psychologists, the cognitive abilities of many species remain unexplored. A strong emphasis on studying group-living species, such as primates, canids, cetaceans and corvids (e.g., [10,11,20–24] has led to the neglect of other aspects of sociality as predictors of social cognition, such as pair bonds and cooperative breeding [18]. Although cooperative breeding has emerged as a possible predictor of social cognitive abilities in primates [25,26], and birds [27–29], other groups present important opportunities for study. For example, within rodents, there exists a wide range of social structures (e.g., [30,31], including the monogamous pair bonds of African crested porcupines—a large species of rodent, ranging from 10–15 kg, that inhabits Central and North Africa, as well as Central Italy [32,33]. These porcupines are good candidates for research on cooperative behavior due to their tendency to pair-bond and cohabitate in dens with other mated pairs [33]. Furthermore, partners share in parental duties, such that they alternate cub guarding in the den for the first two months of life [34]. The few existing studies of these species have been restricted to assessments of temporal activity patterns [32,35], observations of home site selection and fidelity to those locations [33,35], and scavenging behavior [36]. A single study of their cognitive abilities found that African crested porcupines could be successfully trained to touch and hold to a target for 30 s using a shaping procedure [37]. Thus, very little is currently known of their cognitive capacities, particularly in regard to social cognition. Our ultimate goal was to test their capacity to engage in cooperation in an experimental task, but, because porcupines have relatively poor vision [38], and little is known about their capacity to coordinate their actions, we needed to first test the feasibility of presenting them with an experimental task requiring behavioral coordination. To do so, we piloted the popular loose string paradigm (e.g., [12,39] with a single mated pair of African crested porcupines.

Various paradigms have been employed to reveal a species' capacity to cooperate and the underlying cognitive mechanisms (e.g., simultaneous handle pulling: [40]; synchronized button pressing: [22]). Individuals can learn to engage in cooperation by learning associations between their own behaviors within the presence of a partner without understanding the essential role of the partner; thus, experimental studies are necessary to probe the mechanisms underlying their performance. One method for doing so, the loose-string task, a popular cooperation paradigm, involves two individuals pulling two ends of a rope attached to an out of reach platform baited with food. The rope is typically looped through a hole attached to the platform that allows the rope to come loose if only one individual pulls. Thus, two individuals must pull to access the platform, or it becomes inaccessible. Some previously tested species have excelled at this task (e.g., capuchins, *Cebus apella*: [41]; elephants, *Elephas maximus*: [42]; domestic dogs, [43]; wolves, *Canis lupus*: [44,45]), although others have shown difficulty in understanding the role of the partner (e.g., African gray parrots, *Psittacus erithacus*: [46]; keas, *Nestor notabilis*: [47]; rooks: [17]; domestic dogs, [44]; chimpanzees, [12]). Although the ecological relevance of the task for many tested species may be questioned, it is important to test the capacity for animals to learn to perform behaviors that have not been extensively shaped by natural selection to test their capacity for behavioral flexibility and the generalization of behavior to novel contexts. Typically, subjects are first trained to pull the ropes independently to achieve reward, either by allowing them access to both ropes for animals that use their hands or tying the ropes together for animals that pull using their mouths, beaks or one foot. Often, shaping procedures are implemented to ensure that subjects acquire proficiency with independent pulling before partners are introduced (e.g., [44,48]). When partners are introduced, the two may solve the task by both pulling backwards at the same time, although each individual may differ in the speed and force of their pulls. For instance, one of the elephants assessed by [42] solved the task by stepping on its side of the rope while its partner pulled the platform within reach. Subsequent phases introduce a partner with only one end of the rope being available to each partner, thus necessitating that both partners pull simultaneously or pull for short distances alternately. In this phase, researchers assess whether the subjects look to each other to coordinate their actions or solicit pulling behavior from the partner. This allows researchers to determine whether subjects appreciate the role of their partner. Additional tests can be constructed in such a way that one partner's access to the rope is delayed, allowing researchers to assess whether the actor waits to pull until the partner is in position. Thus, this unique paradigm allows a test of the capacity of the subject to understand essential components of cooperation.

Tolerance (i.e., expressed as the ability to eat from the same food source within proximity of one another) has also been identified as directly impacting the results of these studies. The level of tolerance in a dyad predicts chimpanzee spontaneous cooperation and highly tolerant bonobos (*Pan paniscus*) cooperate more successfully than chimpanzees on highly monopolizable rewards [49,50]. Testing a familiar mated pair of porcupines maximized our likelihood of observing cooperative behavior.

We presented two African crested porcupines with the standard loose string task. We were unable to provide them additional planned opportunities to learn about the role of the partner similar to [46], who tested African gray parrots. However, our results serve as a pilot test of the capacity of porcupines to participate in a task requiring coordination and tolerance. The first phase ensured the individuals would pull a rope and that they associated pulling with a reward. In this phase, we assessed how quickly the porcupines interacted with the apparatus, pulled on the apparatus, and completed the task. In the second phase, the subjects were required to pull the rope simultaneously to move the apparatus within reach. In this phase, we assessed their ability to complete the task successfully, the speed at which they did so, and any soliciting behaviors that might have occurred. To probe their understanding, we evaluated whether the likelihood of pulling became more closely synchronized with the partner's initiation of pulling over sessions. This pilot test serves to improve our limited understanding of the social cognition of the African crested porcupine, but future testing is needed to assess their ability to learn to understand the role of their partner.

2. Materials and Methods

2.1. Participants

Two adult African crested porcupines, one female, Lady Gaga, and one male, Bedhead, were tested. These porcupines were housed at The Creature Conservancy, a nonprofit educational sanctuary in Ann Arbor, Michigan in the United States. This pair had been bonded for the duration of their time at this sanctuary. The porcupines had participated in some husbandry training prior to this study, including target training. They previously participated in a study investigating behavioral flexibility through presentation of a multi-access box and in a study investigating their ability to track the number of responses required in a particular spatial location (Vonk, unpublished). They were housed together in an indoor enclosure with intermediate access to an outdoor habitat depending on the weather (see Figure 1). The porcupines could choose not to participate in the study at any time.

2.2. Materials

The loose string paradigm [12] requires an out of reach tray baited with food. For this study, the tray was built out of a square piece of wood covered by metal sheets and was similar to the apparatus used by Heaney et al. [51]. In the first phase, the rope, made out of non-toxic manila and sisal, was fed around the back of the tray and kept in place by a U-shaped metal ridge attached to the back. The bait used for the porcupines depended on the food available to the researchers at The Creature Conservancy, but it was typically either sweet potatoes or apples. These foods were typical to the diet of the porcupines, but the specific items utilized in research were provided in addition to their daily diet. The food was placed in front of a metal barrier at the very front of the tray to prevent it from sliding backward when the porcupines attempted to grasp it with their mouths (Figure 2). This design was modified after four sessions of the second phase to include extending arms that contained the food, which would protrude into the porcupines' habitat to allow them to retrieve the food more easily. A metal loop was also attached in the center of the tray to feed the rope through to ensure the tray would move in a straight trajectory and the rope would easily come loose (Figure 3).

Figure 1. Porcupine indoor habitat including testing area.

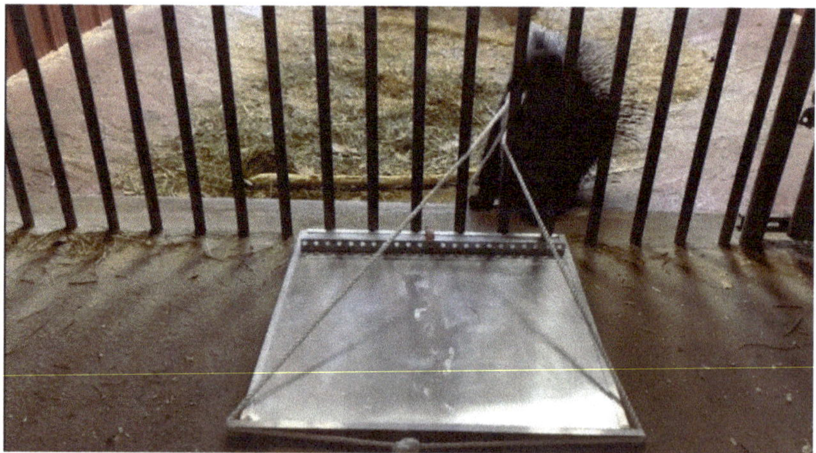

Figure 2. Apparatus used in Phase 1 and first 4 sessions of Phase 2.

Figure 3. Apparatus used in the last 8 sessions of Phase 2.

2.3. *Procedure*

2.3.1. Phase 1: Individual Pulling

To ensure the subjects were able to pull a rope to gain access to a reward, a training phase was implemented. Sessions consisted of 10 trials each, and subjects were given a maximum time limit of five minutes for each trial before the apparatus was removed (Some sessions consisted of fewer trials if the porcupines stopped participating (1 session for Bedhead, 1 session for Lady Gaga), or more than 10 trials if there was extra food (no more than 13 trials per session, this occurred in 5 sessions for Lady Gaga). Note that trials ranged from 5–10 within sessions in the original [12] study. If there were fewer than 10 trials, the session was not counted toward the criteria for changing training or phases. If there were more than 10 trials, only the first 10 trials were counted toward the criteria for changing training or phases.). Typically, one session per subject occurred per day. The porcupines were separated to test each subject individually through luring one individual into the outer portion of the enclosure with the food available to the experimenters at the Creature Conservancy (e.g., almonds, corn, sweet potatoes, bananas), while distracting the target subject with the same food in the inside portion of their enclosure similar to the procedure of [12]. Once one individual was successfully lured outside, the experimenter closed off the opening to the outer enclosure by sliding a metal door into place. This door was secured to ensure the other porcupine could not access this indoor room during testing. Once separated, training with the desired subject began. Previous studies have also trained subjects to pull the rope individually before interacting with a partner. In the original version of the task with chimpanzees, the chimpanzees were trained to pull both ends of the rope at the same time, or the rope would be pulled out of the apparatus and the trial would be a failure [12]. This is feasible with primates that use their hands with human-like dexterity. However, with other species that pull ropes with their mouths, beaks or one foot (e.g., domestic dogs, [44]; rooks, [48]), it is not feasible to have the subjects pull two ropes simultaneously. Porcupines use their mouths to manipulate objects, which we had witnessed in other studies (Vonk, unpublished). Thus, here, the two ends of the rope attached to the apparatus were tied together so that one animal alone could pull the apparatus forward (as in [44,51]. When training started, this rope was attached to the inaccessible apparatus so that only the rope was accessible to the porcupine through the gates in front of the enclosure.

At the start of the trial, the experimenter positioned the apparatus approximately 20 cm from the bars separating the test area from the porcupine's habitat and extended the rope through the bars into the habitat. The trial lasted for five minutes or until the porcupine obtained the reward, whichever occurred first. If five minutes passed with no interaction with the rope, the trial was considered unsuccessful, and the apparatus was reset for the next trial. If the porcupine could not be lured back for the next trial, the session was ended. Once the porcupine obtained the reward, the apparatus was pulled back to the starting position and rebaited and the rope was placed back within the porcupine's habitat for the next trial without additional delay. The experimenter remained behind the bars on the outside of the enclosure on either side of the apparatus. The subject was required to pull the apparatus until it reached the front bars of their enclosure to access the reward. If the porcupine successfully pulled the apparatus to within reach but did not immediately take the reward, their attention was directed to the food or the experimenter handed them the food. This occurred more often with the female than the male, due to her being slower to find and retrieve the reward. Handing the food directly to the porcupines was necessary to ensure they received a reward soon following a correct response to reinforce the desired behavior and motivate continued participation. As these are program animals, it was essential to reduce frustration by ensuring they received rewards for performing desired actions. Once the porcupine pulled the apparatus flush, the experimenter would say "Yes!", the cue used by trainers at this facility to indicate the animal had reached criteria, and then the experimenter would pull the ropes outside of the enclosure.

This phase continued until both individuals reached criterion. The criterion required that the subjects responded correctly on 8/10 trials for two consecutive sessions on two different testing days without any prompting from the experimenter. If one individual reached criterion before the other, that individual received individual refresher sessions before moving on to simultaneous pulling to maintain criterion level performance up until Phase 2 could be implemented.

If the subject responded correctly on fewer than 5/10 trials within a session, shaping was introduced in the next session. For shaping, the experimenter rewarded the porcupine with small pieces of the desired food as soon as the porcupine engaged in the desired behavior (biting the rope and pulling, even if the rope did not move the desired distance). Once the porcupine had done so, and been rewarded three times (i.e., on three consecutive trials), the experimenter did not offer a reward until the porcupine pulled the apparatus to the desired distance on the fourth trial. If the porcupine needed to be lured back to the rope, a less desirable food was placed near it (e.g., corn). If the porcupine reached a criterion of five out of ten correct trials, that subject was presented with a regular training session without shaping procedures for the next session. If the porcupine exhibited less than five out of ten correct trials in that session, it continued to receive shaping on the subsequent session. Once both porcupines reached the final training criterion without shaping, both porcupines were moved on to Phase 2

Immediately before starting Phase 2, both porcupines received a reminder session of Phase 1 in which they were required to achieve success on 8/10 trials without prompting.

2.3.2. Phase 2: Simultaneous Pulling

This phase included 12 sessions of approximately 10 trials each. Three sessions in Phase 2 consisted of 11 trials, 1 session consisted of 9 trials, and 1 session consisted of 7 trials. The same method as Phase 1 was utilized in this phase to determine which trials would have been considered toward the criterion. Two experimenters were present in this phase, each positioned on opposite sides of the apparatus. The apparatus was baited and made inaccessible to the porcupines. The tips of the extending arms were approximately 2 cm from the back of the bars. This meant that the apparatus had to be pulled approximately 20 cm to become flush with the bars. The rope was attached to the apparatus and untied so that there were two accessible ends. During this phase, both porcupines were required to be within 9 m of the front of their enclosure, within 30 cm of each other positioned side to side, and facing the apparatus to ensure they reached the rope at similar times before the rope was made accessible on each trial. Corn was used to lure them nearer to the dividing bars, if necessary. When the porcupines were in position, each experimenter placed one of the rope ends into the enclosure simultaneously so that each rope end was an approximately equal distance from its respective porcupine on each side of the apparatus. Once the ropes were placed in the habitat, the porcupines had five minutes to obtain the rewards.

If only one subject pulled and the rope became fully detached from the apparatus so that the other end of the rope became inaccessible to the other porcupine, the rope was removed, and the apparatus was returned to its starting position and rebaited, and the trial was scored as unsuccessful. The experimenters did not provide any guidance or cueing and remained in position looking straight ahead at each other in profile to the porcupines during the trial. A trial was also considered unsuccessful when there was no response after five minutes, meaning neither porcupine had interacted with the rope via touching or biting. If the porcupines were successful, meaning that they had pulled the tray forward far enough that the baited cups on the extending arms were accessible to them through the bars of the enclosure, they received their reward. If one porcupine had not found its reward by the time the other porcupine was finishing its reward, the researchers would attempt to direct the porcupine to their reward or move the reward to the porcupine, if necessary to prevent stealing.

The apparatus was modified as described above after the first four sessions. Two extendable arms with cups for food were attached to the front of the apparatus, and a metal loop was attached near the front of the tray to thread the rope through (see Figure 3). The modifications were made for two reasons. First, the male could sometimes pull the apparatus close enough to receive rewards without the rope being detached even though the female was not pulling her end of the rope in synchrony. Second, the female porcupine continued to have difficulty obtaining food in this phase, which led to the male having an opportunity to consume her food. The modifications ensured the rope would come loose when pulled by only one individual and improved the female's ability to find the food. Once both subjects had consumed their rewards on successful trials, the apparatus was returned to its starting position and rebaited and the ropes were placed in the habitat to commence the next trial. The Supplementary Material Video S1 depicts a portion of a trial in this phase.

If the porcupines had reached the criterion of eight out of ten successful trials on four sessions across four testing days, they would have progressed to a planned delayed arrival phase.

2.3.3. Video Coding

One coder coded all trials from video for the following behaviors: success, latency to pull, latency for each to receive a reward (i.e., pulling apparatus flush with enclosure), first to pull the rope, and (in Phase 2 only), soliciting behaviors of one porcupine toward the other. For soliciting behaviors, we asked coders to identify any behaviors that a porcupine engages in that could be soliciting towards the other (i.e., behavior to elicit cooperation). We avoided specifying behaviors as we found no previous literature relevant to this species of porcupine, and we did not want to introduce bias in the coder's decisions. For the trial to be counted as a success in Phase 1, the porcupines were required to engage in the desired pulling behavior (i.e., biting the rope and pulling backward) without engaging in any other undesired behavior first (e.g., biting on the rope and pulling upwards while holding the rope with the front paws, scratching front paws on top of the rope more than once). In Phase 2, a trial was considered a success if both porcupines pulled the apparatus flush with the bars and accessed their reward. A second coder coded a randomly determined 20% of trials in each phase.

3. Results

3.1. Reliability

In Phase 1, reliability for success was represented by Cohen's Kappas; (κ = 1.000). Pearson correlations were conducted for the reliability between coders for latencies; (latency to pull: $r = 0.998$, $p < 0.001$; latency to reward: $r = 0.998$, $p < 0.001$). In Phase 2, reliability for the following behaviors was represented by Cohen's Kappas; (first to pull: $\kappa = 0.344$; Lady Gaga soliciting behaviors towards Bedhead: $\kappa = 1.000$; Bedhead soliciting behaviors towards Lady Gaga: $\kappa = 1.000$). Pearson correlations were conducted for the reliability between coders for latencies; (Bedhead latency to pull: $r = 0.475$, $p = 0.007$; Lady Gaga latency to pull: $r = 0.816$, $p < 0.001$; latency to reward: $r = 0.938$, $p < 0.001$). The data of the primary coder were used for analyses, although we acknowledge a low level of agreement between the two coders for first to pull and Bedhead's latency to pull in Phase 2. This is likely due to Bedhead's more animated behaviors surrounding the rope, which made it difficult to determine precisely when he was pulling as defined in our coding instructions (e.g., pulling backward rather than upward).

3.2. Phase 1: Individual Pulling

The purpose of this phase was to ensure that the porcupines were capable of consistently performing the basic action required to cooperate in the later phases. Bedhead required 63 trials of individual training to reach criterion and Lady Gaga required 144 trials of individual training. Both subjects pulled on the rope in 100% of trials. For Bedhead, the average latency to interact with the apparatus was 7.42 s, the average latency to pull the rope was 10.03 s, and the average latency to receive the reward was 18.38 s. For Lady Gaga, the average latency to interact with the apparatus was 14.32 s, the average latency to pull the rope was 18.23 s, and the average latency to receive the reward was 26.31 s. Overall, Lady Gaga's response time was slower than Bedhead's response time. There was no evidence that they learned to pull more quickly over time (Figure 4).

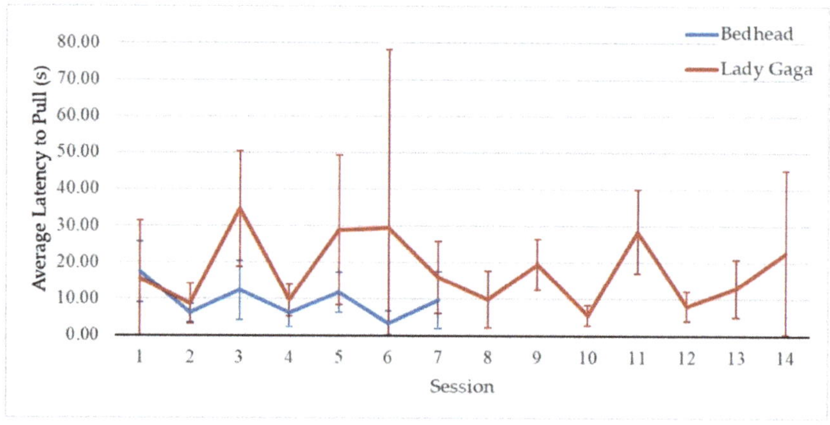

Figure 4. Average latency until first pull for each porcupine across sessions in Phase 1. Error bars depict standard deviations.

3.3. Phase 2: Simultaneous Pulling

This phase consisted of 12 sessions, the first 4 sessions (41 trials) with the first version of the apparatus, and the last 8 sessions (78 trials) with the updated version of the apparatus. However, the first 4 sessions were not included in analyses as those sessions allowed for success without cooperation. Thus, these initial sessions provided additional experience for the porcupines but did not contribute toward demonstration of cooperative behavior. In this phase, Bedhead pulled on the rope in 98.72% of trials and Lady Gaga pulled on the rope in 93.59% of trials. The average latency before Bedhead pulled on the rope was 2.40 s, and the average latency before Lady Gaga pulled on the rope was 2.62 s. Across all trials where at least one porcupine pulled, Bedhead was the first to pull in 39.74% of trials, Lady Gaga was the first to pull on the rope in 21.79% of trials, and they pulled on the rope simultaneously in 38.46% of trials. The average latency until the first pull for each porcupine across all sessions is shown in Figure 5. There is no evidence of learning in that the porcupines did not become quicker at pulling the rope over time or increasingly likely to pull simultaneously. The average latency until success was achieved was 7.54 s.

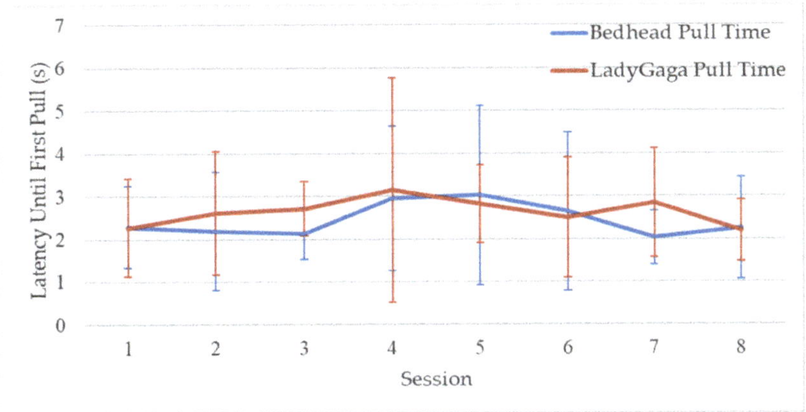

Figure 5. Average latency until first pull for each porcupine across sessions in Phase 2. Error bars depict standard deviations.

A Chi-square goodness of fit test was conducted to compare the frequency of the porcupines pulling simultaneously versus pulling separately across all sessions. All categories were expected to be equal. Overall, the porcupines were significantly more likely to pull individually ($N = 48$) than in a coordinated fashion ($N = 30$), $\chi^2(1) = 4.15$, $p = 0.04$. The latency until successful completion of the task is shown in Figure 6. Again, there is no evidence of learning across the sessions. As another indicator of what the porcupines might have understood about the need to coordinate their pulling, we examined the average latency for each porcupine to pull after the other porcupine pulled first. These data appear in Figure 7. Values of zero indicate that the porcupines pulled at the same time. There does not appear to be an increased likelihood to pull simultaneously with increased sessions. Figure 8 demonstrates the percentage of successful trials across Phase 2, which indicates that the porcupines did not become more likely to succeed over time. We also examined the possibility of soliciting behaviors between the porcupines. However, there were no recorded instances of such behavior.

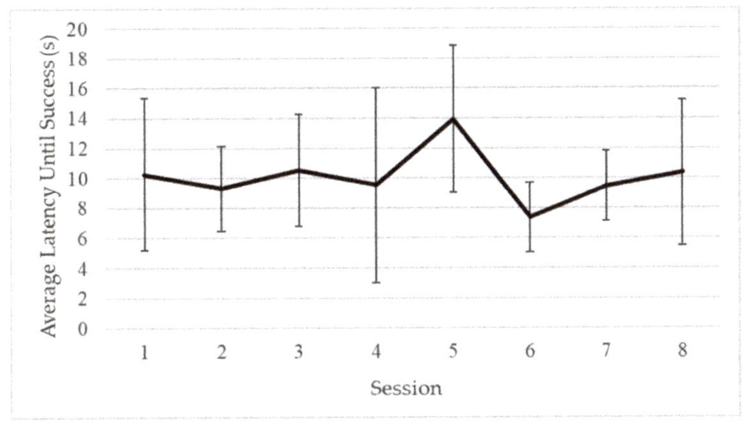

Figure 6. Time until success across sessions in Phase 2. Error bars depict standard deviations.

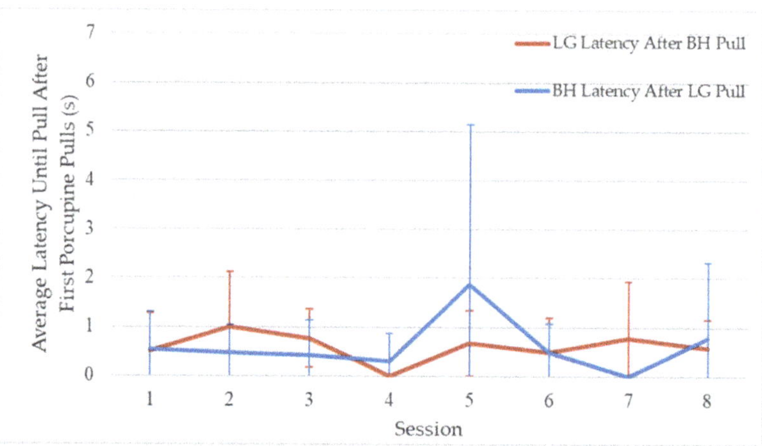

Figure 7. Latency for partner to pull after the initiating partner pulls in Phase 2. Error bars depict standard deviations. BH = Bedhead, LG = Lady Gaga.

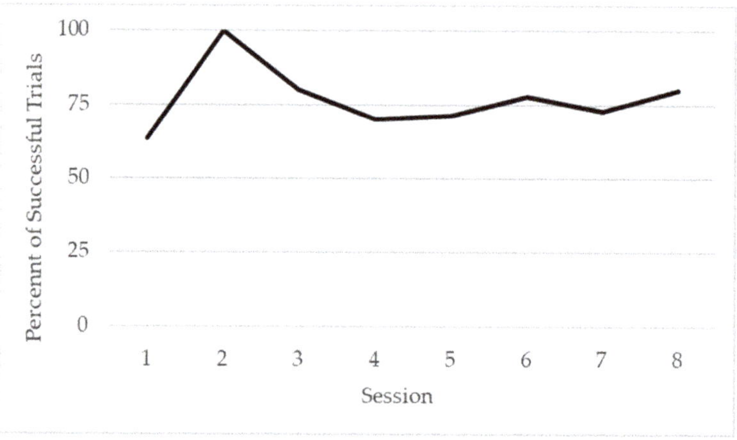

Figure 8. Percentage of Successful Trials across Phase 2.

4. Discussion

Other species widely considered good candidates for cooperative behavior, such as domestic dogs [44], chimpanzees [12] and keas [47,52] have struggled to succeed in the loose string task, although other members of the same species have succeeded [43,50,51]. The successful chimpanzees may have succeeded because the rope was longer in [50], which meant that they did not have to completely synchronize their behavior to succeed, which was also the case in the current study although it was necessary for both partners to pull in both studies. Even successful rooks [48] appeared to synchronize their pulling to external cues, rather than to their partner's behavior. Hirata and Fuwa's [12] chimpanzees also failed to look at their partner in the first 30 trials of the task or to engage in soliciting behaviors with their conspecific, although they did show soliciting behavior when paired with human partners. Although the porcupines tested here did not show soliciting behaviors to engage their partner, they did pull the ropes together often enough to succeed on 111 of the 119 trials presented in Phase 2. Whereas chimpanzees appeared to learn to be successful over time, the porcupines' behavior did not appear to change across sessions.

As Hirata and Fuwa [12] noted, cross species comparisons are fundamental to determining the evolutionary roots of cooperative behavior. Here, we used this popular cooperative task for the first time with African crested porcupines—a species overlooked in studies of cooperation to date. Our findings confirm that the cooperative capabilities of the African crested porcupines can be assessed utilizing the loose string task. The porcupines were very likely to engage with the task, as both porcupines pulled on 100% of trials in the first phase and over 95% of trials in the second phase. There were individual differences in learning, as Bedhead reached the criteria for the second phase much more quickly than did Lady Gaga. Bedhead was also quicker to interact with the rope, pull on the rope, and to receive the reward compared to Lady Gaga, on average. This trend continued in Phase 2, as Bedhead was the first to pull on the rope in 46.28% of trials. No conclusions can be drawn regarding sex differences given the very small sample size. What can be confirmed is that the porcupines did learn to pull two ends of a rope to access separate rewards in most trials in the second phase, and the average time to complete the task decreased in the second phase by over 9 s for Bedhead and over 17 s for Lady Gaga compared to the average time of completion of each porcupine in the first phase. This increased speed in completing the task might be taken as a sign that the porcupines approached the task intentionally with the understanding that they could receive the rewards if the partner was in place.

As with previous implementations of the loose string task, we were able to show that porcupines could succeed in the task, and we assessed the extent to which they adjusted their own behavior to account for the behavior of their partner. Although the porcupines achieved some level of success, indicating that this task was appropriate for testing cooperation in porcupines, it did not appear that the porcupines improved their coordination on the task over time by monitoring their partner's actions. Specifically, the latency until the first pull for each porcupine, the latency until success, and the latency to pull after the first porcupine pulled did not decrease across sessions in Phase 2. We would have expected the porcupines to solve the task more quickly as they learned to intentionally coordinate their pulling as soon as the partner was in position. The conclusion that the porcupines were not attending to their partner is supported by the lack of soliciting behaviors from either porcupine toward the other and that the porcupines were significantly more likely to pull individually rather than together. Later phases were originally planned to provide porcupines the opportunity to learn to coordinate their behavior with their partner, but testing was unfortunately terminated due to COVID-19. Other species tested with delayed partner arrival conditions (i.e., when one partner has access to the rope before the other partner has arrived), have demonstrated at least some understanding of the necessity of a partner, for example, in kea [51], elephants [42], capuchins [41], domestic dogs [43], and wolves [44]. Although we were unable to assess the porcupines' understanding of the partner's role under a delayed partner arrival condition, we did not observe the signatures of this understanding in Phase 2, consistent with other researchers' observations with chimpanzees (e.g., [12]). Thus, our data suggest that porcupines may not spontaneously take the partner's role into account in the loose string task. Future work is needed to determine whether they can do so in other contexts.

Capuchins successfully cooperated more often when in visual contact with their partner [41]. The limited eyesight of the porcupines may hinder their ability to monitor their partner while completing this task, which may explain their lack of soliciting behaviors and synchronized pulling. However, other cues (e.g., auditory cues) to the partner's presence were available and the partners were positioned at a distance where they would have been visible to each other. The porcupines did not appear to decrease their time until the first pull across Phase 1 or increasingly synchronize their pulling actions across Phase 2, unlike wolves that improved their performance across sessions within every condition tested [44]. Thus, the porcupines may not integrate the feedback necessary for improved coordination across multiple trials. This may be due to the fact that porcupines do not forage cooperatively as hunting species may do. These difficulties in coordination are interesting given that previous research has pointed at the importance of strong social bonds for cooperation in species such as wolves and chimpanzees [13,44], and the porcupines were a mated pair. Male-female dyads were found to perform better than same sex dyads in ravens, *Corvus corax* (i.e., a pair bonding species; [53]). It would be interesting to test different types of dyads in a larger sample of porcupines in future work.

Some challenges with this style of apparatus became apparent for this species. First, the poor eyesight of the African crested porcupine proved to be problematic in ensuring they were able to take the reward from the apparatus in a timely manner. The male was generally able to find the bait relatively quickly, possibly due to his quick responding, but this proved more difficult for the female. Thus, her understanding of the association between pulling and access to the food via the movement of the apparatus may have been hindered, as she may have understood the operant contingencies of the task (she received a reward after pulling the rope), but not the causal contingencies (pulling caused the reward to move closer to her) [3,54]. Allowing her extra time to find the reward would have given the male enough time to then take the reward meant for the female, which might have led to her not interacting with the apparatus in the future. Furthermore, these animals frequently participate in training procedures at the Creature Conservancy. Thus, it may be problematic for her future training if she learned that she does not receive a reward for her efforts, or if the timing of the reward is so delayed that she does not associate it with the task at hand. Thus, the reward was handed to her soon after she pulled, even though doing so might interrupt her causal understanding of the task. This is similar to the procedure used with dolphins (*Tursiops truncatus*) that were tossed a fish when they simultaneously pushed two buttons that were not connected to the delivery of food [22]. Thus, researchers have considered coordination of causally arbitrary actions as evidence for cooperation in previous research, mitigating against the concern that our methods could not evoke cooperation. Handing the reward to the porcupines on only successful trials would still allow the porcupines to learn the necessity of pulling synchronously for reward, albeit via association rather than by functional understanding (see also [43,55]).

However, there are other limitations of the current study. With only one male and one female porcupine, it is difficult to generalize to other captive members of the species, let alone their wild counterparts who have additional agency in mate selection. In particular, these porcupines were housed in a notably different environment from African crested porcupines in the wild, and this pair did not select each other as mates even though they were a mated pair. The Creature Conservancy also involves these porcupines in husbandry training, like target training; thus, they are likely more experienced with training procedures than other members of their species. However, any information that can enhance our limited understanding of porcupine cognition is of value, given its scarcity.

This research demonstrates that African crested porcupines are a promising species for the study of cooperative behaviors as they are capable of interacting with an apparatus that requires pulling as well as being capable of pulling together when given simultaneous access. They were also tolerant enough to receive rewards simultaneously and this did not inhibit participation in the task. However, given that the male was able to pull the apparatus mostly on his own in the first four sessions of Phase 2 and sometimes took the female's reward, it is possible that the female's performance was impacted by motivation. That is, a lack of motivation from sometimes not being rewarded for her effort, albeit this did not occur often and was remedied by the apparatus change. It would have been ideal to test her with other less forceful and dominant partners. Social tolerance has been cited as a necessary precursor to developing cooperative behaviors [43,56] and important for success on cooperative tasks in multiple species [17,49,50,53,57]. Future research is necessary to test porcupines' ability to learn about the role of their partner. However, given that this was the first experimental test of cooperation in this taxon, the results contribute to the ongoing understanding of the breadth of species exhibiting cooperation behaviors. This new information improves our understanding of their cognition, as there is currently no other research on their ability to problem solve.

Future Directions

To understand whether African crested porcupines can understand the role of their partner, future studies could implement a delayed partner arrival phase (e.g., [17,42,44,46,51]), which would have made the apparatus immediately available to one partner, while the other partner was just released from a distant location, requiring the individual closest to the apparatus to wait until the partner had reached the apparatus. This phase would test whether the individual understands that the partner is necessary to pull the apparatus forward and receive the reward and can inhibit their own pulling behavior in their absence. Due to the extended time until the partner arrives, it is possible that the animals may engage with the apparatus due to frustration rather than a misunderstanding of the necessity of a partner. The last phase of our experiment would have attempted to address this problem. In the planned covered rope phase, a randomly selected piece of the rope, out of the two sides available to the porcupines, would have been covered with a moveable blocker. This blocker would need to be removed before the porcupine closest to this rope end could access the rope and pull. This would allow the subject to facilitate the partner's response through allowing access to the rope.

Because porcupines burrow to den, it is possible that a method that would allow the porcupines to dig may be a more intuitive paradigm for this species. In the first phase especially, the porcupines were likely to attempt to scratch at the rope rather than pull utilizing their mouth. In future studies, an apparatus that would provide a benefit only if both porcupines dug together may be a better test of their cooperative abilities. We hope that the current results will encourage other researchers to probe the origins of cooperative behavior in other understudied species.

Supplementary Materials: The following supporting information can be downloaded at: https://www.mdpi.com/article/10.3390/jzbg3030034/s1. Video S1: An example of a successful test trial. Bedhead and Lady Gaga pull their ropes simultaneously and each receive rewards.

Author Contributions: All four authors contributed substantially to this article. Conceptualization, J.T. and J.V.; methodology, J.T. and J.V.; apparatus construction, J.T. and Z.K.B.; formal analysis, J.T. and J.V.; data curation, J.T., J.L.V.; writing—original draft preparation, J.T.; writing—review and editing, J.T., J.V., J.L.V. and Z.K.B.; supervision, J.V. All authors have read and agreed to the published version of the manuscript.

Funding: This research was partially funded by Oakland University's Provost Graduate Student Research Award.

Institutional Review Board Statement: The animal study protocol was approved by the Institutional Animal Care and Use Committee of Oakland University (protocol code 2021–1126 and 25 June 2021).

Data Availability Statement: As the data reported here are preliminary, we have not made these data publicly available.

Acknowledgments: This work would not have been possible if not for the help of The Creature Conservancy, their staff, and their animals. We would also like to thank Bridget Benton and Eric Sienkowski who contributed to data coding. Thank you to Todd Shackelford for his insightful comments on this manuscript.

Conflicts of Interest: The authors declare no conflict of interest.

References

1. Creel, S.; Creel, N.M. Communal hunting and pack size in African wild dogs, *Lycaon pictus*. *Anim. Behav.* **1995**, *50*, 1325–1339. [CrossRef]
2. Bossema, I.; Benus, R.F. Territorial defence and intra-pair cooperation in the carrion crow (*Corvus corone*). *Behav. Ecol. Sociobiol.* **1985**, *16*, 99–104. [CrossRef]
3. Boesch, C.; Boesch, H. Hunting behavior of wild chimpanzees in the Tai National Park. *Am. J. Phys. Anthropol.* **1989**, *78*, 547–573. [CrossRef] [PubMed]
4. Schoech, S.J. Physiology of helping in Florida scrub-Jays: When these birds are young, they delay reproduction and help others raise their offspring. The hormone prolactin may influence that cooperation. *Am. Sci.* **1998**, *86*, 70–77. [CrossRef]

5. Stander, P.E. Cooperative hunting in lions: The role of the individual. *Behav. Ecol. Sociobiol.* **1992**, *29*, 445–454. [CrossRef]
6. Bronstein, J.L. The scope for exploitation within mutualistic interactions. In *Genetic and Cultural Evolution of Cooperation*; Hammerstein, P., Ed.; MIT Press: London, UK, 2003; pp. 185–202.
7. Noë, R. Cooperation experiments: Coordination through communication versus acting apart together. *Anim. Behav.* **2006**, *71*, 1–18. [CrossRef]
8. Hamilton, W.D. The genetical evolution of social behaviour II. *J. Theor. Biol.* **1964**, *7*, 17–52. [CrossRef]
9. Trivers, R.L. The evolution of reciprocal altruism. *Q. Rev. Biol.* **1971**, *46*, 35–57. [CrossRef]
10. Brosnan, S.F.; Silk, J.B.; Henrich, J.; Mareno, M.C.; Lambeth, S.P.; Schapiro, S.J. Chimpanzees (*Pan troglodytes*) do not develop contingent reciprocity in an experimental task. *Anim. Cogn.* **2009**, *12*, 587–597. [CrossRef]
11. Hamann, K.; Warneken, F.; Greenberg, J.R.; Tomasello, M. Collaboration encourages equal sharing in children but not in chimpanzees. *Nature* **2011**, *476*, 328. [CrossRef]
12. Hirata, S.; Fuwa, K. Chimpanzees (*Pan troglodytes*) learn to act with other individuals in a cooperative task. *Primates* **2007**, *48*, 13–21. [CrossRef] [PubMed]
13. Melis, A.P.; Hare, B.; Tomasello, M. Chimpanzees recruit the best collaborators. *Science* **2006**, *311*, 1297–1300. [CrossRef] [PubMed]
14. Rekers, Y.; Haun, D.B.; Tomasello, M. Children, but not chimpanzees, prefer to collaborate. *Curr. Biol.* **2011**, *21*, 1756–1758. [CrossRef]
15. Suchak, M.; Eppley, T.M.; Campbell, M.W.; de Waal, F.B. Ape duos and trios: Spontaneous cooperation with free partner choice in chimpanzees. *PeerJ* **2014**, *2*, e417. [CrossRef] [PubMed]
16. Yamamoto, S.; Tanaka, M. Do chimpanzees (*Pan troglodytes*) spontaneously take turns in a reciprocal cooperation task? *J. Comp. Psychol.* **2009**, *123*, 242–249. [CrossRef]
17. Seed, A.M.; Clayton, N.S.; Emery, N.J. Cooperative problem solving in rooks (*Corvus frugilegus*). *Proc. R. Soc. Lond. B Biol. Sci.* **2008**, *275*, 1421–1429. [CrossRef]
18. Vonk, J.; Vincent, J.; O'Connor, V. It's hard to be social alone: Cognitive complexity as transfer within and across domains. *Comp. Cogn. Behav. Rev.* **2021**, *16*, 33–67. [CrossRef]
19. Gräfenhain, M.; Behne, T.; Carpenter, M.; Tomasello, M. Young children's understanding of joint commitments. *Dev. Psychol.* **2009**, *45*, 1430. [CrossRef]
20. Baciadonna, L.; Cornero, F.M.; Emery, N.J.; Clayton, N.S. Convergent evolution of complex cognition: Insights from the field of avian cognition into the study of self-awareness. *Learn. Behav.* **2021**, *49*, 9–22. [CrossRef]
21. Clayton, N.S. Ways of thinking: From crows to children and back again. *Q. J. Exp. Psychol.* **2015**, *68*, 209–241. [CrossRef]
22. Jaakkola, K.; Guarino, E.; Donegan, K.; King, S.L. Bottlenose dolphins can understand their partner's role in a cooperative task. *Proc. R. Soc. B Biol. Sci.* **2018**, *285*, 20180948. [CrossRef] [PubMed]
23. Lea, S.E.; Osthaus, B. In what sense are dogs special? Canine cognition in comparative context. *Learn. Behav.* **2018**, *46*, 335–363. [CrossRef] [PubMed]
24. Wittmann, M.K.; Lockwood, P.L.; Rushworth, M.F. Neural mechanisms of social cognition in primates. *Annu. Rev. Neurosci.* **2018**, *41*, 99–118. [CrossRef] [PubMed]
25. Burkart, J.M.; Hrdy, S.B.; Van Schaik, C.P. Cooperative breeding and human cognitive evolution. *Evol. Anthropol. Issues News Rev.* **2009**, *18*, 175–186. [CrossRef]
26. Burkart, J.M.; Van Schaik, C.P. Cognitive consequences of cooperative breeding in primates? *Anim. Cogn.* **2010**, *13*, 1–19. [CrossRef]
27. Duque, J.F.; Stevens, J.R. Voluntary food sharing in pinyon jays: The role of reciprocity and dominance. *Anim. Behav.* **2016**, *122*, 135–144. [CrossRef]
28. Horn, L.; Scheer, C.; Bugnyar, T.; Massen, J.J. Proactive prosociality in a cooperatively breeding corvid, the azure-winged magpie (*Cyanopica cyana*). *Biol. Lett.* **2016**, *12*, 20160649. [CrossRef]
29. Leete, J.A.; Vonk, J. In mixed company: Two macaws are self-regarding in a symbolic prosocial choice task. *Behav. Ecol. Sociobiol.* **2022**, *76*, 1–14. [CrossRef]
30. Clarke, F.M.; Faulkes, C.G. Dominance and queen succession in captive colonies of the eusocial naked mole–rat, *Heterocephalus glaber*. *Proc. R. Soc. Lond. Ser. B Biol. Sci.* **1997**, *264*, 993–1000. [CrossRef]
31. Lacey, E.A.; Ebensperger, L.A.; Wolff, J.O.; Sherman, P.W. Social structure in octodontid and ctenomyid rodents. In *Rodent Societies: An Ecological and Evolutionary Perspective*; Wolff, J.O., Sherman, P.W., Eds.; University of Chicago Press: Chicago, IL, USA, 2007; pp. 403–415.
32. Corsini, M.T.; Lovari, S.; Sonnino, S. Temporal activity patterns of crested porcupines *Hystrix cristata*. *J. Zool.* **1995**, *236*, 43–54. [CrossRef]
33. Monetti, L.; Massolo, A.; Sforzi, A.; Lovari, S. Site selection and fidelity by crested porcupines for denning. *Ethol. Ecol. Evol.* **2005**, *17*, 149–159. [CrossRef]
34. Mori, E.; Menchetti, M.; Lucherini, M.; Sforzi, A.; Lovari, S. Timing of reproduction and paternal cares in the crested porcupine. *Mamm. Biol.* **2016**, *81*, 345–349. [CrossRef]
35. Viviano, A.; Amori, G.; Luiselli, L.; Oebel, H.; Bahleman, F.; Mori, E. Blessing the rains down in Africa: Spatiotemporal behaviour of the crested porcupine *Hystrix cristata* (Mammalia: Rodentia) in the rainy and dry seasons, in the African savannah. *Trop. Zool.* **2020**, *33*. [CrossRef]

36. Coppola, F.; Guerrieri, D.; Simoncini, A.; Varuzza, P.; Vecchio, G.; Felicioli, A. Evidence of scavenging behaviour in crested porcupine. *Sci. Rep.* **2020**, *10*, 12297. [CrossRef] [PubMed]
37. Fernandez, E.J.; Dorey, N.R. An examination of shaping with an African Crested Porcupine (*Hystrix cristata*). *J. Appl. Anim. Welf. Sci.* **2021**, *24*, 372–378. [CrossRef]
38. Woods, C.A. Erethizon dorsatum. *Mammal Species* **1973**, *29*, 1–6. [CrossRef]
39. Crawford, M.P. The cooperative solving of problems by young chimpanzees. *Comp. Psychol. Monogr.* **1937**, *14*, 1–88.
40. Chalmeau, R. Do chimpanzees cooperate in a learning task? *Primates* **1994**, *35*, 385–392. [CrossRef]
41. Mendres, K.A.; de Waal, F.B. Capuchins do cooperate: The advantage of an intuitive task. *Anim. Behav.* **2000**, *60*, 523–529. [CrossRef]
42. Plotnik, J.M.; Lair, R.; Suphachoksahakun, W.; de Waal, F.B. Elephants know when they need a helping trunk in a cooperative task. *Proc. Natl. Acad. Sci. USA* **2011**, *108*, 5116–5121. [CrossRef]
43. Ostojić, L.; Clayton, N.S. Behavioural coordination of dogs in a cooperative problem-solving task with a conspecific and a human partner. *Anim. Cogn.* **2014**, *17*, 445–459. [CrossRef] [PubMed]
44. Marshall-Pescini, S.; Schwarz, J.F.; Kostelnik, I.; Virányi, Z.; Range, F. Importance of a species' socioecology: Wolves outperform dogs in a conspecific cooperation task. *Proc. Natl. Acad. Sci. USA* **2017**, *114*, 11793–11798. [CrossRef] [PubMed]
45. Möslinger, H.; Kotrschal, K.; Huber, L.; Range, F.; Virányi, Z. Cooperative string-pulling in wolves. *J. Vet. Behav. Clin. Appl. Res.* **2009**, *4*, 99. [CrossRef]
46. Péron, F.; Rat-Fischer, L.; Lalot, M.; Nagle, L.; Bovet, D. Cooperative problem solving in African grey parrots (*Psittacus erithacus*). *Anim. Cogn.* **2011**, *14*, 545–553. [CrossRef]
47. Schwing, R.; Reuillon, L.; Conrad, M.; Noë, R.; Huber, L. Paying attention pays off: Kea improve in loose-string cooperation by attending to partner. *Ethology* **2020**, *126*, 246–256. [CrossRef]
48. Scheid, C.; Noë, R. The performance of rooks in a cooperative task depends on their temperament. *Anim. Cogn.* **2010**, *13*, 545–553. [CrossRef]
49. Hare, B.; Melis, A.P.; Woods, V.; Hastings, S.; Wrangham, R. Tolerance allows bonobos to outperform chimpanzees on a cooperative task. *Curr. Biol.* **2007**, *17*, 619–623. [CrossRef]
50. Melis, A.P.; Hare, B.; Tomasello, M. Engineering cooperation in chimpanzees. *Anim. Behav.* **2006**, *72*, 275–286. [CrossRef]
51. Heaney, M.; Gray, R.D.; Taylor, A.H. Keas perform similarly to chimpanzees and elephants when solving collaborative tasks. *PLoS ONE* **2017**, *12*, e0169799. [CrossRef]
52. Schwing, R.; Jocteur, E.; Wein, A.; Noë, R.; Massen, J.J. Kea cooperate better with sharing affiliates. *Anim. Cogn.* **2016**, *19*, 1093–1102. [CrossRef]
53. Massen, J.J.; Ritter, C.; Bugnyar, T. Tolerance and reward equity predict cooperation in ravens (*Corvus corax*). *Sci. Rep.* **2015**, *5*, 15021. [CrossRef] [PubMed]
54. Jacobs, I.F.; Osvath, M. The string-pulling paradigm in comparative psychology. *J. Comp. Psychol.* **2015**, *129*, 89. [CrossRef] [PubMed]
55. Seed, A.M.; Jensen, K. Large-scale cooperation. *Nature.* **2011**, *472*, 424–425. [CrossRef] [PubMed]
56. Stringham, S.F. Salmon fishing by bears and the dawn of cooperative predation. *J. Comp. Psychol.* **2012**, *126*, 329–338. [CrossRef] [PubMed]
57. Drea, C.M.; Carter, A.N. Cooperative problem solving in a social carnivore. *Anim. Behav.* **2009**, *78*, 967–977. [CrossRef]

Journal of
Zoological and Botanical Gardens

MDPI

Communication

Digit Entrapment Due to Plastic Waste in a Verreaux's Eagle Owl (*Bubo lacteus*)

Lindsay Thomas *, **Phillipa Dobbs** and **Samantha Ashfield**

Twycross Zoo, A444 Burton Road, Junction 11, Atherstone CV9 3PX, UK
* Correspondence: lindsay.thomas@twycrosszoo.org

Abstract: Plastic waste has become a hot topic in sustainability and conservation, helped in part by popular documentaries which have highlighted the issue to the general public. Much of the current literature focuses on the effect of microplastics in the marine environment, with very little information on macroplastic interactions or the terrestrial environment. In this report, the management of digit constriction due to macroplastic debris in a Verreaux's eagle owl (*Bubo lacteus*) is presented, and the role of zoos in decreasing littering behaviour both within the collection and in the wider global context is discussed.

Keywords: Verreaux's eagle owl; *Bubo lacteus*; macroplastic; litter; digit constriction

Citation: Thomas, L.; Dobbs, P.; Ashfield, S. Digit Entrapment Due to Plastic Waste in a Verreaux's Eagle Owl (*Bubo lacteus*). *J. Zool. Bot. Gard.* **2022**, *3*, 442–447. https://doi.org/10.3390/jzbg3030033

Academic Editors: Kris Descovich, Caralyn Kemp and Jessica Rendle

Received: 30 May 2022
Accepted: 12 September 2022
Published: 15 September 2022

1. Introduction

The issue of plastic waste and its impact on wildlife species has become one of the most pressing environmental issues of recent times, in part due to popular documentaries such as *Blue Planet II* which have highlighted the issue to the general public [1,2]. The primary focus for both the public and policymakers has been marine microplastics, with relatively little focus on terrestrial and macroplastic pollution [3,4].

The effects of plastic pollution on wildlife species have been widely reported in the literature, from invertebrates [5] to megavertebrates [6], on both land [7] and in the oceans [8]. While microplastics can have a combination of chemical and physical effects on organisms which interact with them, macroplastics have primarily physical effects [7]. Birds appear particularly at risk of negative interactions with macroplastic debris, with ingestion and entrapment being the two most commonly reported negative interactions [4,9]. Accumulation of litter, including macroplastic debris, in nests is also very common and may have negative, positive or net neutral implications [4].

Although anecdotal interactions with litter are not uncommon in zoo-housed species, there is little published information available [10]. Whilst wildlife species residing in zoological collections are protected from most anthropogenic threats, for example, habitat degradation and climate change, they are also brought into close contact with humans and the risks associated therein, including exposure to anthropogenic litter. Furthermore, these species frequently demonstrate cryptic behaviour which can make it difficult to assess the need for intervention where an interaction has occurred.

This report describes the management of a case of digit entrapment due to macroplastic debris in a zoo-housed Verreaux's eagle owl (*Bubo lacteus*) and discusses the wider issues of littering and how zoos can influence this behaviour amongst visitors.

2. Case Description

A male 14-months-old Verreaux's eagle owl was presented by keeping staff after blood was noticed around one of his digits. On examination, a white plastic cable was found embedded in the tissue around digit three on the left foot. Oral analgesia (meloxicam 0.2 mg/kg; Metacam, Boehringer Ingelheim, Berkshire, UK) and an antibiotic (enrofloxacin

10 mg/kg; Baytril, Bayer, Reading, UK) was prescribed by the locum vet who was on call and anaesthesia for further investigation and cable removal was scheduled for the following day.

Anaesthesia was induced with 8% sevoflurane in oxygen via a facemask. The animal was intubated with a size 3.5 mm endotracheal tube and anaesthesia maintained on 4–6% sevoflurane via a T-piece circuit. On examination, the animal was found to be in good body condition with no abnormalities other than the cable tie encircling digit three of the left foot. The tissue underlying the cable tie was necrotic and the surrounding tissue was inflamed and oedematous. Butorphanol was given by intramuscular injection at a dose of 1 mg/kg (Tobugesic, Zoetis UK, Surrey, UK) and radiographs were taken prior to removal of the cable tie (Figure 1).

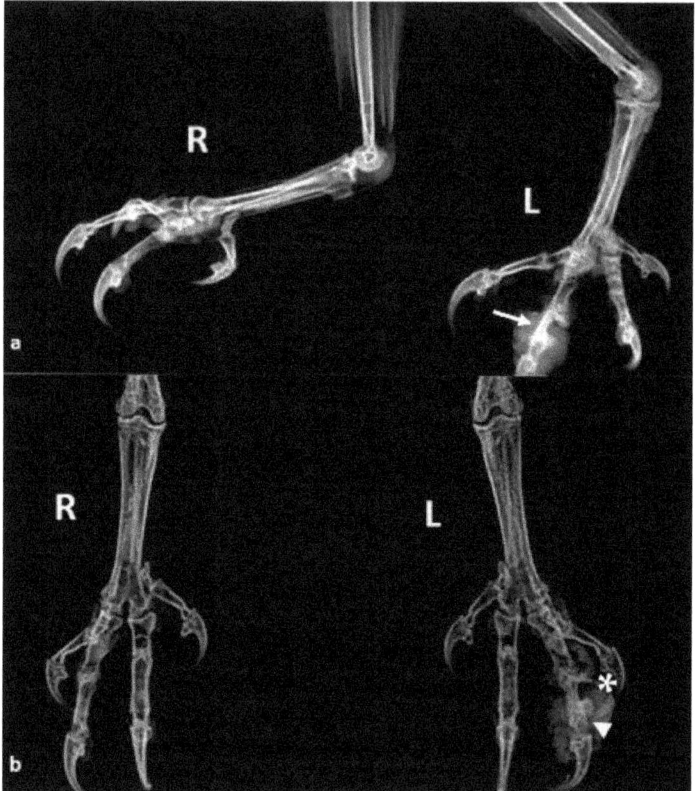

Figure 1. Verreaux's eagle owl feet pretreatment. Lateral (**a**) and dorsoplantar (**b**) radiographs taken prior to removal of the cable tie showing changes to phalanx 2 of digit three in the left foot. There is narrowing of the bone (white arrow) compared with the bone of the contralateral foot and osteolysis is present (white arrowhead). Osteophyte formation can be appreciated at both ends of the bone, particularly the proximal end (white asterisk).

Lateral and dorsoplantar views of both feet showed no abnormalities of the right foot; however, digit three of the left foot showed narrowing of the central part of phalanx 2 with some evidence of osteolysis and some osteophyte formation at the ends of phalanx 2, particularly the proximal end.

A low-speed dental burr (Cocoon spray dental unit, Eickmeyer, Surrey, UK) was used to carefully remove the embedded cable tie while flushing was performed with room temperature Hartmann's solution (Aqupharm No.11, Animalcare UK, York, UK) to avoid overheating and thermal necrosis of the surrounding tissue. Once the cable tie was removed, the wound was flushed with 100 mL of additional fluid and laser treatment was performed with a class 3B laser (Xp mobile, Omega laser systems, Essex, UK). Very little tissue remained overlying P2 once the cable tie had been removed, so post-removal radiographs were performed to ensure the bone had remained intact (Figure 2).

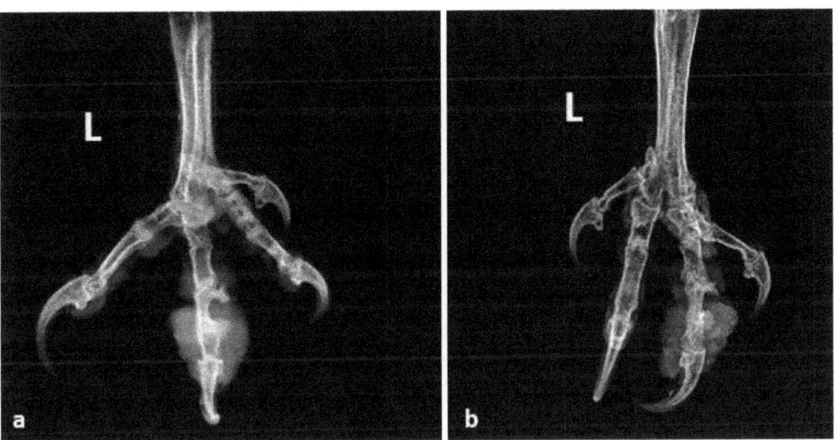

Figure 2. Verreaux's eagle owl feet immediately following removal of the cable tie. Lateral (**a**) and DP (**b**) radiographs taken following removal of the cable tie to ensure no fractures had occurred during the removal process.

Postoperatively, the animal was given oral meloxicam 0.5 mg/kg (Loxicom, Norbrook, Corby, UK), tramadol 10 mg/kg (Tramadol oral drops, MercuryPharma, Croydon, UK), marbofloxacin 12 mg/kg (Marbocare, Animalcare UK, York, UK), and prophylactic itraconazole 10 mg/kg (Itraconazole capsules, Glenmark Pharmaceuticals Europe Ltd., Watford, UK) while receiving antibiotic treatment. Medication was administered in food to minimise stress from handling; however, due to the feeding ecology of this species, this did mean that the medication could be administered only once daily. For this reason, the antibiotic was switched from enrofloxacin, which should be administered twice daily in most avian species, to marbofloxacin, to reduce the risk of antimicrobial resistance with inappropriate dosing intervals. He was kept separate from his mate to allow accurate monitoring of medication intake.

The animal was monitored closely in the postoperative period to ensure the digit swelling was reducing, primarily through the use of binoculars to allow visual assessment of the wound without stressing him through regular capture events. Appetite was also monitored closely, as stress from both capture events and foot injuries have the potential to decrease food intake in owls (J. Mihr, personal communication). Food pieces were counted in and out, with appetite remaining stable throughout the treatment period. Every fourteen days, he was captured for physical examination of the affected digit. Six weeks postoperatively, anaesthesia was repeated using the same protocol and repeat radiographs were taken of both feet (Figure 3), as well as standard right lateral and ventrodorsal whole-body views.

Radiographs revealed no abnormalities of the right foot. On the left foot, phalanx 2 of digit three had increased in width and now exceeded that seen on the same bone in the right foot. The osteolysis which had been apparent six weeks previously appeared to have resolved, and while osteophytes were still present at both ends of phalanx 2 and, in particular, the proximal end, they appeared less obvious than previous. The tissue deficit was largely resolved by this point, and clinically there was no evidence of any further infection, necrotic tissue or pain.

The osteolysis of phalanx 2 in digit three on the left foot had resolved. Osteophytes were still appreciable, particularly at the proximal end of the bone, but the width of the bone had increased to exceed that of the contralateral foot.

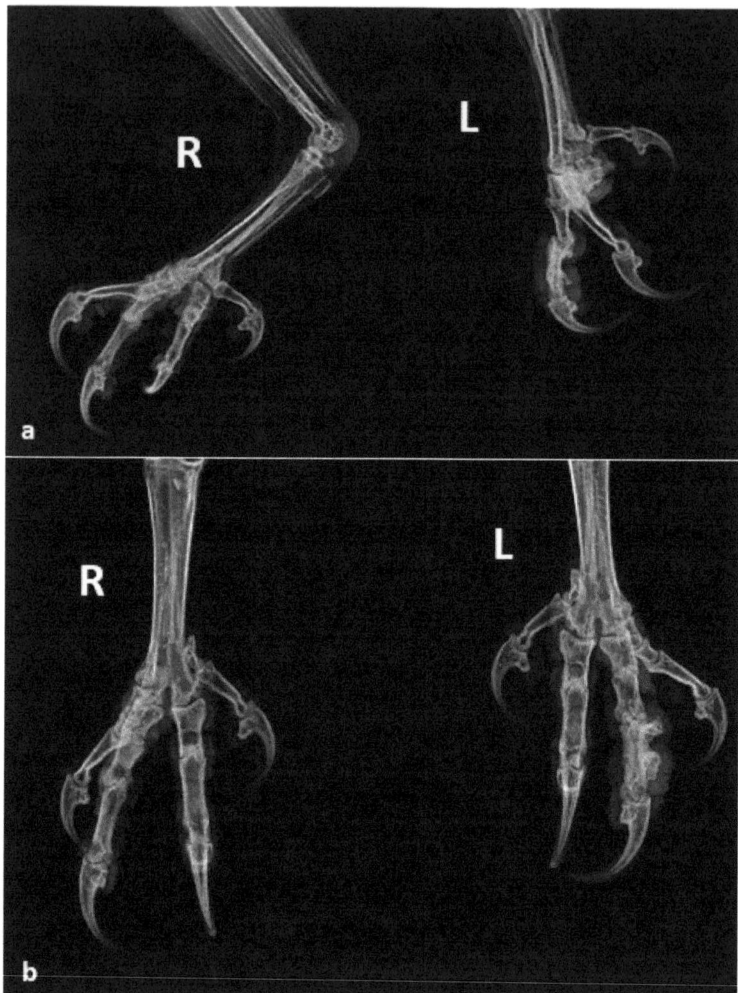

Figure 3. Verreaux's eagle owl feet six weeks post-treatment. Lateral (**a**) and DP (**b**) follow-up radiographs taken six weeks post-removal of the cable tie.

3. Discussion

Constriction injuries in captive birds are most commonly seen as a result of improperly placed identification rings, which can cause significant trauma, and, in some cases, even result in the loss of affected limbs [11]. Digit constriction in captive avian species is most commonly reported as a result of fibres wrapping around the digit, low humidity leading to annular ring formation and circumferential wounds leading to scab/fibrotic tissue formation [12]. Entrapment and constriction injuries have previously been reported in free-living wildlife species, for example, plastic strings incorporated into nests were reported to cause entrapment and degeneration of leg bones in juvenile white storks presented to two wildlife rehabilitation centres in Poland [13]. However, to the author's knowledge, this is the first published report of a constriction injury in a zoo-housed wildlife species due to anthrogenic litter.

The source of the cable tie was unknown in this case, as it did not match the type used by collection staff and no cable ties were in use within the enclosure that the animal inhabited. Possible sources include accidental inclusion in wood chip bedding material used in the enclosure or littering by a member of the public. Other cases of interactions with athropogenic litter have previously been observed in this collection. A metal coin was found in the ventriculus on postmortem examination of a Humboldt penguin (*Spheniscus humboldti*) but was not a contributing factor to mortality. The source of the coin was thought to be visitors throwing money into the penguin pool, as this had been observed on multiple occasions despite signage discouraging this behaviour. In another case, a De Brazza monkey (*Cercopithecus neglectus*) was found to have a plastic bottle cap impacted into a cheek pouch. The source of the bottle cap was, again, thought to be littering by members of the public.

Littering is a major global issue with widespread consequences for both human and non-human health and welfare [14]. There are many published studies available investigating a variety of factors which influence littering behaviour, including situational factors, such as the presence of litter bins or the amount of pre-existing litter in an area, and societal/psychological factors, such as the behaviour of other individuals or the presence of signage discouraging littering behaviour [14].

Signs have been shown to be effective deterrents for antisocial behaviour such as littering [14]; however, littering continues to be a problem within the collection despite the presence of signage. Research has shown that including a brief explanation as to why a behaviour is prohibited and the addition of 'watching eyes' images can increase the efficacy of signage but will not completely eliminate the antisocial behaviour [15] and may increase the incidence of other 'displacement' behaviours which may also be antisocial [16]. The presence of litter bins has been shown to decrease littering; however, the design and positioning of the litter bin as well as associated signage may have a significant impact on their use [17,18]. Review of current signage and litter bins may aid in decreasing incidences of littering within the collection.

Anthropogenic litter can impact wildlife species via various routes. Senko et al. (2020) reviewed studies published between 1969 and 2020 reporting the effects of plastic pollution of marine megafauna and highlighted nine 'pathways' for interactions between marine megafauna and plastic pollution, including entanglement, ingestion and increased exposure to contaminants [19]. Bletter and Mitchell 2021 documented 90 individual cases of encounters between macroplastic waste and freshwater and terrestrial species and noted that plastic entanglement was the second most common encounter, the most common being the use of plastic for nesting material [4]. Entanglements may have negative effects by reducing mobility and, therefore, the ability to ingest food or escape predation, or, as in this case, leading to physical injury via constriction [4].

In addition to decreasing on-site littering and, therefore, the potential negative effects on collection animals, zoos can also play an important role in educating the public on the wider issues of littering and plastic pollution. Mellish et. al. (2019) reported a positive change in attitudes towards balloon litter and the use of balloons at outdoor events after visitors viewed an exhibit, with or without an accompanying presentation, on this issue [20]. However, few other studies have specifically looked at the effect and outcome of targeted pollution education programmes, despite studies suggesting that the majority of zoo visitors are willing to engage with learning during their visit [21].

4. Conclusions

Zoos have an important role to play in highlighting the dangers of anthropogenic litter and the effect this can have on wildlife species, both in captivity and in the wild. The cryptic behaviour displayed by many wildlife species housed in zoos can make management of cases such as the one discussed here difficult, and so risk reduction by both decreasing littering on an individual collection basis and by educating the public on the wider risks associated with litter and plastic pollution is of paramount importance.

Funding: This research received no external funding.

Institutional Review Board Statement: Not applicable.

Data Availability Statement: Not applicable.

Acknowledgments: The authors would like to give thanks to the bird team at Twycross zoo, without whose care this case would have had a very different outcome.

Conflicts of Interest: The authors declare no conflict of interest.

References

1. Males, J.; Van Aelst, P. Did the Blue Planet set the Agenda for Plastic Pollution? An Explorative Study on the Influence of a Documentary on the Public, Media and Political Agendas. *Environ. Commun.* **2020**, *15*, 40–54. [CrossRef]
2. Bailey, I. Media coverage, attention cycles and the governance of plastics pollution. In *Environmental Policy and Governance*; Wiley: Hoboken, NJ, USA, 2022; pp. 1–13. [CrossRef]
3. Malizia, A.; Monmany-Garzia, A.C. Terrestrial ecologists should stop ignoring plastic pollution in the Anthropocene time. *Sci. Total Environ.* **2019**, *668*, 1025–1029. [CrossRef] [PubMed]
4. Blettler, M.C.; Mitchell, C. Dangerous traps: Macroplastic encounters affecting freshwater and terrestrial wildlife. *Sci. Total Environ.* **2021**, *798*, 149317. [CrossRef] [PubMed]
5. Sussarellu, R.; Suquet, M.; Thomas, Y.; Lambert, C.; Fabioux, C.; Pernet, M.E.; Le Goïc, N.; Quillien, V.; Mingant, C.; Epelboin, Y.; et al. Oyster reproduction is affected by exposure to polystyrene microplastics. *Proc. Natl. Acad. Sci. USA* **2016**, *113*, 2430–2435. [CrossRef] [PubMed]
6. Fossi, M.C.; Panti, C.; Baini, M.; Lavers, J.L. A review of plastic-associated pressures: Cetaceans of the Mediterranean Sea and eastern Australian shearwaters as case studies. *Front. Mar. Sci.* **2018**, *5*, 173. [CrossRef]
7. De Souza Machado, A.A.; Kloas, W.; Zarfl, C.; Hempel, S.; Rillig, M.C. Microplastics as an emerging threat to terrestrial ecosystems. *Glob. Chang. Biol.* **2018**, *24*, 1405–1416. [CrossRef] [PubMed]
8. Li, W.C.; Tse, H.F.; Fok, L. Plastic waste in the marine environment: A review of sources, occurrence and effects. *Sci. Total Environ.* **2016**, *566*, 333–349. [CrossRef] [PubMed]
9. Battisti, C.; Staffieri, E.; Poeta, G.; Sorace, A.; Luiselli, L.; Amori, G. Interactions between anthropogenic litter and birds: A global review with a 'black-list'of species. *Mar. Pollut. Bull.* **2019**, *138*, 93–114. [CrossRef] [PubMed]
10. Malepa, P.L. Visitor Perceptions and Awareness of Litter at the Johannesburg ZOO. Doctoral Dissertation, University of South Africa, Pretoria, South Africa, 2014.
11. Montesinos, A. Basic techniques. In *BSAVA Manual of Avian Practice*; BSAVA Library: Gloucester, UK, 2018; pp. 215–231. [CrossRef]
12. Guzman, D.S. Avian soft tissue surgery. *Vet. Clin. Exot. Anim. Pract.* **2016**, *19*, 133–157. [CrossRef] [PubMed]
13. Kwieciński, Z.; Kwiecińska, H.; Botko, P.; Wysocki, A.; Jerzak, L.; Tryjanowski, P. Plastic strings as the cause of leg bone degeneration in the White Stork (Ciconia ciconia). In *White Stork Study in Poland: Biology, Ecology and Conservation*; Bogucki Wydawnictwo Naukowe: Poznań, Poland, 2006.
14. Chaudhary, A.H.; Polonsky, M.J.; McClaren, N. Littering behaviour: A systematic review. *Int. J. Consum. Stud.* **2021**, *45*, 478–510. [CrossRef]
15. Ernest-Jones, M.; Nettle, D.; Bateson, M. Effects of eye images on everyday cooperative behavior: A field experiment. *Evol. Hum. Behav.* **2011**, *32*, 172–178. [CrossRef]
16. Parker, E.N.; Bramley, L.; Scott, L.; Marshall, A.R.; Slocombe, K.E. An exploration into the efficacy of public warning signs: A zoo case study. *PLoS ONE* **2018**, *13*, 0207246. [CrossRef]
17. Portman, M.E.; Behar, D. Influencing beach littering behaviors through infrastructure design: An in situ experimentation case study. *Mar. Pollut. Bull.* **2020**, *156*, 111277. [CrossRef]
18. Johannes, H.P.; Maulana, R.; Herdiansyah, H. Prevention of Littering through Improved Visual Design. *Environ. Res. Eng. Manag.* **2021**, *77*, 86–98. [CrossRef]
19. Senko, J.F.; Nelms, S.E.; Reavis, J.L.; Witherington, B.; Godley, B.J.; Wallace, B.P. Understanding individual and population-level effects of plastic pollution on marine megafauna. *Endanger. Species Res.* **2020**, *43*, 234–252. [CrossRef]
20. Mellish, S.; Pearson, E.L.; McLeod, E.M.; Tuckey, M.R.; Ryan, J.C. What goes up must come down: An evaluation of a zoo conservation-education program for balloon litter on visitor understanding, attitudes, and behaviour. *J. Sustain. Tour.* **2019**, *27*, 1393–1415. [CrossRef]
21. Roe, K.; McConney, A. Do zoo visitors come to learn? An internationally comparative, mixed-methods study. *Environ. Educ. Res.* **2015**, *21*, 865–884. [CrossRef]

Journal of
Zoological and
Botanical Gardens

MDPI

Article

An Approach to Assessing Zoo Animal Welfare in a Rarely Studied Species, the Common Cusimanse *Crossarchus obscurus*

Danielle Free [1], William S. M. Justice [1], Sarah Jayne Smith [1], Vittoria Howard [2] and Sarah Wolfensohn [2,*]

1 Marwell Wildlife, Winchester SO21 1JH, Hampshire, UK
2 School of Veterinary Medicine, University of Surrey, Guildford GU2 7AL, Surrey, UK
* Correspondence: s.wolfensohn@surrey.ac.uk

Abstract: Objective welfare assessments play a fundamental role in ensuring that positive welfare is achieved and maintained for animals in captivity. The Animal Welfare Assessment Grid (AWAG), a welfare assessment tool, has been validated for use with a variety of both domestic and exotic species. It combines both resource- and animal-based measures but relies heavily on knowledge of the species to effectively assess welfare. Many zoo species are understudied in the wild due to their cryptic nature or habitat choice; therefore, the published literature needs to be supported with captive behavioural observations and zoo records. Here we adapted previously published AWAG templates to assess the welfare of *Crossarchus obscurus*. A total of 21 factors were identified, and the final template was used to retrospectively score the welfare of two male and two female *C. obscurus* at Marwell Zoo, UK, validating the use of this process for preparing a welfare assessment for a species where the published literature is scarce.

Keywords: cusimanse; *Crossarchus obscurus*; zoo; behaviour; welfare; AWAG; welfare assessment; carnivore; evidence-based; understudied

Citation: Free, D.; Justice, W.S.M.; Smith, S.J.; Howard, V.; Wolfensohn, S. An Approach to Assessing Zoo Animal Welfare in a Rarely Studied Species, the Common Cusimanse *Crossarchus obscurus*. *J. Zool. Bot. Gard.* **2022**, *3*, 420–441. https://doi.org/10.3390/jzbg3030032

Academic Editors: Kris Descovich, Caralyn Kemp and Jessica Rendle

Received: 16 June 2022
Accepted: 19 August 2022
Published: 24 August 2022

1. Introduction

For most UK zoos, maintaining positive animal welfare is not only important from a moral and ethical perspective but from a legislative perspective as well. It also underpins the zoo's ability to fulfil their education and conservation aims, as laid out in the Zoo Licensing Act, 1981. Therefore, there is increasing pressure on animal caregivers to be able to demonstrate that their animals are experiencing positive welfare. There is currently no single definition for 'animal welfare'; however, it is largely agreed within the scientific community that it involves the reflection of physical and psychological health as perceived by the animal itself—the state of the animal and what it then experiences as a result [1–3]. Probably the most well-known welfare concept, the Five Freedoms, developed by the UK Farm Animal Welfare Committee in 1979 [4] to monitor and improve the welfare of livestock, became the key checklist for assessing the welfare of animals, domestic or exotic, across all industries, globally. However, tools used to monitor animal welfare have adapted as more and more has been understood about the factors that feed into an animal's welfare state. The Five Freedoms focuses on resources (such as, food, water, shelter, and veterinary care) with the implication that if these are provided, negative welfare states improve. It is now, however, generally accepted that to understand an animal's welfare state, it is necessary to include animal-based measures (i.e., behaviour and physical/physiological factors) within the assessment. As a result, there is no single method for assessing animal welfare, but a variety, adapted for different species, contexts, and resource levels.

The Animal Welfare Assessment Grid (AWAG) is one such tool, combining both resource- and animal-based measures, i.e., the effects of environment, physical and psychological well-being and procedural and management events, on welfare. Welfare is context specific and is a subjective experience; therefore, although a group of animals may share an enclosure, receive the same nutrition and live in the same social group, they could experience very different welfare states. The AWAG objectively examines the welfare of the individual animal at key points throughout its life, taking into account the duration as well as the intensity of any suffering and produces both a numeric and visual presentation of the animal's overall quality of life. By using a template to score four parameters (Physical, Psychological, Environmental, Procedural), this system develops a matrix based on data collected as an intrinsic part of husbandry records. Within each parameter various factors are scored to assess the level contributing to welfare and the factors for each parameter can be modified to suit different types of animal husbandry systems and so be relevant for the specific context. Thus, it can identify key events which impact on welfare, and by providing a whole-life assessment of an animal's welfare with a temporal approach, the AWAG allows those caring for animals to plan or intervene with targeted and timely refinements that can improve, or prevent the deterioration of, an animal's quality of life. AWAG templates have already been produced and validated at Marwell Zoo (MZ), UK, with a variety of exotic species, including various primates [5], giraffe *Giraffa camelopardalis*, scimitar horned oryx *Oryx dammah* and cheetah *Acinonyx jubatus* [6], as well as various bird species [5] and Western lowland gorillas *Gorilla gorilla gorilla* [7] at other organisations. This study is the first to use the AWAG for a small, exotic carnivore species, and, where previous studies have either focussed on broad taxonomic groups or species where plentiful information is available, this study shows how the AWAG can be adapted for a species where relatively little published information is available.

1.1. *Crossarchus obscurus*

Crossarchus obscurus, also known as the common cusimanse (or kusimanse), is a mongoose species native to equatorial western Africa [8–10]. Whilst the species is frequently seen locally [8,9,11], due to its cryptic nature and habitat preference, detailed research of the species in the wild has proved difficult, and with few individuals to study in captivity, this species remains understudied. *C. obscurus* was classified as 'Least Concern' by the International Union for Conservation of Nature (IUCN) in 2015, with an unknown population trend [12]. Current population size is unknown.

Compared to other mongoose species, *C. obscurus* are stocky in appearance, seeming unkempt due to a combination of fine, pale but dense underfur and dark, coarse outer fur (Figure 1). They are opportunistic omnivores and will take advantage of whatever food is available, although their diet consists primarily of invertebrate species and fruits and berries, depending on the season [9,10,13,14]. They will occasionally eat small vertebrates and bird or lizard eggs and have also been reported co-operatively hunting larger species such as rats [9,11,15,16].

Figure 1. *Crossarchus obscurus* or common cusimanse. Marwell Zoo, UK.

C. obscurus is an obligate social carnivore and in the wild has been found living in mixed-sex groups of adults and juveniles, ranging in number from four to 20 individuals [9,10,17]. It has been suggested that the larger groups may be formed of multiple family groups, which consist of a dominant breeding pair and their offspring from both current and previous litters [9,14]. Although it is yet to be confirmed, it appears that these groups generally remain stable: foraging, moving and resting together [13]. As seen exhibited by other social mongoose species (e.g., slender-tailed meerkat *Suricata suricatta*), *C. obscurus* communicate with one another using vocalisations such as 'peeping' contact calls and 'shrill' alarm calls (observed directed towards humans in the wild [13]), supporting the theory that they live in stable, cooperative groups [13,17]. They also communicate via olfaction, with both males and females using anal and cheek gland secretions to communicate [9,10,17,18]. These scents may be deposited by anal dragging, anal tapping, alternate cheek rubbing, in a 'handstand' position, lifting a hind leg or in addition to kicking the hind legs and urinating [13,17,18]. Scents will be deposited on objects in the environment, the ground, on faeces and on conspecifics and are believed to advertise information on identity, status and possibly ownership [13,17,18]. Similar to the meerkat, *C. obscurus* are a highly complex and cooperative species [17].

As with other members of the *Herpestidae* family, *C. obscurus* are a predominantly diurnal species, active from sunrise to sunset, with this activity punctuated by periods of rest. Research has shown that outside of these hours they are inactive, most likely sleeping, within shelters [13]. Species-typical behaviour is also similar to that of other mongoose species, including group foraging, various forms of locomotion (walking, running and trotting), climbing, resting/sleeping, scent marking, digging, hunting, foraging, stalking prey, sniffing, drinking and predator defence (for example, mobbing or head-darting towards a predator, piloerection, alarm calling and hiding) [9,13,17,19]. Social behaviours include allogrooming (grooming a conspecific, helping to maintain social bonds), mating, aggression, bundling (huddling close to conspecifics to maintain warmth), scent marking and play [9,13,17]. Self-directed and comfort behaviours include scratching, stretching, yawning and autogrooming (grooming itself) [9].

Although *C. obscurus* will climb when foraging, they are more commonly seen active at ground level [9]. During a 2001 study on wild *C. obscurus* in Sierra Leone, Olson [13] found that groups showed a preference for resting overnight in trees and recorded them from a height of 5 m up to 25 m. The only resting places that deviated from this were ~8 m high tree stumps found in more open habitats. The groups changed resting site almost every night, possibly to avoid predation. Other literature has suggested that they also take shelter in underground burrows, either dug themselves or by another species, and hollow logs or fallen trees [9]. There is little information on home range size for this species; however, Olson [13] calculated it to vary from 20 ha to 30 ha based on the three groups that were studied. Olson [13] also found that they will cover ~$^1/_4$–$^1/_3$ of their home range within a day, a minimum distance of 1036–1714 m travelled.

C. obscurus appears to be ecologically versatile, primarily inhabiting dense rainforest habitats but also found living in riparian and logged forest, open grassland, fallow, agricultural fields and plantations, up to an altitude of 1500 m above sea level [9,12,13,20]. It has been suggested that this tolerance of varied habitat may be a result of the changing availability of food resources [11]. Temperatures across equatorial western Africa vary from cool nights of 10 °C to hot day temperatures typically around the mid-20s °C but up to low 30s °C with the region receiving a high level of rain annually, <2000 mm over a >5-month rainy season. The rainforest habitats are usually cooler than savannah habitats and are high in humidity [21].

1.2. *Crossarchus obscurus* in Captivity

The Association of Zoos and Aquariums (AZA) has produced the 'Mongoose, Meerkat and Fossa (*Herpestidae/Eupleridae*) Care Manual' [22], which includes a compilation of expert knowledge on the management of *C. obscurus* in captivity. This manual identifies the environmental parameters for captive *C. obscurus*. Basing their recommendations on what is currently known about the species' wild environment, they suggest a temperature of 20–25 °C, with an indoor area of 22–25 °C. Whilst *C. obscurus* can tolerate cooler temperatures, something they would sometimes experience in the wild, the AZA recommend that the temperature should not dip below 13 °C and stipulate that a heated indoor area must be available if these low temperatures are expected. Humidity level is more difficult to specify as there is a lack of information on levels in the *C. obscurus'* wild habitat. Humidity is also likely to fluctuate in the wild depending on environmental factors such as density of vegetation and weather conditions [21]. The AZA do, however, suggest a humidity level of ~55–60% as the minimum. They also recommend the provision of shade and shelter from the elements, sunlight or heat sources for basking and a photoperiod resembling that experienced in the wild habitat (12 h of daylight/12 h of dark) [22].

Enclosure complexity or 'naturalness' should be species-specific, enabling the animals to exhibit an extensive array of behaviours from their natural repertoire with minimal input from animal carers, so that behavioural needs can be met outside of staff working hours, thus providing for the animal's overall lifetime experience [23]. The enclosure complexity should physically reflect *C. obscurus'* wild habitat with extensive vegetative cover at varying heights. Living trees and/or dead branching and logs should be provided to offer the opportunity to climb, forage in various environments and rest at height, as seen in wild *C. obscurus* (the AZA recommends at least 1.22 m above the ground [22]). These will also provide the opportunity for scent marking behaviour. The provision of visual barriers that the animals can use to take themselves out of view of both conspecifics (particularly important for subordinate individuals) and people (both familiar and unfamiliar) are vital for improving welfare. Species that inhabit dense environments, such as *C. obscurus*, have been noted to exhibit a greater negative response to the presence of people than species that would inhabit more open environments [24]. Providing animals with the opportunity in their environment to escape the view of visitors can reduce fear and stress, thus improving welfare [25]. Provision of a variety of substrates is important for enabling digging and foraging, two behaviours that are likely to play a key role in positive welfare for this species. *C. obscurus* also have non-retractable claws whose length can be managed through the provision of digging opportunities rather than requiring veterinary intervention [22]. The provision of resources required to fulfil the species' evolutionary and biological needs will result in positive affects (Figure 2).

Figure 2. Outside habitat for *C. obscurus*, MZ. Environmental features include a stream, a variety of planting, a mesh roof and branching/logs.

For a gregarious species such as *C. obscurus*, being able to maintain good social interactions with conspecifics is key to the individual's welfare state. There are records of

C. obscurus that were housed in isolation experiencing stress, which was noted leading to apathy and self-harm [13,22]. All veterinary procedures requiring general anaesthesia, inevitably resulted in the isolation of an individual from the rest of the group. The period of isolation varies depending on the length of the procedure and the recovery time immediately after the procedure, which itself is affected by the length of the procedure and the physiological health of the individual [26]. As *C. obscurus* are a highly sociable species [17], periods of isolation may have a negative impact on the welfare of the isolated individual and potentially the rest of the group, with the risk that the isolated individual could be rejected on reintroduction. Duration of separation should be kept to a minimum [22]. Captive *C. obscurus*, similar to other members of the *Herpestidae* family, are also prone to developing hypercholesterolaemia [22]. As this condition is not reported in wild individuals, it is hypothesised that it is linked to a captive diet and one dietary trial led to improved blood cholesterol levels [27,28].

Crossarchus obscurus at Marwell Zoo

Two males and two females (a non-breeding sibling group, containing surgically castrated males) are currently housed at MZ, where they arrived in July 2017, aged one year. This group is of a similar size but differs in that it lacks the multigenerational composition seen in wild social groups. Contrary to Goldman's species' description [9], aggression between *C. obscurus* appears fairly common in captivity, possibly as a result of inappropriate social groupings. Behavioural information gathered from MZ's animal records confirmed that it is frequently exhibited by this group where it seems to occur predominantly around food, although it mostly comprises vocalisations and pushing, rather than physical attacks [29].

Relevant behavioural knowledge from MZ's *C. obscurus* was derived from data collected during a behavioural study on the four individuals. This utilised an ethogram developed from a literature search and confirmed that they exhibit a similar activity budget to their wild counterparts, with, on average, 63% of daylight hours spent active (max. 66%, min. 58%), 37% spent resting (max. 42%, min. 34%) and no sleeping behaviour seen. Little difference in behaviour was seen between either the individuals or sexes (males active 61.5%, resting 38.5%; females active 64.5%, resting 35.5%). At 5–6 years of age, with a captive life expectancy, on average, of 8–10 years [14], activity levels would not yet be expected to be affected by age. These observations also indicate a preference for resting at height, particularly when sleeping, as would be expected from what is known of the species' wild behaviour.

Common veterinary procedures historically performed in this group of animals, as identified from medical records, include the administration of medication, manual restraint for minor procedures or the application of topical treatmen, and general anaesthesia for blood samples and skin biopsies. The welfare impact of these procedures was considered to be less if they did not require restraint or general anaesthesia. Whilst the general anaesthesia of small mammals, induced using inhalation medication such as isoflurane, is associated with a negative impact on animal welfare due to the need for pre-operation starvation and the marked breath-holding behaviour that it can cause during induction [30] and the recovery phase, it may have a lower impact on overall welfare compared with manual restraint, depending on the duration of restraint and whether habituation to restraint has occurred. However, in this case, manual restraint occurred infrequently in this group and none of the individuals appeared to experience long-term negative effects when it did occur; therefore, manual restraint was considered to have a lower impact on welfare than general anaesthesia, an example of welfare being dependent on context.

Further clinically relevant information was gathered from the medical records of the animals. This highlighted certain pathological conditions exhibited by all individuals of this group, that appear to occur more frequently in captive *C. obscurus*, in particular, alopecia and poor coat quality (outside of the natural seasonal shedding [9], linked also to behaviour) and skin disease resulting in dry flaky skin and sores. After considerable veterinary investigation, the underlying cause remains inconclusive, although one individual was identified as having skin allergies that led to more frequent and severe lesions compared to the other three (Figure 3). This condition is now managed using a steroid medication. There are no records of this affecting *C. obscurus* in the wild.

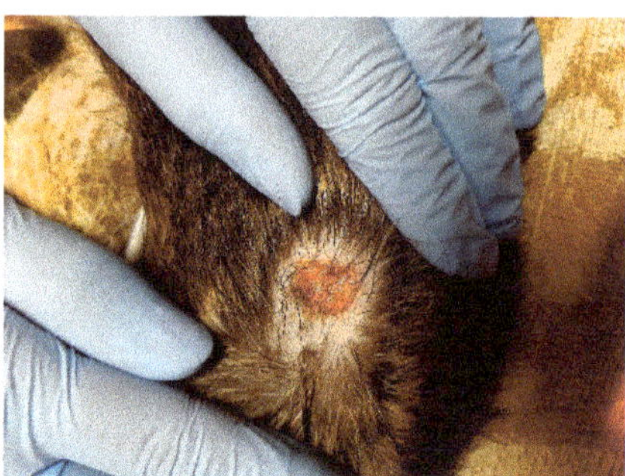

Figure 3. Visual record of a sore on one *C. obscurus* at MZ. Photograph taken while the animal was under general anaesthetic, MZ, 2019.

Abnormal behaviours, such as overgrooming (both auto- and allo-) and barbering, self–mutilation, excessive scratching, repetitive pacing and circling behaviours have been recorded for various mongoose species in captivity [13]. These behaviours can be used as animal-based indicators of potentially compromised welfare [31–33]. Evidence gathered via camera traps during a study in 2019 indicated that barbering, over-grooming and/or excessive scratching were contributing to hair loss, possibly in response to pathological issues. In addition to indicating an attempt to cope with an aversive situation, the hair loss itself can have a negative impact on welfare, affecting communication with both con- and contraspecifics (when threatened piloerection will occur along with an arched back to appear larger [9]) and the animal's ability to thermoregulate and protect the skin from the sun and injuries. No other abnormal behaviours have, to date, been exhibited by this group of individuals.

Since 2017 there have been changes in the group's diet with the replacement of items of a higher fat content (e.g., chicks and mice) with more invertebrates, plus the addition of crayfish/crab and whitebait to determine whether this would impact coat condition, as was found by Totten [27]. The reduction in large single item feeds for invertebrate feeds that stimulate active foraging behaviours also led to reduced aggression among the group at feed times. A higher-than-expected cholesterol level was found during a veterinary exam on one individual in early 2019.

The indoor area of the *C. obscurus* enclosure at MZ is heated by a bar heater and fluctuates in temperature from 13.5 to 30 °C (Figure 4), depending on the season, with heat lamps provided during cooler weather. Data on enclosure usage by this group identified that above the bar heater is the area most used by all individuals (on average, 32% of their time). Behavioural data also showed a positive correlation between use of this area and poor weather. This indicates that temperature is an important environmental factor for these individuals and likely the species, considering the environment it evolved in, and excessive use of this area in addition to an increase in bundling behaviour can indicate a potential compromise to welfare. As a link between sub-optimal humidity and skin and hair issues has been found in rodents [34], it was speculated that low humidity level, along with the behaviour of sitting above the radiator, was a contributor to the skin issues in this group. Humidity levels were therefore increased by changing the substrate from wood shavings to bark mulch that could be dampened regularly.

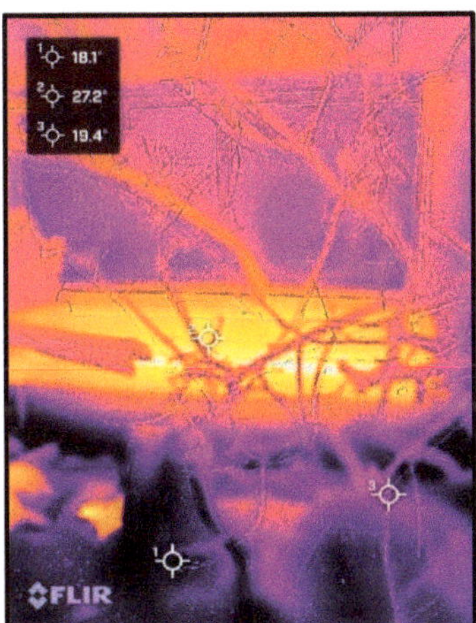

Figure 4. Image from the inside area of the *C. obscurus* enclosure. FLIR thermal image of bar heater reading 27.2 °C (as indicated by the white crosshairs numbered 2) and floor temperature reading 18.1–19.4 °C (as indicated by the white crosshairs numbered 1 and 3).

This *C. obscurus* group typically show little behavioural response to unfamiliar people, as supported by data on their enclosure use where they were found to spend 51% of the observed time in areas of the enclosure that are adjacent to guest viewing windows. It is probable that resources in these areas attract the *C. obscurus*, for example, one area was particularly suitable for digging, one of the most common behaviours observed, and gave the best view of familiar people approaching the enclosure, whilst the other area was on top of a heater. Even so, the presence of unfamiliar people in proximity did not deter them from using these areas. Individual and species-specific responses will lead to variation in behaviour exhibited, as will habituation and previous positive or negative experiences with people.

To assess and monitor the welfare of these individuals, previously published AWAG templates were adapted utilising the limited species-specific knowledge that was available from the literature supplemented with data from zoo records and direct observations to better reflect this species, these individuals and their specific context. As this system relies on knowledge of the species in order to effectively assess welfare, it is more difficult to put together a template for species where there is relatively little information available. Hence, this paper shows one possible approach to creating a template when information is scarce.

2. Parameters of the AWAG

2.1. Physical

Wolfensohn, et al. [35] published the following factors under this parameter: general condition, clinical assessment, activity level/mobility, presence of injury, not eating/drinking. Justice, et al. [5] adapted the template to include 'faecal consistency', as it is a commonly used indicator of gastrointestinal health and diet suitability in zoological institutions, and exclude 'presence of injury', instead including this under 'clinical assessment'. For this study, these animal-based measures were largely kept the same, using relevant species knowledge available from the literature.

The factor 'clinical assessment' was split to allow the scoring of skin condition separately from other clinical signs. 'Body condition score' (BCS), where subjective visual assessments of muscle and fat, typically scored on a 1–5 scale with 3 representing optimum condition, are used to determine whether an animal is a healthy weight, replaced general condition. Change in an animal's BCS can indicate the presence of underlying health conditions and can be used to monitor the progression of the disease and the success of veterinary intervention [36] and husbandry practices [37]. Extremes in body condition can predispose an individual to disease [38,39]. This measurement is used because although weight is important to monitor, it is not always easy to assess its relevance to health depending on the size and age of an individual. Whilst the coarse fur of this species may impede a consistently reliable result, BCS is non-invasive and quick to carry out, both important factors when considering the practicalities of assessing welfare in a zoo setting [40,41]. BCS is also validated by hands-on physical examination and weighing when the opportunity arises (i.e., during veterinary care). 'Faecal consistency' was retained in this template but as *C. obscurus* is a midden utilising species, faecal consistency was scored for the group.

See Table 1 for the full list of factors and 1–10 criteria for the parameter: Physical.

2.2. Psychological

Psychological factors scored in a previous study by Wolfensohn, et al. [35] comprised the following animal-based measures: stereotypy, self-harming, unusual grooming; response to catching events; hierarchy upset/dispute, aggression/bullying; alopecia score; use of enrichment; and aversion to 'normal' events. These were revised by Justice, et al. [5] for use in a zoo context to consist of: abnormal behaviours; response to catching event; hierarchy upset/dispute, aggression/bullying; use of enrichment; aversion to 'normal' events and training. Although there are some similarities, for this study the factors scored under this parameter were significantly adapted.

'Abnormal behaviour' was retained, and the criteria adapted to include alopecia that may be occurring as a result of barbering, over-grooming and/or excessive scratching. Videos indicated that a proportion of these behaviours were occurring overnight; hence, the criteria for this factor were adapted to allow scoring to be carried out on behaviour seen as well as the extent of hair loss as a proxy for these behaviours occurring out of sight. The 'Rule of Nines', as used by the emergency services to quickly assess the total body surface area of burns victims [42], was adapted for *C. obscurus* (Figure 5), to help zookeepers objectively quantify hair loss. Excessive scent-marking of conspecifics was also considered as a possible cause of hair loss, with areas of 'wet' fur occurring in some of the same locations. This was encapsulated in this factor as a possible result of an abnormal frequency of the behaviour.

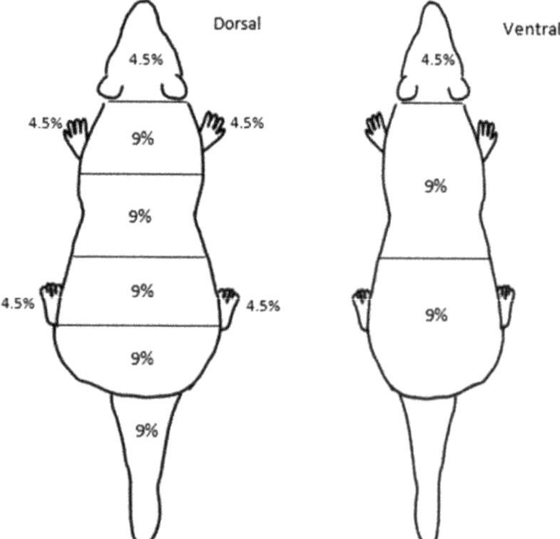

Figure 5. The 'Rule of Nines' schematically depicted for *C. obscurus*. The 'Rule of Nines' can be used to calculate a rough percentage of hair loss to support welfare assessment.

Table 1. Factors scored within the physical parameter.

	Body Condition Score	Food Intake	Activity Levels (Based on Wild Behaviour). Diurnal Species.	Faecal Consistency	Clinical Assessment (Excluding Skin Condition)	Skin Condition
Score						
1	3—Ideal	Eating normally. No signs of increased/decreased hunger. All food is consumed.	Normal balance between activity and resting. Little to no sleeping seen during the day. Activity begins around sunrise and continues throughout day until sunset or drop in temperature, with periods of resting.	Normal	Nothing observed	Skin looks healthy, no dryness or flaking. No wounds.
2	More/less than 3 but not quite 2.5/3.5	Food intake slightly lower than normal for one day OR animal reported hungry	Slight increase/decrease in activity not related to normal daily variation	Loose (below 10% of total) or clumpy (below 30%)	Mild clinical signs that show no impact on the animal's ability to perform normal behaviours. Full recovery expected	Small area/s of dry, flaky skin visible.
3	2.5/3.5—Slightly over/under	Food intake significantly lower than normal for one day OR reported hungry for 2–3 days	Moderate increase/decrease in activity (no obvious cause)-showing full recovery within 8 h	Loose (below 20% of total) or clumpy (below 40%)	Mild clinical signs having a short term impact on the animal's ability to perform normal behaviours. Full recovery expected.	Large area/s of dry, flaky skin visible.
4	More/less than 2.5/3.5 but not quite 2/4	Food intake slightly lower than normal for 2 days (lower than 80%) OR reported hungry for 4–5 days	Moderate increase/decrease in activity (no obvious cause)-showing full recovery within 12 h	Loose (below 30% of total) or clumpy (below 50%)	Mild clinical signs having a longer term impact on the animal's ability to perform normal behaviours. Full recovery expected.	Healing/scabbed sore/s
5	2/4—Over or under	Food intake significantly lower for 2 days (lower than 50%) OR reported hungry for 6–7 days	Significant increase/decrease in activity (no obvious cause)—showing full recovery within 12 h	Loose (below 40% of total) or clumpy (below 60%)	Moderate clinical signs having limited impact on the animal's ability to perform normal behaviours. Full recovery expected	One small open sore
6	More/less than 2/4 but not quite 1.5/4.5	Food intake slightly lower than normal for 3 days (lower than 80%) OR reported hungry for 8–9 days	Significant increase/decrease in activity (no obvious cause)—not showing full recovery to normal or recovery taking over 12 h.	Loose (below 50% of total) or clumpy (below 70%)	Moderate clinical signs having limited impact on the animal's ability to perform normal behaviours. Recovery potential unknown.	One medium OR multiple small open sores OR multiple medium/large wounds scabbed and beginning to heal
7	1.5/4.5—Very over/under	Food intake significantly lower than normal for 3 days (lower than 50%) OR reported hungry for 10–11 days	Inactive but does get up to eat/drink/defecate OR very active throughout the day with not much rest	Loose (below 60% of total) or clumpy (below 80%)	Moderate clinical signs with medium to long term impact on animals ability to perform normal behaviours. Recovery potential unknown.	Multiple medium open sores OR one large open sore OR open wound on face, ear/s, groin or foot/feet
8	More/less than 1.5/4.5 but not quite 1/5	No sign animal has eaten for 1 day OR reported hungry for 12–13 days	Very minimal movement/signs of hyperactivity	All clumpy or all loose	Severe clinical signs but with short term impact and expected recovery OR moderate to severe signs with long term impact on animal's welfare and little chance of recovery.	One large open wound present for >1 week <4 weeks

Table 1. *Cont.*

	Body Condition Score	Food Intake	Activity Levels (Based on Wild Behaviour). Diurnal Species.	Faecal Consistency	Clinical Assessment (Excluding Skin Condition)	Skin Condition
Score						
9	1/5—Very Thin/Obese	No sign animal has eaten for 2 days or reported hungry for 14–15 days	Completely hyperactive OR inactive	All watery diarrhoea	Severe or chronic clinical signs that are having serious negative impact on the animal's ability to perform normal behaviours	Multiple open wounds, various locations on body for <4 weeks OR one large open wound present for >4 weeks.
10	0—Emaciated/Starving OR 6—Morbidly Obese	No sign animal has eaten for 3 days OR reported hungry for >15 days	Animal causing itself harm through inactivity or hyperactivity OR completely recumbent	Presence of abnormal elements e.g., mucus or blood	Severe clinical signs that are rendering the animal recumbent/unable to carry out any normal behaviour	Multiple open wounds, various locations on body, present for >4 weeks

The presence of humans in and around the animals' enclosure is an inevitable consequence of living in a captive environment. Research across various species has shown that human presence can elicit a variety of responses, although most research to date has focussed on the negative impacts on animal welfare [25,43]. It has also been evidenced that animals are able to distinguish between familiar and unfamiliar people [44]. Therefore, the factor 'response to unfamiliar people' was added to this parameter, with scoring criteria to reflect the range of impacts on welfare and response to familiar people, i.e., zookeepers, encapsulated under the factor 'response to normal events'. The term 'unfamiliar people' was used instead of visitor or guest as this was felt to be more encompassing of the array of people that may have an impact on the animals (for example, contractors or service providers).

The factor 'aversion to normal events' was changed to 'response to normal events' to allow the inclusion of scores that would reflect a positive impact on welfare, in line with contemporary welfare science [45]. The criteria for this factor were adapted to include anticipatory behaviours (as was included by Brouwers and Duchateau [7]), such as increased activity ahead of scheduled feed times, as these can indicate the level of importance an animal attributes to a positive event, with those positive events that occur less frequently potentially resulting in the greater duration or intensity of anticipatory behaviour. Whilst anticipation is associated with dopamine production and thus in limited duration may indicate positive welfare at the moment when it is occurring, the behaviour may be more useful as an indicator that outside of that time, the animal's overall welfare is suffering due to a lack of something or the lack of ability to do something that the animal considers important [46–48]. Affiliative behaviours, including play, grooming and bundling, are positive social interactions, likely resulting in positive emotional states for the individuals involved [49]; therefore, 'hierarchy upset' was changed to 'social interaction', to enable these positive impacts on welfare to be taken into account. As food-related aggression is commonly seen in mongoose species in captivity [22], it was important that this was also captured in this factor's criteria.

'Response to catching event' was replaced with 'response to restricted access to part of the enclosure' as catching occurs infrequently for these individuals. When it is required, it is usually for veterinary care; therefore, it was incorporated in the procedural parameter instead. Restricting the *C. obscurus* group to part of the enclosure occurs more often, for example they would be shut into the house for landscaping of the outside area. This group rarely shows any negative response to being shut out of the house, or being shut in the house for short periods of time. However, they show signs of frustration if shut inside the house for an extended period; therefore, it is important to assess each occurrence individually.

The term 'enrichment' in everyday animal care seems to have become synonymous with the provision of resources, e.g., novel objects, which are aimed at reducing indicators of poor welfare, such as abnormal behaviour, or to stimulate positive, but mostly short-term, changes in behaviour, instead of focussing on enabling and encouraging species-specific behavioural repertoires. As a non-invasive, accessible and thus practical tool, assessing behaviour is the most used method for evaluating animal welfare [41,50]. Both the specific behaviours exhibited, and overall behavioural diversity can be used as animal-based indicators of welfare. Behavioural diversity is classically compared to the species-typical wild behaviours. Several studies have found when the amount of abnormal or stereotypical behaviour displayed is high, behavioural diversity is generally low and vice versa (see Miller, et al. [51] for examples) and although stereotypes may develop as a coping mechanism that helps to improve welfare, their presence can indicate a suboptimal situation that the animal is attempting to cope with [33]. Although information on wild behaviour and behavioural diversity is limited for this and numerous other captive species, what is known can provide a benchmark that captive animal behaviour can be compared to. For these reasons, the factor 'use of enrichment' was adapted and renamed 'species-typical behaviours' for this template. Finally, 'training' was removed from the template as the group do not receive any training at present.

See Table 2 for the full list of factors and 1–10 criteria for the parameter: Psychological.

Table 2. Factors scored within the psychological parameter.

Score	Abnormal Behaviour (Overgrooming, Barbering, Increased Scratching—For Hair Loss (Use 'Rule Of Nines')	Response to Presence of Unfamiliar People (e.g., Guests, Contractors, Keepers from Other Sections)	Response to Normal Events	Response to Restricted Access to Part of the Enclosure	Social Interaction (with Conspecifics)	Species-Typical Behaviours—Either Observed Occurring or Evidence of
1	No hair loss, coat is in healthy condition, looking bright and glossy. No abnormal behaviour or frequency of behaviour observed.	Actively seeks out familiar and unfamiliar people. Appears to prefer parts of enclosure close to people.	Actively seeks out keeper during normal events, exhibiting positive behaviours.	No restricted access to part of the enclosure	Animals are interacting with one another positively. Exhibiting affiliative behaviours e.g., bundling, frequently	Animal is observed performing all expected natural, primarily positive, behaviours (e.g., group foraging, exploration, scent marking, walking, running, trotting, climbing, resting, sleeping, bundling, digging, hunting, foraging with nose, foraging in vegetation, egg smashing, stalking, sun bathing, sniffing, drinking, aggression, allogrooming, mating, scratching, stretching, play, predator defence and hiding) at expected rate with no abnormal behaviours.
2	No hair loss but coat is looking dull, greasy or wet AND/OR limited time spent, low frequency, distractable	Completely habituated to presence of unfamiliar people, no concern shown. No preference for particular parts of enclosure noticeable.	Well habituated to keeper interactions/daily events, no response	Animal comes in easily with call and no further intervention required. Animal remains calm until released.	Animals are interacting with one another positively. Exhibiting affiliative behaviours e.g., bundling, occasionally	Animal is observed performing most natural behaviours as expected in captivity (may not include, for example, mating, egg smashing or predator defence) at expected rate with no abnormal behaviours.
3	Hair is thinning in patches but there are no bald areas AND/OR limited time spent, low frequency, distractable	Preference to avoid areas close to viewing areas but does use rest of enclosure. No signs of stress observed.	Anticipatory behaviour begins on seeing keeper heading to enclosure but stops immediately on keeper arrival. Well habituated to keeper interactions/daily events.	Animal well trained and/or habituated and comes in easily but intervention/enticement is required. Animal remains calm until released.	Animal interacts with conspecifics without affiliative behaviours or fear/stress/aggression (exhibited/received)	Animal is observed performing a wide variety of positive natural behaviours including: foraging, stalking/hunting, digging, locomotion, social, self-maintenance, climbing, resting, at expected rate with no abnormal behaviours.
4	1–5% hair loss AND/OR moderate time spent, medium frequency, distractable	Spends most of the day away from viewing areas or hiding out of sight, utilises enclosure when zoo is closed. No signs of stress observed	Well habituated to most keeper interactions/daily events but mild stress seen for single specific interaction.	Animal shows some reluctance to come in and mild signs of stress. Takes between 5–15 minutes to get the animal inside/attempt abandoned. OR animal is showing signs of stress as a result of attempts to shut conspecific in. AND/OR mild signs of stress when shut in.	Animal has opportunity to interact with conspecific but choses not to. No fear/stress/aggression (exhibited/received) noted.	Animal is observed only performing fundamental behaviours such as feeding, resting, drinking, urinating, defecating but no negative or abnormal behaviours.
5	>5–15% hair loss AND/OR moderate time spent, higher frequency, no damage done and distractable OR 2–4 and not distractable	Preference to avoid viewing areas/hide out of sight but does use rest of enclosure for some of the day. Shows mild signs of stress when unfamiliar people around but recovers when they leave area.	Shows some mild signs of stress during normal daily interactions but recovers as soon as interaction is over AND/OR short-term anticipatory behaviour in lead up to keeper arrival for a single routine event.	Animal very reluctant to come in and shows moderate signs of stress. Single attempt required but takes over 15 minutes/abandoned after 15 mins.	Animal shows mild fear/stress/aggression (exhibited/received) in interactions with conspecifics which is temporary e.g., around food. Majority of interactions are normal.	Animal is observed performing a wide variety of positive natural behaviours including: foraging, stalking/hunting, digging, locomotion, social, self-maintenance, climbing, resting, at expected rate with some negative or abnormal behaviours.

Table 2. *Cont.*

	Abnormal Behaviour (Overgrooming, Barbering, Increased Scratching—For Hair Loss (Use 'Rule Of Nines')	Response to Presence of Unfamiliar People (e.g., Guests, Contractors, Keepers from Other Sections)	Response to Normal Events	Response to Restricted Access to Part of the Enclosure	Social Interaction (with Conspecifics)	Species-Typical Behaviours—Either Observed Occurring or Evidence of
Score						
6	>15–25% hair loss AND/OR significant time spent, higher frequency, not distractable	Spends most of the day away from viewing areas/hide out of sight and show mild signs of stress when unfamiliar people are around but recovers when they leave area.	Shows some moderate signs of stress during normal daily interactions AND/OR short-term anticipatory behaviour in lead up to keeper arrival for all routine events.	Animal very reluctant to come in and shows moderate signs of stress. Multiple attempts required/abandoned after multiple attempts. AND/OR moderate signs of stress when shut in.	Animal shows moderate fear/stress/aggression (exhibited/received) in interactions with conspecifics that is temporary.	Animal is observed primarily performing fundamental behaviours such as feeding, resting, drinking as well as a few positive behaviours (e.g., digging, climbing or self-maintenance) with an increase of negative or abnormal behaviour.
7	>25–35% hair loss AND/OR significant time spent, higher frequency, not distractable	Avoid viewing areas/hide out of sight and show moderate signs of stress when unfamiliar people around but recovers when they leave area.	Shows significant stress behaviour during normal daily interactions AND/OR considerable time spent in anticipatory behaviour in lead up to keeper arrival for a single routine event.	Animal very reluctant to come in and showing severe signs of stress/fear. Significant time (over half an hour) and/or multiple attempts to get in/abandoned.	Animals shows moderate fear/stress/aggression (exhibited/received) in interactions with conspecifics OR no opportunity to interact with conspecific (gregarious species)	Animal is observed performing fundamental behaviours such as feeding, resting, drinking but spending some time on negative or abnormal behaviours.
8	>35–45% AND/OR majority of time spent, high frequency, not distractable.	Significant signs of stress when unfamiliar people are around but recover within an hour when they leave area.	Shows significant stress/fear behaviour during normal daily interactions AND/OR considerable time spent in anticipatory behaviour in lead up to keeper arrival for all routine events.	Animal very reluctant to come in and showing severe signs of stress/fear. Significant time spent (over 1 hour). AND/OR showing severe signs of stress when shut in.	Animal shows severe fear/stress/aggression (exhibited/received) in most interactions with conspecifics.	Animal is observed performing fundamental behaviours such as feeding, resting, drinking but spending greater proportion of time exhibiting negative or abnormal behaviours.
9	>45–55% AND/OR majority of time spent, very high frequency, not distractable.	Significant signs of stress when unfamiliar people are present. Takes up to 8 h to recover when they leave area.	Animal is stressed and aggressive during normal daily routine events	Animal extremely difficult to get in and showing aggressive behaviour in response to attempts	Animal shows fear/stress/aggression (exhibited/received) in all interactions with conspecifics.	Animal is observed performing fundamental behaviours such as feeding, resting, drinking but spending most of their time exhibiting negative or abnormal behaviours.
10	>55% hair loss AND/OR majority of time spent, very high frequency, not distractable.	Significant signs of stress AND/OR aggression/self harm in response to presence of unfamiliar people.	Animal is self-harming as a result of a normal or routine event.	Animal harming itself and/or conspecifics as a result of being shut in or attempts at shutting in.	Animal is aggressive and either self harming or harming conspecific.	Complete lack of natural behaviour observed, overwhelming abnormal or negative behaviour exhibited (e.g., aggression, hiding, pacing, self-directed or escape behaviours)

2.3. Environmental

Wolfensohn, et al. [35] used housing, group size, provision of 3D enrichment, provision of manipulable enrichment and contingent events, under the environmental parameter. These were refined by Justice, et al. [5] for the zoological environment where 'furnishing/enclosure design' replaced 'housing', enrichment was moved to the psychological parameter and 'nutrition' and 'access' (to enclosure) were added. The factors under this parameter were further adapted for this template.

The factors included in this parameter are predominantly resource-based, assessing what has been provided/is available to the animals and thus what the animals could be experiencing as a result. Nevertheless, this does not consider whether the animals are utilising these resources oor the resulting affects. 'Enclosure' and 'enclosure complexity' are scored separately under this factor to account for the suitability of the enclosure parameters to provide physical comfort and fulfil biological functions, and complexity to provide for the species' behavioural needs that may not be captured under the factor 'species-typical behaviour' when not observed. The factors 'group size/structure' and 'contingent events' were retained, whist the impact of reduced 'access' was incorporated into the psychological parameter, under the factor 'response to restricted access to part of the enclosure', to change it from a resource- to an animal-based indicator. As it is not possible to regularly test cholesterol level without an invasive veterinary procedure, this measure was not included in the AWAG, so the only changes made to the criteria for 'nutrition' were to include the presence of a variety of tastes, textures and smells that increase the pleasurable experience of eating, leading to positive welfare [52].

See Table 3 for the full list of factors and 1–10 criteria for the parameter: Environmental.

2.4. Procedural

In Wolfensohn, et al. [35], restraint, sedation, effect of intervention and change in daily routine were included in the procedural parameter. Justice, et al. [5] adapted 'effect of intervention' to focus on veterinary procedures specifically, and two factors were added for birds: 'time bird restrained before/during procedure' and 'visitor score'. Although they may not occur frequently, veterinary procedures are likely to be some of the most stressful events a zoo animal will experience during its lifetime; therefore, it is vital that the negative impact on welfare caused by veterinary procedures is considered.

'Isolation' was included as a separate factor in this study due to the greater welfare impact it could have on this highly social species. 'Vet procedure' and the 'impact of vet procedure' were both included to cover the effect of the procedure itself on welfare as well as the effect on welfare in the lead up to and following the procedure, including manual restraint and changes in husbandry. The factor 'changes in daily routine' was removed as a factor as it was felt that changes to food intake and enclosure would be captured elsewhere in the template. 'Visitor score' was included under the psychological parameter instead and the focus was placed on the animals' response to unfamiliar people rather than assuming increasing group size and noise level has a negative impact on the welfare of these individuals [5].

See Table 4 for the full list of factors and 1–10 criteria for the parameter: Procedural.

Table 3. Factors scored within the environmental parameter.

	Enclosure (Species Specific, e.g., Size, Lighting, Shelter, Ventilation, Temperature, Drainage, Noise Levels, Substrate etc.)	Enclosure Complexity (Species Specific, e.g., Planting, Water Bodies, Food, Shelter, Choice, Hiding Places, Furniture, Sunlight/Heat Lamp) Plus Opportunities Provided by Keepers	Group Size/Structure (Based on Wild Size and Composition)	Contingent Events (e.g., Animal Movement, Enclosure Changes, Building Works, Visitor Event)	Nutrition
Score					
1	Enclosure mirrors the species' wild habitat preference (tropical rainforest, transitional forest, logged forest, agricultural land), size (20–30 ha) with access to temperature >22 degrees C, >60% humidity, 12/12 photoperiod, shelter from inclement weather and sun, guest viewing less than 360 degrees, off show area, ventilation, UVB, low noise level, adequate drainage and substrate.	Enclosure complexity is reflective of the wild environment, including waterbody/ies, suitable substrates for foraging and digging burrows, dense vegetation plus some open areas, climbing opportunities, ability to sleep at height as a group in a variety of locations, and be able to rest as a group elsewhere, variety of weather conditions. All natural behaviours can be expressed.	Group size is reflective of natural wild group size (4–20 individuals) and suitable group structure (dominant breeding pair and young from current or previous litters, both adult males and females, groups appear fairly stable). Stocking density is appropriate for the enclosure.	None	Diet available is optimally suited to the species-specific needs (nutritional, physiological and behavioural (natural acquisition and manipulation)) and the individual. Diet includes a variety of tastes, textures and smells.
2	Enclosure is smaller than wild territory but mirrors other elements	All natural behaviours can be expressed with little reliance on keepers	Group size and structure is similar to wild. No stress observed and natural behaviours seen from all of the group.	External works/visitor event with minimal disturbance	Diet provided is suited to the species-specific needs (nutritional, physiological and behavioural (natural acquisition and manipulation)) and the individual but regularly lacks variety.
3	Enclosure is smaller than wild territory and lacks one other element.	All natural behaviours can be expressed with considerable reliance on keepers	Group size and structure are dissimilar to the wild but no stress observed and natural behaviours seen from all of the group. Stress behaviours NOT seen when separated.	Enclosure move to familiar enclosure with no other events taking place	Diet provided has a slightly reduced suitability to species and/or individual needs AND/OR lacks variety.
4	Enclosure is smaller than wild territory and lacks 2 other elements.	Most natural behaviour can be expressed with minimal reliance on keepers	Group size and structure are not like wild but no stress observed and natural behaviours seen from all of the group. Some stress behaviours seen when separated.	External works/visitor event with some disturbance including visitor event outside of usual opening times but during daylight.	Diet provided has reduced suitability to the individual needs
5	Enclosure is smaller than wild territory and lacks 3 other elements.	Most natural behaviours can be expressed with considerable reliance on keepers	Group size is reflective of natural wild group size and suitable group structure. Stocking density is slightly high for the enclosure (e.g., presence of young)	Enclosure move to a completely new enclosure OR significant change to the existing furniture of the enclosure.	Diet provided has reduced suitability to the species needs

Table 3. *Cont.*

Score	Enclosure (Species Specific, e.g., Size, Lighting, Shelter, Ventilation, Temperature, Drainage, Noise Levels, Substrate etc.)	Enclosure Complexity (Species Specific, e.g., Planting, Water Bodies, Food, Shelter, Choice, Hiding Places, Furniture, Sunlight/Heat Lamp) Plus Opportunities Provided by Keepers	Group Size/Structure (Based on Wild Size and Composition)	Contingent Events (e.g., Animal Movement, Enclosure Changes, Building Works, Visitor Event)	Nutrition
6	Enclosure is smaller than wild territory and lacks 4–5 other elements.	Some natural behaviours can be expressed with considerable reliance on keepers	Group size and structure is similar to wild but environmental pressures cause stress/aggressive behaviours	External works/visitor event taking place with definite disturbance, including visitor event taking place after sunset.	Diet provided lacks behavioural requirements for the species and individual.
7	Enclosure is smaller than wild territory and lacks 6–7 other elements.	Enclosure complexity and keeper intervention are minimal, preventing the expression of numerous natural behaviours.	Group size and structure not completely like wild. Moderate stress behaviours observed either when together OR separated.	Introduction of new unfamiliar animal to group.	Diet provided lacks physiological requirements for the species and individual
8	Enclosure is smaller than wild territory and lacks 8–9 other elements.	Enclosure complexity and keeper intervention are minimal, preventing the expression of most natural behaviours	Group size and structure not completely like wild. Significant stress behaviours observed either when together OR separated.	Prolonged external works with definite disruption.	Diet provided lacks nutritional requirements for the species and individual
9	Enclosure is smaller than wild territory and lacks 10–11 other elements.	Enclosure complexity and keeper intervention is very limited, preventing the expression of almost all natural behaviours.	Group structure very different to wild group and inappropriate for species (e.g., solitary) and/or high degree of overstocking.	New enclosure and new animals introduced at the same time.	Diet provided lacks 2 requirements for the species and individual.
10	Enclosure is smaller than wild territory and lacks 12+ other elements	The options are not available in the enclosure nor provided additionally for the animal to express natural behaviours	Group structure very different to wild group and dangerous for species OR harmful degree of overstocking.	Multiple events happening at the same time (e.g., new enclosure, new group and external works)	Diet provided lacks all requirements for the species and the individual.

Table 4. Factors scored within the procedural parameter.

	Isolation (From Conspecifics)	Vet Procedures (e.g., Daily Medication, Routine Vaccinations, Sedation, Anaesthesia)	Impact of Vet Procedures and/or Catch up for Other Purpose (e.g., Stress/Fear)	Sedation/Anaesthesia
Score				
1	Not isolated	No vet procedure occurred	No vet procedure/catch up occurred	No sedation/anaesthesia
2	Isolated for less than 2 h	Minor procedure performed with minimal effect on animal (e.g., delivery of oral medication in food).	Procedure can be performed easily with no stress or aggressive behaviour (e.g., delivery of oral medication in food).	Mild sedation (e.g., sedated not asleep/recumbent). Calm induction and recovery. Rapid return to normal feeding and behaviour
3	Isolated for less than 6 h	Minor procedure, short term low impact effecting animal (e.g., parasite spot-on treatment)	Animal does not show anticipatory stress/fear behaviour before the procedure (e.g., triggered by arrival of vet or change in husbandry) but some mild stress/fear shown afterwards. Recovery from stress takes less than 8 h	Deeper sedation (e.g., asleep) with calm induction and recovery. Rapid return to normal feeding and behaviour.
4	Isolated for less than 12 h	Minor procedure, medium term low impact effecting animal (e.g., nail clipping).	Animal does not show anticipatory stress/fear behaviour before the procedure but mild stress/fear shown afterwards. Recovery from stress takes less than 12 h	Sedation with stressful induction and/or recovery but rapid return to normal feeding and behaviour after procedure
5	Isolated for >12 <24 h	Moderate procedure with short or medium term moderate impact effecting animal.	Animal shows mild anticipatory stress/fear behaviour before and stress/fear after procedure but recovers from stress within 4 h.	Sedation with stressful induction and/or recovery and/or effects on normal feeding and behaviour for a few hours after procedure.
6	Isolated for >24 <48 h	Moderate procedure with longer term moderate impact effecting animal.	Animal shows moderate anticipatory stress/fear behaviour before and moderate stress/fear behaviour after procedure but recovers from stress within 8 h.	Sedation with stressful induction and/or recovery and/or up to 12 h for normal feeding and behaviour to return after procedure.
7	Isolated for more than 2 days	Moderate procedure with longer term serious impact effecting animal.	Animal shows moderate anticipatory stress/fear behaviour before and severe stress/fear behaviour after procedure but recovers from stress within 12 h	Sedation with stressful induction and/or recovery and/or over 12 h for normal feeding and behaviour to return after procedure.
8	Isolated for more than 1 week	Severe procedure with short or medium term moderate impact effecting animal	Animal shows severe anticipatory stress/fear behaviour before and severe stress/fear behaviour after procedure and takes up to 24 h to recover.	Sedation with stressful induction and/or recovery and/or over 24 h for normal feeding and behaviour to return after procedure.
9	Isolated for more than 2 weeks	Extensive procedure with significant impact on animal and short term pain despite appropriate treatment and analgesia (e.g., tail amputation)	Animal shows severe anticipatory stress/fear behaviour before and aggressive behaviour after procedure.	Sedation with highly stressful induction and moderate to long term effects on normal feeding and behaviour after the procedure
10	Isolated for more than 1 month.	Extensive procedure with significant impact on animal and long term pain despite appropriate treatment and analgesia (e.g., tail amputation)	Animal shows severe anticipatory stress/fear behaviour before and aggressive behaviour after procedure. Animal continues to be aggressive to keepers more than 24 h after the procedure	Sedation with highly stressful induction and prolonged effects on normal feeding and behaviour after the procedure

3. Application of the Template to Real Data

The final template derived from the information above consisted of 21 animal- and resource-based welfare indicators, each scored 1–10, with 1 being best possible welfare state and 10 being the worst, based on incrementally defined criteria (Tables 1–4). To calculate a welfare score, the individual factor scores for each parameter were averaged, resulting in four separate scores that were plotted on a two-dimensional grid then linked to form a minimum convex polygon (Figure 6). The resulting area of the polygon provided the cumulative welfare assessment score (CWAS) for that period.

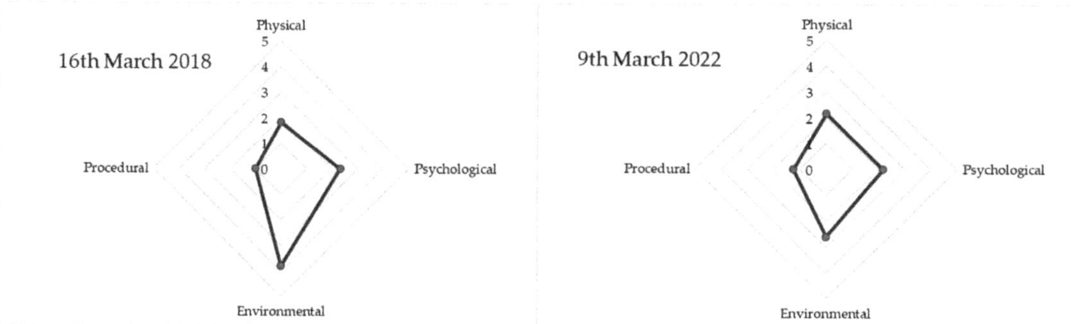

Figure 6. Visual depiction of welfare scores for one individual on two separate days. Reduction in the area of the polygon (from CWAS 9.39 on 16 March 2018 to 8.16 on 9 March 2022) indicates the potential improvement to welfare resulting primarily from a change in habitat.

In order to validate this approach, the adapted template was used to retrospectively assess welfare using the animal care team's daily animal records, which are based on their direct observations of the animals at least twice daily (see Justice, et al. [5] for further details of methods). All assessments were undertaken by MZ's experienced Animal Behaviourist in conjunction with the veterinary team, to maintain consistency of scoring. The data were analysed using dedicated cloud-based software (AWAG, Reuben Digital).

The impact of management decisions on welfare was assessed using the template to score welfare for one individual on a single day in 16 March 2018 to compare with a score for the same individual when living in a different habitat in 9 March 2022 (Figure 6). The primary difference in score between these two periods related to the environmental parameter, reflecting the improvement in suitability of the enclosure for the species (e.g., improved substrate, ventilation, humidity and temperature) and its complexity (e.g., presence of a waterbody, greater climbing/height opportunities).

The template was also used to retrospectively assess the welfare of all four individual *C. obscurus* for the period 9 March 2022 to 24 March 2022. Over this period, the CWAS across the group varied between 6 and 9 (the increase in score indicating reduced welfare), from a total possible score of 200 (Figure 7). The peaks and troughs in scores can be linked to specific incidences noted in the zoo records, as has been highlighted in Figure 7, providing evidence that this tool is sensitive enough to pick up these nuances in welfare state even where changes in score remain small. It also supports the addition of a separate factor for 'skin condition' in addition to 'clinical assessment'. As shown, the main welfare determinants over this period were skin condition, abnormal behaviour and social interaction, all examples where welfare is context dependent.

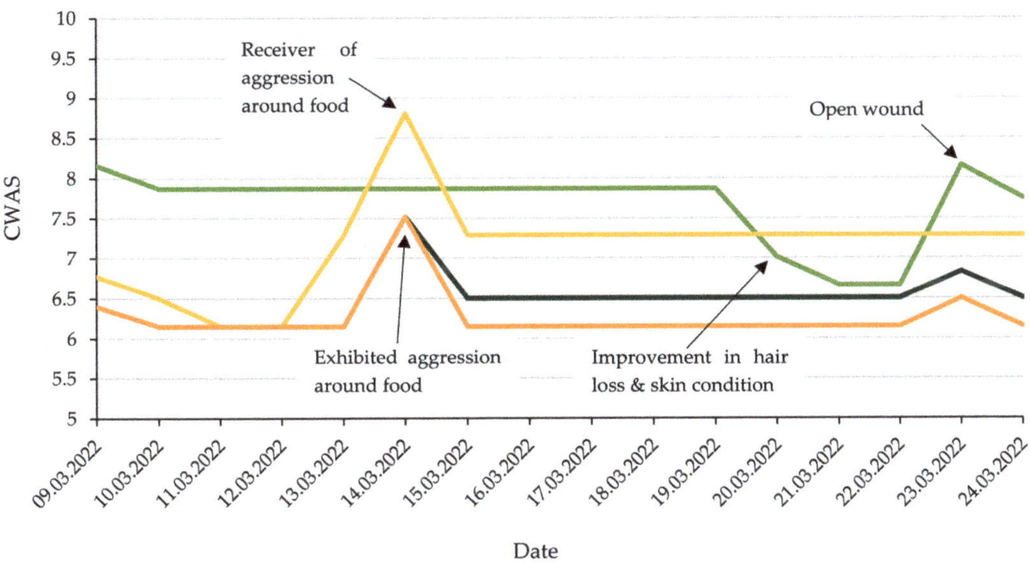

Figure 7. Cumulative welfare assessment scores (CWAS) for the four *C. obscurus* for the dates 9 March to 24 March 2022. Changes in CWAS indicate a likely reduction or improvement in welfare and have been highlighted alongside specific incidences that occurred on the day. Please note, the y-axis was adjusted from a minimum of 0 and maximum of 200 to emphasise the changes in CWAS which highlight discrete events that impact welfare.

4. Discussion

The aim of this study was to demonstrate how a welfare assessment can be created for a species with scarce published information available. Using the specifically designed AWAG template to retrospectively score daily animal records, this approach has been validated. In addition, it has highlighted the benefits of using behavioural observations and zoo records to provide context-dependent information to support the information gathered from the literature.

Reviewing the available literature is a key step in the process of designing a welfare assessment and will save the researcher both time and resources by removing the need to gather this information first-hand. However, when dealing with cryptic species, as many zoo-housed animals are, one of the limitations faced is the lack of published literature, resulting in welfare assessments based, sometimes, on only one or two wild observations. When sources are limited, the information presented must be considered in the original context and the relevance to captivity not exaggerated. For the purpose of welfare assessments, in some cases it may be better to avoid comparison to the wild environment and instead focus on how the captive environment provides for the needs of the species, placing more emphasis on animal-based factors. Some behaviours relevant to welfare may not have been observed in the wild, for example no evidence could be found of wild *C. obscurus* sunbathing, a behaviour commonly seen in other mongoose species [17], yet it is a behaviour seen exhibited by captive *C. obscurus* [27,29] and provision of access to sunlight is recommended in the AZA guidelines [22]. As evidenced here, direct observations of captive individuals can be used to support information gathered from the literature. They provide the opportunity to site individual health and behaviour within context, which is vital for an accurate understanding of that individual animal's welfare. However, care should be taken not to extrapolate generalisations from observations based on small sample sizes.

A limitation of this template in its current form is the inclusion of multiple resource-based indicators to assess welfare. Whilst utilising both resource- and animal-based indicators can provide a greater holistic understanding of welfare, good husbandry and care, or 'inputs', do not necessarily result in good welfare. Although resource-based factors can be used as a proxy for what the animal might be experiencing, only by assessing animal-based factors is it possible to ascertain the animal's likely mental state in response to the provided resources. The list of validated welfare measures is long; therefore, to produce a practical assessment, welfare measures need to be chosen depending on the context. Resource-based factors are often quantifiable, non-invasive, quick to assess and easily replicated, and having been used and validated with various species in different contexts they remain popular. Animal-based factors are still being developed for welfare assessment and there is currently a lack of information on affective states in many zoo-housed species. At present, this multi-faceted approach is valuable in the absence of being able to obtain all the evidence from animal-based factors.

Management decisions in captive environments should be based on scientifically validated evidence, preferably collected over time. For this study, only 16 days of CWAS were assessed to validate the methods; however, continuous monitoring over time is more likely to accurately reflect the impact of life stage or seasonal change on welfare compared with point-in-time audits [23]. One of the key advantages to using the AWAG is that welfare can be rapidly and easily scored, recorded and reviewed at regular frequencies, enabling continuous assessment over the animal's lifetime. This permits prompt identification of changes to specific contexts where welfare may be compromised, allowing the necessary adjustments to be made and their impact to be monitored.

Whilst this template has been successfully validated with *C. obscurus* at MZ, it is important to highlight that much of the data presented have been gathered from a small sample of individuals of the same age, and care should be taken if extrapolating this information to other individuals of a different age. Several suggested changes also resulted from the trial. For this species there was no clear link between 'faecal consistency' and welfare, so further evaluation of this factor as an indicator of welfare may be necessary. Similarly, the previously discussed limitations relating to the use of the wild environment as a benchmark suggest the factor definition for environmental complexity should be re-evaluated. The factor definitions for 'nutrition' could also be adapted to incorporate a time component as food items that may improve welfare short-term (consider the dopamine hit from eating sugary foods) but lead to decreased welfare in the long-term need to be accounted for. Finally, future score definitions could place more focus on animal-based factors, for example, preference testing, cognitive bias and Qualitative Behaviour Assessment.

5. Conclusions

Species-specific knowledge is a crucial part of developing the AWAG template's relevance for use with zoo species. This study demonstrated the development of the AWAG for an understudied species, *C. obscurus,* for which there is little published literature, using behavioural observations and zoo records to place that information within the specific environmental, social and individual context. Limitations of the methods, such as utilising resource-based factors, have been addressed, and future changes to this specific template have been highlighted. However, retrospectively scoring the welfare of the *C. obscurus* group at MZ validated the use of this tool for identifying factors that may have impacted animal welfare (in this instance, aggression, possibly the result of unnatural social dynamics; alopecia and skin lesions; and the environment). Consequently, using this methodology the AWAG demonstrated how the environmental changes likely improved animal welfare based on the features and complexity of the wild environment in which the species evolved (e.g., improved substrate, ventilation, humidity and temperature, presence of a waterbody and greater climbing/height opportunities). The AWAG is a flexible continuous welfare monitoring tool using scoring templates that can, and should, be regularly reviewed and updated with the latest knowledge as it becomes available, supporting the development of evidence-based management practices that promote the welfare of captive wild animals.

Author Contributions: Conceptualisation, D.F., W.S.M.J.; Methodology, D.F., W.S.M.J.; Validation, D.F., W.S.M.J., S.W.; Formal Analysis, D.F.; Investigation, D.F., V.H.; Data Curation, D.F., V.H.; Writing—Original Draft Preparation, D.F., S.J.S., V.H.; Writing—Review and Editing, D.F., W.S.M.J., S.W., S.J.S.; Project Administration, D.F. All authors have read and agreed to the published version of the manuscript.

Funding: The publication of this work was supported by the University of Surrey.

Institutional Review Board Statement: Ethical review and approval were waived in this study as there was no manipulation to the study animal or their environment.

Informed Consent Statement: Not applicable.

Data Availability Statement: The data presented in this study are available on request from the corresponding author.

Acknowledgments: The authors would like to thank Marwell Zoo's Carnivore Team; work experience students for their support. We also wish to thank the anonymous reviewers for their helpful feedback.

Conflicts of Interest: The authors declare no conflict of interest.

References

1. Broom, D.M. Animal welfare defined in terms of attempts to cope with the environment. *Acta Agric. Scand. Sect. A Anim. Sci. Suppl.* **1996**, *27*, 22–28.
2. Bracke, M.B.; Spruijt, B.M.; Metz, J.H. Overall animal welfare assessment reviewed. Part 1: Is it possible? *Neth. J. Agric. Sci.* **1999**, *47*, 279–291. [CrossRef]
3. Buller Blokhuis, H.; Jensen, P.; Keeling, L. Towards Farm Animal Welfare and Sustainability. *Animals* **2018**, *8*, 81. [CrossRef]
4. Farm Animal Welfare Council. *Farm Animal Welfare in Great Britain: Past, Present and Future*; Farm Animal Welfare Council: London, UK, 2009.
5. Justice, W.S.M.; O'Brien, M.F.; Szyszka, O.; Shotton, J.; Gilmour, J.E.M.; Riordan, P.; Wolfensohn, S. Adaptation of the animal welfare assessment grid (AWAG) for monitoring animal welfare in zoological collections. *Vet. Rec.* **2017**, *181*, 143. [CrossRef]
6. Wolfensohn, S.; Shotton, J.; Bowley, H.; Davies, S.; Thompson, S.; Justice, W.S.M. Assessment of welfare in zoo animals: Towards optimum quality of life. *Animals* **2018**, *8*, 110. [CrossRef] [PubMed]
7. Brouwers, S.P.; José, M.; Duchateau, H.M. Feasibility and validity of the Animal Welfare Assessment Grid to monitor the welfare of zoo-housed gorillas Gorilla gorilla gorilla. *J. Zoo Aquar. Res.* **2021**, *9*, 208–217. [CrossRef]
8. Angelici, F.M.; di Vittorio, M. Common Cusimanse Crossarchus obscurus in Ghana and Flat-headed Cusimanse C. platycephalus in Nigeria: A tentative comparison between habitat parameters affecting their distribution. *Small Carniv. Conserv.* **2013**, *48*, 96–100.
9. Goldman, C.A. Crossarchus obscurus. *Mamm. Species* **1987**, *290*, 1–15. [CrossRef]
10. Kingdon, J. *The Kingdon Field Guide to African Mammals*; Academic Press: London, UK, 1997.
11. Djagoun, C.A.M.S.; Akpona, H.A.; Sinsin, B.; Mensah, G.A.; Dossa, N.F. Mongoose species in southern Benin: Preliminary ecological survey and local community perceptions. *Mammalia* **2009**, *73*, 27–32. [CrossRef]
12. Angelici, F.; do Lin Sanh, E. Crossarchus Obscurus, Common Cusimanse. The IUCN Red List of Threatened Species™. 2015. Available online: https://doi.org/10.2305/IUCN.UK.2015-4.RLTS.T41595A45205532.en (accessed on 13 June 2022). [CrossRef]
13. Olson, A.L. The behavior and ecology of the long-nosed mongoose, Crossarchus obscurus. Ph.D. Thesis, University of Miami, Coral Gables, FL, USA, 2001.
14. Nowak, R.M.; Walker, E.P. Carnivora; Viverridae: Cusimanses. In *Walker's Mammals of the World*, 6th ed.; Johns Hopkins University Press: Baltimore, MD, USA, 1999; Volume 1, p. 1168.
15. Hunter, L. *Carnivores of the World*, 1st ed.; Princeton University Press: Princeton, NJ, USA, 2011.
16. Jennings, A.; Veron, G. *Mongooses of the World*; Whittles Publishing, Ltd.: Dunbeath, UK, 2019.
17. Estes, R.D. Genets, Civets, and Mongooses: Family Viverridae. In *The Behaviour Guide to African Mammals: Including Hoofed Mammals, Carnivores, Primates*; University of California Press, Ltd.: London, UK, 1991; pp. 278–322.
18. Decker, D.M.; Ringelberg, D.; White, D.C. Lipid components in anal scent sacs of three mongoose species (Helogale parvula, Crossarchus obscurus, Suricata suricatta). *J. Chem. Ecol.* **1992**, *18*, 1511–1524. [CrossRef]
19. Struhsaker, T.T.; McKey, D. Two Cusimanse Mongooses Attack a Black Cobra. *J. Mammal.* **1975**, *56*, 721–722. [CrossRef]
20. Djagoun, C.A.M.S.; Gaubert, P. Small carnivorans from southern Benin: A preliminary assessment of diversity and hunting pressure. *Small Carniv. Conserv.* **2009**, *40*, 1–10.
21. Bush, E.R.; Jeffery, K.; Bunnefeld, N.; Tutin, C.; Musgrave, R.; Moussavou, G.; Mihindou, V.; Malhi, Y.; Lehmann, D.; Ndong, J.E.; et al. Rare ground data confirm significant warming and drying in western equatorial Africa. *PeerJ* **2020**, *8*, e8732. [CrossRef]

22. AZA Small Carnivore TAG. *Mongoose, Meerkat, & Fossa (Herpestidae/Eupleridae) Care Manual*; Association of Zoos and Aquariums: Silver Spring, MD, USA, 2011.
23. Brando, S.; Buchanan-Smith, H.M. The 24/7 approach to promoting optimal welfare for captive wild animals. *Behav. Processes* **2018**, *156*, 83–95. [CrossRef]
24. Queiroz, M.B.; Young, R.J. The Different Physical and Behavioural Characteristics of Zoo Mammals That Influence Their Response to Visitors. *Animals* **2018**, *8*, 139. [CrossRef]
25. Sherwen, S.L.; Hemsworth, P.H. The Visitor Effect on Zoo Animals: Implications and Opportunities for Zoo Animal Welfare. *Animals* **2019**, *9*, 366. [CrossRef] [PubMed]
26. Van Zeeland, Y.; Schoemaker, N. Current anaesthetic considerations and techniques in rabbits Part I: Pre-anaesthetic considerations and commonly used analgesics and anaesthetics. *EJCAP* **2014**, *24*, 19–30.
27. Totton, J. Cholesterol in the Common Cusimanse (*Crossarchus obscurus*). An Intake Study: Observing the Effects of a Diet Change. Master's Thesis, Zoo Conservation Biology Research Project, Plymouth University, Plymouth, UK, 2016.
28. Totten, J.; Plowman, A. *A Potentially Cholesterol-Reducing Diet is Palatable and Practical for Cusimanse Crossarchus obscurus*; Whitley Wildlife Conservation Trust (Paignton Zoo Environmental Park); Biological Sciences, Plymouth University: Plymouth, UK, 2017.
29. Arnold, C. *Personal Communication*; Marwell Zoo: Winchester, UK, 2022.
30. Flecknell, P.A.; Cruz, I.J.; Liles, J.H.; Whelan, G. Induction of anaesthesia with halothane and isoflurane in the rabbit: A comparison of the use of a face-mask or an anaesthetic chamber. *Lab. Anim.* **1996**, *30*, 67–74. [CrossRef]
31. Greening, L. Stereotypies and other abnormal behavior in welfare assessment. In *Encyclopedia of Animal Behavior*, 2nd ed.; Choe, J., Ed.; Academic Press: Gloucester, UK, 2019; Volume 1, pp. 141–146.
32. Department for Environment Food and Rural Affairs. *Zoos Expert Committee Handbook*; Department for Environment Food and Rural Affairs: Bristol, UK, 2012.
33. Wechsler, B. Coping and coping strategies: A behavioural view. *Appl. Anim. Behav. Sci.* **1995**, *43*, 123–134. [CrossRef]
34. Wolfensohn, S.; Lloyd, M. *Handbook of Laboratory Animal Management and Welfare*, 4th ed.; Wiley: Hoboken, NJ, USA, 2013.
35. Wolfensohn, S.; Sharpe, S.; Hall, I.; Lawrence, S.; Kitchen, S.; Dennis, M. Refinement of welfare through development of a quantitative system for assessment of lifetime experience. *Anim. Welf.* **2015**, *24*, 139–149. [CrossRef]
36. Freeman, L.M.; Lachaud, M.P.; Matthews, S.; Rhodes, L.; Zollers, B. Evaluation of Weight Loss over Time in Cats with Chronic Kidney Disease. *J. Vet. Intern. Med.* **2016**, *30*, 1661–1666. [CrossRef] [PubMed]
37. Morfeld, K.A.; Meehan, C.L.; Hogan, J.N.; Brown, J.L. Assessment of Body Condition in African (*Loxodonta africana*) and Asian (*Elephas maximus*) Elephants in North American Zoos and Management Practices Associated with High Body Condition Scores. *PLoS ONE* **2016**, *11*, e0155146. [CrossRef] [PubMed]
38. Teng, K.T.; McGreevy, P.D.; Toribio, J.A.L.M.L.; Raubenheimer, D.; Kendall, K.; Dhand, N.K. Associations of body condition score with health conditions related to overweight and obesity in cats. *J. Small Anim. Pract.* **2018**, *59*, 603–615. [CrossRef] [PubMed]
39. Hut, P.R.; Hostens, M.M.; Beijaard, M.J.; van Eerdenburg, F.J.C.M.; Hulsen, J.H.J.L.; Hooijer, G.A.; Stassen, E.N.; Nielen, M. Associations between body condition score, locomotion score, and sensor-based time budgets of dairy cattle during the dry period and early lactation. *J. Dairy Sci.* **2021**, *104*, 4746–4763. [CrossRef] [PubMed]
40. Schiffmann, C.; Clauss, M.; Hoby, S.; Hatt, J.M. Visual body condition scoring in zoo animals—Composite, algorithm and overview approaches. *J. Zoo Aquar. Res.* **2017**, *5*, 1–10. [CrossRef]
41. Binding, S.; Farmer, H.; Krusin, L.; Cronin, K. Status of animal welfare research in zoos and aquariums: Where are we, where to next? *J. Zoo Aquar. Res.* **2020**, *8*, 166–174.
42. Moore, R.A.; Waheed, A.; Burns, B. Rule of Nines. In *StatPearls [Internet]*; StatPearls Publishing: Treasure Island, FL, USA, 2021.
43. Hosey, G. A preliminary model of human-animal relationships in the zoo. *Appl. Anim. Behav. Sci.* **2008**, *109*, 105–127. [CrossRef]
44. Martin, R.A.; Melfi, V. A Comparison of Zoo Animal Behavior in the Presence of Familiar and Unfamiliar People. *J. Appl. Anim. Welf. Sci.* **2016**, *19*, 234–244. [CrossRef]
45. Mellor, D.J.; Beausoleil, N.J. Extending the 'Five Domains' model for animal welfare assessment to incorporate positive welfare states. *Anim. Welf.* **2015**, *24*, 241–253. [CrossRef]
46. Sharma, A.; Phillips, C.J.C. Avoidance distance in sheltered cows and its association with other welfare parameters. *Animals* **2019**, *9*, 396. [CrossRef]
47. Watters, J. Searching for behavioral indicators of welfare in zoos: Uncovering anticipatory behavior. *Zoo Biol.* **2014**, *33*, 251–256. [CrossRef] [PubMed]
48. Ward, S.J.; Sherwen, S.; Clark, F.E. Advances in applied zoo animal welfare science. *J. Appl. Anim. Welf. Sci.* **2018**, *21*, 23–33. [CrossRef] [PubMed]
49. Boissy, A.; Manteuffel, G.; Jensen, M.B.; Moe, R.O.; Spruijt, B.; Keeling, L.J.; Winckler, C.; Forkman, B.; Dimitrov, I.; Langbein, J.; et al. Assessment of positive emotions in animals to improve their welfare. *Physiol. Behav.* **2007**, *92*, 375–397. [CrossRef] [PubMed]
50. Jones, N.; Sherwen, S.L.; Robbins, R.; McLelland, D.J.; Whittaker, A.L. Welfare Assessment Tools in Zoos: From Theory to Practice. *Vet. Sci.* **2022**, *9*, 170. [CrossRef]
51. Miller, L.J.; Vicino, G.A.; Sheftel, J.; Lauderdale, L.K. Behavioral Diversity as a Potential Indicator of Positive Animal Welfare. *Animals* **2020**, *10*, 1211. [CrossRef]
52. Mellor, D.J.; Beausoleil, N.J.; Littlewood, K.E.; McLean, A.N.; McGreevy, P.D.; Jones, B.; Wilkins, C. The 2020 Five Domains Model: Including Human–Animal Interactions in Assessments of Animal Welfare. *Animals* **2020**, *10*, 1870. [CrossRef]

Journal of
Zoological and Botanical Gardens

MDPI

Article

Location, Location, Location! Evaluating Space Use of Captive Aquatic Species—A Case Study with Elasmobranchs

Alexis M. Hart [1,*], Zac Reynolds [2] and Sandra M. Troxell-Smith [1,*]

[1] Department of Biological Sciences, Oakland University, Rochester, MI 48309, USA
[2] SEA LIFE Michigan Aquarium, Auburn Hills, MI 48326, USA; zacreynolds313@gmail.com
* Correspondence: alexismhart@outlook.com (A.M.H.); stroxellsmith@oakland.edu (S.M.T.-S.)

Abstract: The space use of captive animals has been reliably used as a tool to measure animal welfare in recent years. However, most analyses of space use focus primarily on terrestrial animals, with very little emphasis placed on the space use of aquatic animals. By comparing the space use of these animals to their natural histories and what would be expected of them physiologically, a general assessment of their overall welfare can be obtained. Using the Zoomonitor program, this study investigated the space use of five elasmobranch species housed in a captive aquatic environment: a blacktip reef shark (*Carcharhinus melanopterus*), a nurse shark (*Ginglymostoma cirratum*), a smooth dogfish (*Mustelus canis*), a bonnethead shark (*Sphyrna tiburo*), and a blacknose shark (*Carcharhinus acronotus*). The exhibit was delineated into five different zones: three represented the animal locations along the X/Y axis ('Exhibit Use'), and two zones were related to the Z-axis ('Depth Use'). The location of each individual on both the X/Y and Z axes was recorded during each observation. Heat maps generated from the Zoomonitor program were used in conjunction with the Spread of Participation Index (SPI) to interpret the data. It was found that while all the individuals used their given space differently, the Exhibit Use was relatively even overall (the SPI values ranged from 0.0378 to 0.367), while the Depth Use was more uneven (the SPI ranged from 0.679 to 0.922). These results mostly reflected what would be expected based on the species' natural histories. However, for the smooth dogfish, the observed Exhibit Use and activity patterns revealed a mismatch between the anticipated and the actual results, leading to further interventions. As demonstrated here, space use results can be utilized to make positive changes to husbandry routines and enclosure designs for aquatic individuals; they are thus an important additional welfare measure to consider for aquatic species.

Keywords: elasmobranch; sharks; space use; ZooMonitor; spread of participation index; animal welfare

Citation: Hart, A.M.; Reynolds, Z.; Troxell-Smith, S.M. Location, Location, Location! Evaluating Space Use of Captive Aquatic Species—A Case Study with Elasmobranchs. *J. Zool. Bot. Gard.* **2022**, *3*, 246–255. https://doi.org/10.3390/jzbg3020020

Academic Editors: Kris Descovich, Caralyn Kemp and Jessica Rendle

Received: 13 March 2022
Accepted: 3 June 2022
Published: 7 June 2022

1. Introduction

Although the space use of animals in captivity has been studied both formally and informally for decades, it has only recently been introduced as an indicator of animal welfare [1–3]. Even space use is typically anticipated for captive animals in a good welfare state, as it suggests that the animals do not actively avoid any areas in their habitat and willingly utilize their enclosure to its fullest potential [4,5]. However, it is also important to note that species' natural history or certain physiological elements can also influence activity level and space use [6–8]. In particular, species' natural history must be considered to ensure that enclosures provide appropriate opportunities for species-typical behaviors [9,10]. Both the specific behaviors of the focal subject(s) and the areas in which they display these behaviors are essential to the evaluation of enclosures. Therefore, an indicator of good enclosure design for an animal is whether the animal uses its enclosure in a way that would be expected for its species.

While space-use evaluations have become more commonplace as metrics of welfare in terrestrial animals, they are not yet widely applied to aquatic species, particularly teleosts and elasmobranchs [11]. Consistent considerations in both applications include establishing a behavioral repertoire of species-specific behaviors and an understanding of natural history. Factors that may uniquely affect aquatic animal space use include the chemical parameters and flow of water, the vibration of pumps and other equipment, and the depth of the environment [12,13]. The influence of many of these factors may not be immediately apparent to caretakers, but may be reflected in the enclosure location choices of aquatic animals. Therefore, the consistent documentation of enclosure use for aquatic species could prove to be even more vital to welfare than its use for terrestrial species.

The reliable documentation and quantification of space use for both aquatic and terrestrial animals can be performed quickly and efficiently when using the right tools. The ZooMonitor program [14] is a web application that allows data to be collected on captive-animal behavior and space use with ease. A project can be created based on research needs, and any focal animals can be entered for behavioral data collection. An image of an enclosure can then be uploaded onto the application, and animal location data can be collected at preset intervals by selecting where the individual was in the enclosure image at any given time [5]. Many enclosure-use studies have utilized the ZooMonitor program to collect data on the space use of a variety of terrestrial zoo animals [14–17]; however, few articles have been published about space use in aquatic environments [18].

The space-use data collected from ZooMontitor can be easily evaluated using a variety of post-occupancy evaluations. Post-occupancy evaluations (POEs) were originally used to determine how effectively space was used in occupied industrial buildings [19,20]. POEs involve assessing the utilization of a given space, as well as interviewing individuals who use the area being evaluated and gaining insight into their level of satisfaction with the space. The insight gained from these evaluations would historically be used to steer architectural changes in how buildings were designed [20]. However, in recent years, POEs have found additional practical applications in studies of animal enclosure use [21]. All POEs usually involve dividing a given space into zones, and then running analyses based on how those zones are used in relation to the entire space [11]. Many different POEs that can be used to interpret how an animal explores its enclosure and interacts with the resources within it [11]. Brereton discusses four main methods that are used to evaluate space use in captive animal species: zone occupancy, Dickens' [22] Spread of Participation Index (SPI), Plowman's [23] Modified Spread of Participation Index, and Vanderploeg and Scavia's [24] Electivity Index. Each of these methods evaluates unique aspects of enclosure use. For example, zone occupancy is used to report the percentage of time a specific zone is in use, whereas SPI is used to determine how evenly a given space is used. Both the Modified SPI and the Electivity Index can be used to determine how resources in space are utilized. The evaluation method chosen ultimately depends on which variables are examined in the study [25]. In addition, while all of these methods have their own merits, very few of them have been used to assess aquatic populations [11].

The lack of quantifiable data on welfare outcomes in aquatic populations has driven the formation of the Association of Zoo and Aquariums' (AZA) Aquatic Collection Sustainability Committee, with the expressed goal of encouraging the proper consideration, documentation, and assessment of welfare indicators in aquatic collections [26]. With that goal in mind, this study aims to raise awareness of the importance of documenting space-use data on captive aquatic species, and of how this information can then be utilized to improve welfare outcomes.

2. Materials and Methods

2.1. Subjects

There were five focal subjects in this study: a female bonnethead shark *(Sphyrna tiburo)*, a female blacknose shark (*Carcharhinus acronotus*), a male blacktip reef shark (*Carcharhinus melanopterus*), a female smooth dogfish (*Mustelus canis*), and a female nurse shark (*Ginglymostoma cirratum*). All individuals resided together with other animals in a mixed-species exhibit.

2.2. Exhibit

Data collection took place at the SEA LIFE Michigan Aquarium. The focal individuals resided in the Ocean Exhibit, which had a volume of 473,000 L and a depth of 7.3 m (Figure 1). This exhibit is designed to mimic conditions of ocean ecosystems, with dissolved oxygen concentration held at 98%, salinity held at 29–30 ppt, temperature held at 24–25 °C, and a photoperiod of 14:10. In addition to the five focal individuals, the Ocean Exhibit is also home to two green sea turtles (*Chelonia mydas*), roughly 250 teleost fish (including golden trevallies (*Gnathanodon speciosus*), tangs (*Naso* sp.), a Goliath grouper (*Epinephelus itajara*), among other assorted tropical marine species, and two dozen other elasmobranchs, including cownose rays (*Rhinoptera bonasus*) and southern stingrays (*Hypanus americanus*). With the exception of the blacktip reef shark, all individuals in this study were the only individuals of their species in this exhibit. The male blacktip reef shark was specifically chosen for data collection, as he was the easiest to reliably differentiate from the other conspecifics.

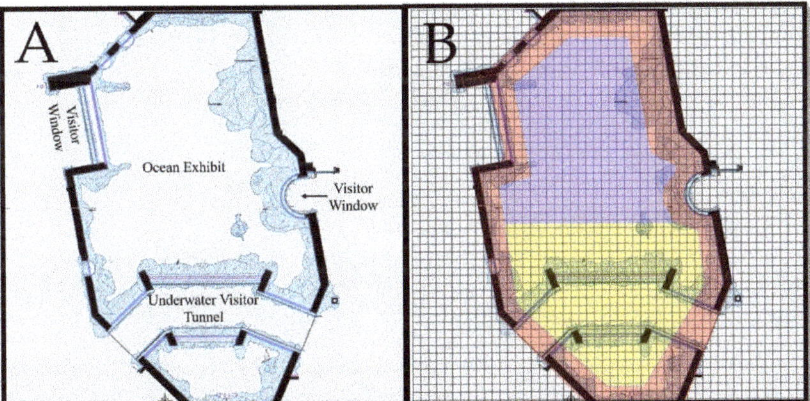

Figure 1. (**A**) The 2-D map of the Ocean Exhibit used to collect the Exhibit Use data. The thick black lines represent the perimeter of the exhibit, whereas the blue shades represent vertical rock formations within the tank. The 'underwater visitor tunnel' was a submerged glass tunnel for visitors to walk through, over which the exhibit animals could swim. (**B**) Map with zones indicated, and the 50 × 50centimeter grid used to ensure the zones were of equal size. Zone 'A' is colored in yellow, Zone 'B' is blue, and Zone 'C' is red.

2.3. Data Collection

Data were collected from 13 December 2018 to 18 June 2019, and data collection sessions were conducted one to two times a week. To ensure well-represented data and attempt to prevent selection biases, every week, a random number generator was used to determine both the weekdays and the time of day that data would be collected. Data were recorded using a tablet with the ZooMonitor program, which allows users to easily input the location of an individual at predetermined intervals [14]. Two observers collected data: one aquarist from SEA LIFE Michigan and one university student studying animal behavior. Prior to data collection, the observers conducted several practice sessions where both observers would record location of an individual simultaneously in order to determine whether results were consistent. Following these practice sessions, an inter-observer reliability test was conducted, which yielded nearly perfect similarity (>90%) in data collection from both observers. Both observers continued to collect data throughout the observation period.

Observational sessions for each individual lasted ten minutes, and focal scan sampling of location was performed at one-minute intervals [27]. If the focal individual was not visible at the one-minute interval, no data were recorded. All focal individuals were recorded once per observation day, and the order in which individuals were observed was also randomized each day of data collection via a random number generator.

2.4. Data Analysis

The data collected from ZooMonitor were used to generate heat maps for all individuals using Microsoft Excel's 3D Map feature. A 2-D map of the exhibit was uploaded into the Zoomonitor program to allow documentation of animal location along the X/Y axis during observations (hereafter referred to as Exhibit Use; Figure 1A). A 50 × 50-centimeter grid was then placed over the exhibit map in order to divide the enclosure into three equal sections of 245 cm^2 (Figure 1B). The area of each zone was calculated by hand using the over-laid grid. The zones were designated as the front of the exhibit (zone 'A', colored in yellow on the map), the back of the exhibit (zone 'B', colored in blue), and the perimeter of the exhibit (zone 'C', colored in red) (Figure 1B). The exhibit was split into equal 'zones' in order to determine whether space use was relatively even overall throughout the exhibit. Even though the zones were equal in size, they were unique in composition. Notably, zone 'A' included the visitor tunnel, which is a large viewing area for guests to walk through. Zone 'B' had far fewer views to offer guests and more open space. Zone 'C' included the rock formations along the perimeter of the exhibit. In addition, the animal depth (Z-axis) was also documented by recording whether the animal was located in the upper 50% or lower 50% of the water column at the time of the observation.

The heat maps generated by Zoomonitor display individual data points as colored dots, and the density of data points at a given location is determined by color [5,28]. Blues and greens indicate a low density of data points, whereas yellows and reds indicate a higher density. The number of data points in each of the three zones for all five individuals was determined following data collection. As all zones chosen were of equal size, if the focal animals used all exhibit space effectively, they were observed in each zone evenly.

The effectiveness of enclosure use for the animals in this study was measured using Dickens' Spread of Participation Index (SPI) [22]. This method of analysis was chosen as it compares evenness of space use for individuals [11]. As our primary goal was to simply determine how the animals in the Ocean Exhibit used their space, and because all three zones chosen for this study were of the same size, this index was determined to be the most appropriate. Moreover, even though space use in aquatic populations is extremely understudied, this method has been used previously to evaluate space use in aquatic habitats [11].

The equation for SPI is as follows:

$$\frac{M(nb - na) + (Fa - Fb)}{2(N - M)}$$

where M is the mean frequency of observations in all pre-determined zones, N is the total number of observations, nb and na is the number of zones with observations less than or greater than M, respectively, and Fb and Fa are the number of observations in those zones [22]. A value of 0 indicated perfectly even space use, whereas a value of 1 indicated highly uneven space use. Two SPI values were calculated: one for the Exhibit Use (XY-axis), and a separate value for Depth Use (Z-axis). The index values ascertained from all individuals were then compared to each other and to what would be expected of the species in its natural environment.

3. Results

In total, 1214 observations were recorded in total for all the individuals, with an average of 243 observations recorded for each individual. There were 30 days of data collection in total, with observation times ranging anywhere from 8AM to 4:15PM. The overall number of data points for each individual in each of the zones is displayed in Table 1. These values were used to calculate the SPIs for all the individuals.

Table 1. Total number of data points in each zone and calculated SPI values for all five focal individuals. The 'Exhibit Use' section compares evenness for the front, back, and perimeter zones (i.e., XY-axis), whereas the 'Depth Use' section compares evenness for the upper and lower water columns (i.e., Z-axis). The ♦ symbol indicates the species with the most even space use, whereas the ♦♦ symbol indicates the species with the most uneven space use.

	Exhibit Use					Depth Use			
	Number of Data Points in Each Zone					Number of Data Points in Each Zone			
Individual	**A**	**B**	**C**	**Total**	**SPI**	**Upper Water Column**	**Lower Water Column**	**Total**	**SPI**
Blacktip reef shark	85 (35.9%)	78 (32.9%)	74 (31.2%)	237	0.0378 ♦	199 (84.0%)	38 (16.0%)	237	0.679 ♦
Bonnethead shark	104 (42.6%)	94 (38.5%)	46 (18.9%)	244	0.217	234 (95.9%)	10 (4.1%)	244	0.922 ♦♦
Blacknose shark	54 (20.4%)	73 (27.7%)	137 (51.9%)	264	0.278	242 (91.7%)	22 (8.3%)	264	0.833
Smooth dogfish shark	72 (31.2%)	33 (14.3%)	126 (54.5%)	231	0.318	221 (95.7%)	10 (4.3%)	231	0.913
Nurse shark	110 (46.2%)	107 (45.0%)	21 (8.8%)	238	0.367 ♦♦	30 (12.6%)	208 (87.4%)	238	0.761

A range of Exhibit Use SPIs was determined for all five individuals, which varied from 0.0378 (indicating very even space use) to 0.367 (indicating less even space use). The nurse shark had the most uneven space use (SPI = 0.367), followed by the smooth dogfish and the blacknose shark. The individuals with the most even Exhibit Use were the bonnethead shark and the blacktip reef shark (SPI = 0.0378; Table 1).

For the Depth Use, the SPI values ranged between 0.679 and 0.922. The individuals that had the most even relative space use for these zones were the blacktip reef shark and the nurse shark. The individuals with the most uneven space use were the blacknose shark, the smooth dogfish, and the bonnethead shark (Table 1).

4. Discussion

Overall, these results display a relatively expected level of space use for all five focal individuals, given their unique natural histories. The non-quantitative view of the heat maps showed that each animal uses their given space uniquely, with some sharks preferring certain areas over others. All five individuals in the study were observed in all three of the aforementioned zones, although the degree to which a zone was utilized varied by individual. For example, while the blacktip reef shark appeared to prefer certain areas of the exhibit (such as the cluster between zones A and C of the exhibit; Figure 2C), the overall Exhibit Use was extremely even (SPI = 0.0378; Table 1).

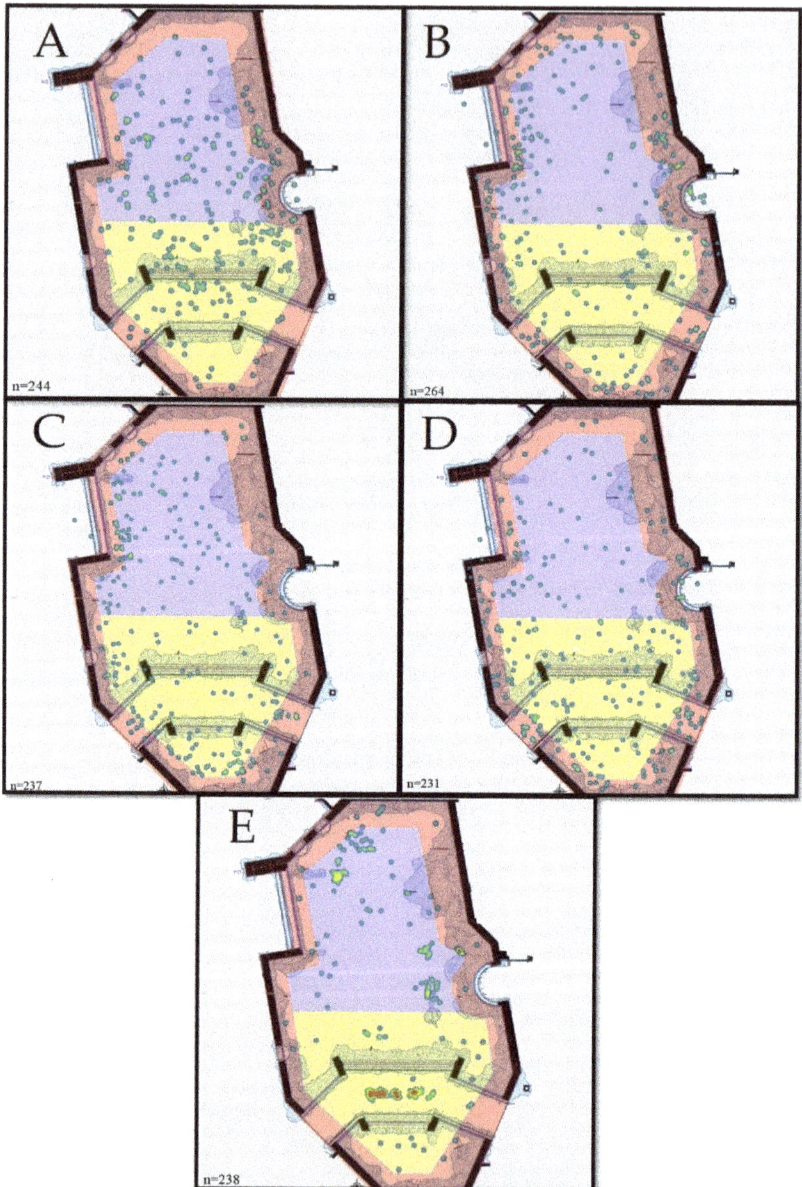

Figure 2. Heat maps generated for the Exhibit Use of all five individuals. The heat maps are labeled as follows: (**A**) bonnethead shark; (**B**) blacknose shark; (**C**) blacktip reef shark; (**D**) smooth dogfish; and (**E**) nurse shark. Zone 'A' is colored in yellow, zone 'B' is blue, and zone 'C' is red. Blue and green dots represent 1–2 data points, whereas reds and yellows represent large clusters of data points (3 or more).

As previously stated, when examining Exhibit Use, it is important to consider the physiology and natural history of an animal. The individual who had the most uneven Exhibit Use, the nurse shark, had an SPI of 0.367 (Table 1). The heat map for this individual (Figure 2E) shows that the nurse shark was indeed observed in all three zones, but preferred very distinct areas within the exhibit (shown by the clusters of red in the figure). Notably, the areas with the greatest concentration of observations of the nurse shark were above the viewing tunnel. Following the data collection, it was discovered that this preferred location was also near a high-flow pump. While several factors could have influenced her preferences, a possible explanation could also lie within the natural history of her species. Importantly, unlike the other shark species in this study (with the exception of the smooth dogfish), nurse sharks are not obligate ram ventilators, meaning they do not have to continually move in order to supply their body with oxygen [29–32]. Nurse sharks instead use a specialized organ, called a buccal pump, to move water over their gills while remaining motionless [31,33]. Thus, nurse sharks are typically considered to be highly sedentary because they have a higher cost for metabolic activity compared to other shark species [8]. With this in mind, it is possible that by positioning herself near the high-flow pump, the nurse shark in this study achieved even greater oxygen exchange with minimal effort. In addition, it was unsurprising that the nurse shark had a high SPI value for depth and was most often found in the lower water column (Table 1), because nurse sharks are primarily a benthic species [33,34] and characteristically spend most of their time resting on the seafloor [29]. These results reinforce the importance of taking natural history elements into account when evaluating space-use results, as uneven space use may not necessarily be a cause for welfare concern in some species.

While both nurse sharks and smooth dogfish are known to be primarily sedentary [8,35], blacktip reef sharks, bonnethead sharks, and blacknose sharks are considered obligate ram ventilators, who must therefore must move continuously in order to receive oxygen [31,36,37]. These three species are therefore typically considered highly exploratory (compared to the smooth dogfish and nurse shark), and, indeed, the SPIs and the heat maps of these species reflected this: all three individuals utilized their exhibits relatively evenly (Table 1). In addition, these three species are generally pelagic, and tend to appear in the open water [34]. It is therefore unsurprising that the data for depth show that these individuals tended to prefer the upper water column.

The notable exception to the species-appropriate Exhibit Use results was the smooth dogfish. Although smooth dogfish are buccal-pumping sharks and can frequently be found at rest [35,38], the smooth dogfish in our experiment appeared to be continuously moving throughout the exhibit and was never observed to be motionless throughout the data collection period. This is uncharacteristic of what would be expected physiologically [30]. In addition, she also had the second-most uneven Exhibit Use. The Depth Use SPI revealed even more startling information: the dogfish again had the second-most uneven space use, but the vast majority of the observations recorded her as being in the upper water column. As smooth dogfish are primarily benthic [39], these results were particularly troubling. Taken together, this information drove the creation of an additional study focused solely on the dogfish, whose space-use and behavior were more thoroughly examined [18]. Interestingly, it was noted that the perimeter of the exhibit (where she spent a disproportionate amount of time) seemed to encourage her to display stereotypical swimming behaviors. This drove a the implementation of a combination of interventions by animal husbandry staff, which ultimately resulted in more even space use, a reduction in stereotypical behaviors, and an increase in species-appropriate behaviors for this dogfish [18]. Without the initial data collection on the dogfish's basic space use, however, this welfare concern, and the resulting improvement in the animal's well-being, would not have been addressed.

Given that the space use of aquatic animals is an understudied aspect of zoo and aquarium welfare, future studies could be performed in many directions. The use of the Dickens' SPI index was appropriate to address the goals of the current study, but its utility is relatively limited, as it only provides an indication of the evenness of space use [25]. Our results indicated that specific areas of the exhibit were in fact preferred by certain individuals. Therefore, one important avenue for future analysis could involve the determination of the biologically relevant aspects of an exhibit (such as pump locations, viewing windows, hiding spaces, etc.) based on species' natural history, and the relation of those variables to individual space use using Plowman's Modified SPI or Vanderploeg and Scavia's Electivity index [23,24]. These indices are useful, as they provide information about whether specific locations (in both the X/Y and Z axes) within an exhibit are over- or under-utilized by animals [25]. Unfortunately, due to limited prior knowledge regarding the exact composition of the exhibit at the time of the data collection, these more inclusive analyses were not run for this dataset. It is also worth noting that since the current study observed five different species, a 'one-size-fits-all' analysis of resource utilization would not have been appropriate. All five species in this study have a unique natural history, and therefore have different biological preferences. These preferences likely result in different resource utilization by each species, which would have impacted how the zones were defined and what the expected values for the time spent in those zones would have been. For researchers interested in incorporating these indices in the future, it may be most effective to focus on a single species at a time for analysis, and to relate the space use to what would be biologically expected for that species based on its natural history. Nonetheless, we encourage researchers to consider utilizing these indices in the future, as they can provide even more detailed information relating to animal preferences and welfare.

This study initially aimed to simply determine whether all the focal individuals properly utilized the exhibit space provided. Even though these results are somewhat limited in their application, they highlight the fact that elasmobranchs *do* utilize their space differently based on their biological context, which was previously only anecdotally noted. By utilizing ZooMonitor software to collect data on the space use of captive animals, caretakers can therefore gain an understanding of the preferences of and potential causes of stress for animals that may not immediately be apparent. As space-use studies in aquatic species are relatively rare in comparison to equivalent studies of their land-dwelling counterparts [11], studies such as these can give important insight into the movement patterns and habitat choices of elasmobranchs and other fishes in captivity. By conducting comprehensive space-use analyses on aquatic populations, animal husbandry professionals can gain insight into changes in activity for individuals, the effects of visitors on space use, areas of the exhibit that show indications of being preferable or uninviting for resident animals, how individuals may alter their space use in relation to one another or seasonally, and many other applications. This information could be particularly useful for species involved in a Species Survival Plan (SSP) [40]. The IUCN states that, currently, 37% of all shark and ray species are endangered to some extent [41], which makes successful captive breeding programs even more essential to the continuation of these species. By learning how elasmobranchs are inclined to use space, animal caretakers can optimize exhibit design and overall welfare for individuals whose genetic diversity is of utmost importance, ultimately supporting conservation interests.

Continued space-use studies on captive aquatic populations can not only positively affect the focal individuals (as in the smooth dogfish in this study), but can also help to respond to the AZA Aquatic Collections Sustainability Committee's call to document, address, and improve the welfare and wellbeing of aquatic populations [26]. We therefore encourage other facilities to take an increased interest in the space use of their aquatic species, and suggest utilizing the ZooMonitor program as an important tool for enhanced aquatic animal management.

Author Contributions: Conceptualization: A.M.H., S.M.T.-S., and Z.R.; methodology: A.M.H. and S.M.T.-S.; formal analysis: A.M.H.; investigation: A.M.H.; data curation: A.M.H.; writing—original draft preparation: A.M.H.; writing—review and editing: Z.R. and S.M.T.-S.; supervision: S.M.T.-S.; All authors have read and agreed to the published version of the manuscript.

Funding: Financial support for this study was provided by an Oakland University Honors College Thesis Grant, awarded to AMH. Funding was provided through reimbursement from the university for travel miles and conferences.

Institutional Review Board Statement: As the data collection was purely observational and no external variables were added to the animals' enclosure, ethical review and approval were waived for this study.

Informed Consent Statement: Not applicable.

Data Availability Statement: Data are available upon request.

Acknowledgments: The authors would like to thank the SEA LIFE Michigan animal husbandry staff for their exceptional care for the animals, as well as their patience and cooperation throughout this study. We would also like to thank three anonymous reviewers for their helpful feedback.

Conflicts of Interest: The authors declare no conflict of interest.

References

1. Goff, C.; Howell, S.M.; Fritz, J.; Nankivell, B. Space use and proximity of captive chimpanzee (*Pan troglodytes*) mother/offspring pairs. *Zoo Biol.* **1994**, *13*, 61–68. [CrossRef]
2. Mallapur, A.; Qureshi, Q.; Chellam, R. Enclosure design and space utilization by Indian leopards (*Panthera pardus*) in four zoos in southern India. *J. Appl. Anim. Welf. Sci.* **2002**, *5*, 111–124. [CrossRef]
3. Rose, P.E.; Brereton, J.E.; Croft, D.P. Measuring welfare in captive flamingos: Activity patterns and exhibit usage in zoo-housed birds. *Appl. Anim. Behav. Sci.* **2018**, *205*, 115–125. [CrossRef]
4. Ross, S.R.; Schapiro, S.J.; Hau, J.; Lukas, K. Space use as an indicator of enclosure appropriateness: A novel measure of captive animal welfare. *Appl. Anim. Behav. Sci.* **2009**, *121*, 42–50. [CrossRef]
5. Wark, J.D.; Cronin, K.A.; Niemann, T.; Shender, M.A.; Horrigan, A.; Kao, A.; Ross, M.R. Monitoring the behavior and habitat use of animals to enhance welfare using the ZooMonitor app. *Anim. Behav. Cognit.* **2019**, *6*, 158–167. [CrossRef]
6. Breton, G.; Barrot, S. Influence of enclosure size on the distances covered and paced by captive tigers (*Panthera tigris*). *Appl. Anim. Behav. Sci.* **2014**, *154*, 66–75. [CrossRef]
7. Troxell-Smith, S.; Miller, L. Using natural history information for zoo animal management: A case study with okapi (*Okapia johnstoni*). *J. Zoo Aquar. Res.* **2016**, *4*, 38–41.
8. Whitney, N.M.; Lear, K.O.; Gaskins, L.C.; Gleiss, A.C. The effects of temperature and swimming speed on the metabolic rate of the nurse shark (*Ginglymostoma cirratum*, Bonaterre). *J. Exp. Mar. Biol. Ecol.* **2016**, *477*, 40–46. [CrossRef]
9. Clark, F.E.; Melfi, V.A. Environmental enrichment for a mixed-species nocturnal mammal exhibit. *Zoo Biol.* **2012**, *31*, 397–413. [CrossRef]
10. Fábregas, M.C.; Guillén-Salazar, F.; Garcés-Narro, C. Do naturalistic enclosures provide suitable environments for zoo animals? *Zoo Biol.* **2012**, *31*, 362–373. [CrossRef] [PubMed]
11. Brereton, J.E. Current directions in animal enclosure use studies. *J. Zoo Aquar. Res.* **2020**, *8*, 1–9.
12. Kolarevic, J.; Baeverfjord, G.; Takle, H.; Ytteborg, E.; Kristin, B.; Reiten, M.; Nergård, S.; Terjesen, B.F. Performance and welfare of Atlantic salmon smolt reared in recirculating or flow through aquaculture systems. *Aquaculture* **2014**, *432*, 15–25. [CrossRef]
13. De Freitas Souza, C.; Baldissera, M.D.; Baldisserotto, B.; Heinzmann, B.M.; Martos-Sitcha, J.A.; Mancera, J.M. Essential oils as a stress-reducing agents for fish aquaculture: A review. *Front. Physiol.* **2019**, *10*, 785. [CrossRef]
14. Ross, M.R.; Niemann, T.; Wark, J.D.; Heintz, M.R.; Horrigan, A.; Cronin, K.A.; Shender, M.A.; Gillespie, K. ZooMonitor (Version 1) (Mobile Application Software). Available online: https://zoomonitor.org (accessed on 10 March 2022).
15. Saiyed, S.T.; Hopper, L.M.; Cronin, K.A. Evaluating the behavior and temperament of African penguins in a non-contact animal encounter program. *Animals* **2019**, *9*, 326. [CrossRef] [PubMed]
16. Fazio, J.M.; Barthel, T.; Freeman, E.W.; Garlick-Ott, K.; Scholle, A.; Brown, J.L. Utilizing camera traps, closed circuit cameras and behavior observation software to monitor activity budgets, habitat use, and social interactions of zoo-housed Asian elephants (*Elephas maximus*). *Animals* **2020**, *10*, 2026. [CrossRef]
17. Huskisson, S.M.; Doelling, C.R.; Ross, S.R.; Hopper, L.M. Assessing the potential impact of zoo visitors on the welfare and cognitive performance of Japanese macaques. *Appl. Anim. Behav. Sci.* **2021**, *243*, 105453. [CrossRef]
18. Hart, A.; Reynolds, Z.; Troxell-Smith, S.M. Using individual-specific conditioning to reduce stereotypic behaviours: A study on smooth dogfish (*Mustelus canis*) in captivity. *J. Zoo Aquar. Res.* **2021**, *9*, 193–199.
19. Preiser, W.F.E.; Rabinowitz, H.Z.; White, E.T. *Post-Occupancy Evaluation*, 1st ed.; Van Nostrand Reinhold: New York, NY, USA, 1988.
20. Hadjri, K.; Crozier, C. Post-occupancy evaluation: Purpose, benefits and barriers. *Facilities* **2009**, *27*, 21–33. [CrossRef]
21. Kelling, A.S.; Gaalema, D.E. Post-occupancy evaluations in zoological settings. *Zoo Biol.* **2011**, *30*, 597–610. [CrossRef]
22. Dickens, M. A statistical formula to quantify the "spread-of-participation" in group discussion. *Speech Monogr.* **1995**, *22*, 28–30. [CrossRef]
23. Plowman, A.B. A note on a modification of the spread of participation index allowing for unequal zones. *Appl. Anim. Behav. Sci.* **2003**, *83*, 331–336. [CrossRef]
24. Vanderploeg, H.A.; Scavia, D. Two electivity indices for feeding with special reference to zooplankton grazing. *J. Fish. Board Can.* **1979**, *36*, 362–365. [CrossRef]
25. Brereton, J.E.; Fernandez, E.J. Which index should I use? A comparison of indices for enclosure use studies. *Anim. Behav. Cognit.* **2022**, *9*, 119–132. [CrossRef]

26. Richard, H. Welfare and Ethics in Aquatic Collections. Available online: https://www.aza.org/connect-stories/stories/animal-welfare-in-aquariums?locale=en (accessed on 10 March 2022).
27. Altmann, J. Observational study of behavior: Sampling methods. *Behaviour* **1974**, *49*, 227–267. [CrossRef]
28. Gehlenborg, N.; Wong, B. Points of view: Heat maps. *Nat. Methods* **2012**, *9*, 213. [CrossRef] [PubMed]
29. Castro, J.I. The biology of the nurse shark, *Ginglymostoma cirratum*, off the Florida east coast and the Bahama Islands. *Env. Biol. Fish.* **2000**, *58*, 1–22. [CrossRef]
30. Dapp, D.R.; Walker, T.I.; Huveneers, C.; Reina, R.D. Respiratory mode and gear type are important determinants of elasmobranch immediate and post-release mortality. *Fish Fish.* **2016**, *17*, 507–524. [CrossRef]
31. Talwar, B.S.; Bouyoucos, I.A.; Brooks, E.J.; Brownscombe, J.W.; Suski, C.D.; Cooke, S.J.; Grubbs, R.D.; Mandelman, J.W. Variation in behavioural responses of sub-tropical marine fishes to experimental longline capture. *ICES J. Mar. Sci.* **2020**, *77*, 2763–2775. [CrossRef]
32. Kelly, M.L.; Spreitzenbarth, S.; Kerr, C.C.; Hemmi, J.M.; Lesku, J.A.; Radford, C.A.; Collin, S.P. Behavioural sleep in two species of buccal pumping sharks (*Heterodontus portusjacksoni* and *Cephaloscyllium isabellum*). *J. Sleep Res.* **2020**, *30*, e13139. [CrossRef]
33. Motta, P.J.; Hueter, R.E.; Tricas, T.C.; Summers, A.P.; Huber, D.R.; Lowry, D.; Mara, K.R.; Matott, M.P.; Whitenack, L.B.; Wintzer, A.P. Functional morphology of the feeding apparatus, feeding constraints, and suction performance in the nurse shark *Ginglymostoma cirratum*. *J. Morph.* **2008**, *269*, 1041–1055. [CrossRef]
34. Mullins, L.L.; Drymon, J.M.; Moore, M.; Skarke, A.; Moore, A.; Rodgers, J.C. Defining distribution and habitat use of west-central Florida's coastal sharks through a research and education program. *Eco. Evo.* **2021**, *11*, 16055–16069. [CrossRef] [PubMed]
35. Schwieterman, G.D.; Winchester, M.M.; Shiels, H.A.; Bushnell, P.G.; Bernal, D.B.; Marshall, H.M.; Brill, R.W. The effects of elevated potassium, acidosis, reduced oxygen levels, and temperature on the functional properties of isolated myocardium from three elasmobranch fishes: Clearnose skate (*Rostrojara aglanteria*), smooth dogfish (*Mustelus canis*), and sandbar shark (*Carcharhinus plumbeus*). *J. Comp. Physiol. B* **2021**, *191*, 127–241.
36. Carlson, J.K.; Parsons, G.R. Respiratory and hematological responses of the bonnethead shark, *Sphyrna tibuto*, to acute changes in dissolved oxygen. *J. Exp. Mari. Biol. Ecol.* **2003**, *294*, 15–26. [CrossRef]
37. Kelly, M.L.; Murray, E.R.P.; Kerr, C.C.; Radford, C.A.; Collin, S.P.; Lesku, J.A.; Hemmi, J.M. Diverse activity rhythms in sharks (Elasmobranchii). *J. Biol. Rhythm.* **2020**, *35*, 476–488. [CrossRef]
38. Larsen, J.; Bushnell, P.; Steffensen, J.; Pedersen, M.; Qvortrup, K.; Brill, R. Characterization of the functional and anatomical differences in the atrial and ventricular myocardium from three species of elasmobranch fishes: Smooth dogfish (*Mustelus canis*), sandbar shark (*Carcharhinus plumbeus*) and clearnose skate (*Raja eglanteria*). *J. Comp. Physiol. B* **2016**, *187*, 291–313. [CrossRef] [PubMed]
39. Gelsleichter, J.; Musick, J.A.; Nichols, S. Food habits of the smooth dogfish, *Mustelus canis*, dusky shark, *Carcharhinus obscurus*, Atlantic sharpnose shark, *Rhizoprionodon terraenovae*, and the sand tiger, *Carcharias taurus*, from the northwest Atlantic Ocean. *Env. Biol. Fish.* **1999**, *54*, 205–217. [CrossRef]
40. Species Survival Plan®Programs. Available online: https://www.aza.org/species-survival-plan-programs (accessed on 14 May 2022).
41. IUCN. The IUCN Red List of Threatened Species. Version 2021-3. 2021. Available online: https://www.iucnredlist.org (accessed on 18 May 2022).

Journal of *Zoological and Botanical Gardens*

MDPI

Article

Behaviour of Zoo-Housed Red Pandas (*Ailurus fulgens*): A Case-Study Testing the Behavioural Variety Index

Caterina Spiezio [1], Mariangela Altamura [2], Janno Weerman [3] and Barbara Regaiolli [1,*]

1 Research & Conservation Department, Parco Natura Viva—Garda Zoological Park, 37012 Bussolengo, Italy; caterina.spiezio@parconaturaviva.it
2 Department of Evolution and Functional Biology, University of Parma, 43121 Parma, Italy; maltamura9319@gmail.com
3 Studbook Keeper Red Panda EEP, Royal Rotterdam Zoological & Botanical Gardens, 3000 AM Rotterdam, The Netherlands; j.weerman@diergaardeblijdorp.nl
* Correspondence: barbara.regaiolli@parconaturaviva.it; Tel.: +39-3346139953

Abstract: The red panda is listed as "endangered" in the IUCN Red List of Threatened Species, due to the rapid population decline. Improving our knowledge on the red panda biology and ethology is necessary to enhance its husbandry and breeding in zoos. Behavioural variety, intended as the presence of a wide array of species-specific behaviour, has been considered a positive welfare index in zoo-housed animals. The aim of this study was to describe the behaviour of two pairs of zoo-housed red pandas, one of them with an offspring, and to investigate the behavioural variability using the Behavioural Variety Index (BVI). Behavioural data from two zoo-living male–female pairs were collected. A continuous focal animal sampling method was used to collect individual and social behaviours of the two pairs. Forty-eight 30 min sessions per subject were carried out. For the BVI, a list of species-specific behaviours previously reported in the red panda was prepared and compared with the behavioural repertoire of the subjects of the study. First, species-specific behaviours were recorded, and no abnormal behaviour was reported. The percentages of time spent on different activities (e.g., routine behaviours, exploratory/territorial behaviours, consumption behaviours, locomotive behaviours, social behaviours, maternal behaviours) were similar to time budgets reported in the red panda, with routine behaviours (resting, comfort and vigilance) being the most performed in both pairs. Moreover, the BVI suggested that each red panda performed on average 73% of the behaviours described in previous literature on this species. In conclusion, studying the behavioural variety of red pandas in zoos can be a useful tool for assessing their welfare as well as improving our knowledge on the behavioural repertoire of a species that is difficult to observe in the wild.

Keywords: red panda; animal welfare; maternal behaviour

Citation: Spiezio, C.; Altamura, M.; Weerman, J.; Regaiolli, B. Behaviour of Zoo-Housed Red Pandas (*Ailurus fulgens*): A Case-Study Testing the Behavioural Variety Index. *J. Zool. Bot. Gard.* **2022**, *3*, 223–237. https://doi.org/10.3390/jzbg3020018

Academic Editors: Kris Descovich, Caralyn Kemp and Jessica Rendle

Received: 31 October 2021
Accepted: 10 May 2022
Published: 13 May 2022

1. Introduction

The red panda (*Ailurus fulgens*) is listed as "Endangered" in the IUCN Red List of Threatened Species, due to the rapid population decline [1]. Given the situation of red pandas in the wild, breeding of this species under human care has become increasingly important to create insurance populations [2]. Improving our knowledge on red panda biology and ethology is necessary to enhance the husbandry standards and breeding success of this species in captivity, as well as for in situ conservation efforts [1]. Past research on red pandas housed in zoological collections has led to improve the husbandry (e.g., diet, housing conditions, group composition) of this species in Europe and to an increase in the total number of births [1]. Moreover, ex situ populations of red panda may allow research opportunities that support wild populations and in situ conservation efforts [1].

Observation of animal behaviour allows us to obtain important information on the psychological and physical health of animals and has become an important non-invasive tool to measure animal welfare [3]. The assessment of positive as well as negative welfare states has become increasingly important in animal welfare science, with the development of indicators to assess and promote positive welfare of each individual, as well as at the species level. Recent studies on animals in zoos focus on the presence of species-specific, natural and normal behaviours, and the absence of abnormal behaviours, such as stereotypes, as an indicator of welfare [3–7]. For example, affiliative social interactions in solitary species such as bears have been found to reduce the performance abnormal behaviours (reviewed in [4]). Behavioural variety in general has been used to assess the welfare of Northern bald ibises in zoos [7].

Behavioural diversity is intended as richness of behaviours, and it would be linked to a positive welfare state because a high behavioural diversity indicates that we are meeting several behavioural needs of the animal [8]. On the contrary, when behavioural diversity is low, and animals have no opportunities to perform their behavioural repertoire, they can become lethargic and develop abnormal behaviours [9]. Thus, behavioural variety could be lost during challenging situations that could characterize controlled environments and human management, and the presence of different normal and natural behaviours performed by each subject could indicate a positive welfare state [3–7]. In particular, behaviours related to positive welfare can be luxury behaviours, such as play, exploration, and allo-grooming, which are the first ones to be lost during challenging situations and require a sufficient welfare level to be performed by the animals [6,10]. In the red panda, a recent study highlighted that personality could impact the welfare of subjects under human care, with behaviours such as locomotion, exploration, and marking, underlining an active and explorative temperament [11].

In the wild, the red panda is a solitary species, with males and females interacting with each other only during the mating season [12,13]. However, in some cases, individuals can move in pairs or small family groups, with males being more territorial than females, protecting and patrolling their home range to a greater extent [13]. This species can be diurnal, crepuscular, or nocturnal, and activity patterns vary across seasons and between sexes, ranging from 45 to 60% of the day depending on temperature and food availability [14,15]. In particular, red pandas are more active in summer, spring, and autumn, whereas in winter, resting periods increase [14,15]. Red pandas are arboreal and are agile climbers, sleeping and resting in trees [14]. Scent-marking is a relevant behaviour in this species, as pandas mark their trails by secretions from foot glands and mark frequently using ano-genital glands, urine, and faeces [14,16]. Females have a gestation of 114–145 days, with births happening in summer, mainly in June and July in the northern hemisphere [16]. Data collected in zoological gardens suggest that cubs stay in the den for 3–4 months and are completely dependent on their mothers [16]. Indeed, the cubs are altricial and nursing is taxing for female survival, requiring high-energy intake and an increased amount of food [16,17]. Before leaving the mother, juveniles show high levels of exploratory behaviour, practice of locomotory and feeding behaviour, and development of social behaviours through play and interactions with conspecifics [18]. In zoos, the bonds between mother and offspring may continue beyond one year of age of the cub, whereas the length is unknown in the wild [16]. In zoos, male red pandas tend to avoid females after parturition and during the denning period. However, after cubs leave the den, male pandas start to gradually interact with them and may become involved in play sessions [16]. In zoological institutions, red pandas are mostly kept in pairs. On some occasions, red pandas are kept in same-sex pairs or small groups [13]. Given that the red panda is solitary, assessing whether group structures and size meet the welfare needs of all individuals, and allowing the performance of species-specific behaviours are of great importance [1,13]. Moreover, first-year mortality of zoo-housed red pandas is high (36% in countries with warm and humid climates during the species breeding season), probably because mothers leave their cubs unattended due to heat stress [19].

Therefore, it is important to monitor the behaviour of red panda mothers, identifying signals and developing strategies to prevent or reduce infant mortality that can help to enhance breeding success of this species under human care.

The aim of this study was to investigate and describe individual and social behaviour of two pairs of zoo-housed red pandas, focusing on the variety of species-specific normal behaviours performed by the subjects. One of the pairs had offspring and special attention was given to maternal behaviour. To describe the behavioural variety of our subjects, we used the Behavioural Variety Index (BVI) developed and described in our previous research [7]. This index provides a measure of behavioural richness of each individual, quantifying the presence of species-specific natural behaviours described for a species in both wild and captive contexts [7]. Discussing behavioural time budgets of zoo-housed individuals based on data collected on the species as well as analysing the BVI of the subjects could be a useful tool that provides some quantitative and qualitative insights into the welfare of the two red panda pairs [7,18–21].

2. Materials and Methods

2.1. Subjects and Area

The study involved two pairs of red pandas housed in two different enclosures at Parco Natura Viva-Garda Zoological Park, an Italian zoological garden, in October and November 2015. The first pair consisted of one female and one male, Ilosha and Ny'ma. The female of this pair, Ilosha, gave birth to two cubs in July 2015. The second pair consisted of one female and one male, Lin and Maituc. At the time of the study, this pair had no offspring, and the female Lin was contracepted with a Suprelorin implant to prevent pregnancy. The two red panda pairs were housed in two different enclosures. The enclosure of Ilosha and Ny'ma consisted of a 260 m^2 outdoor area and a 34 m^2 indoor area; the enclosure of Lin and Maituc consisted of a 358 m^2 outdoor area and a 34 m^2 indoor area. Both outdoor areas contained tall and leafy trees, bushes, rocks, horizontal branches linking different trees together, a water pool, small wooden houses, and artificial nests in large tree trunks. Red pandas were fed with fresh bamboo and fruits once a day. Once a week, the diet was supplemented with fresh meat (quail). In both enclosures, fruit was chopped and provided in bowls (two bowls per enclosure, in feeding points placed on trees, at a height of approximately two meters). Feeding time was at 1 PM, whereas bamboo was provided in the late afternoon (5 PM). Both pairs were involved in an environmental enrichment program and were provided with different kinds of stimulation daily (e.g., food-related enrichments such as hanging fruits, sensory enrichment such as cloth with spices or scents). Water was always available. No human–animal interactions were permitted.

2.2. Procedure and Data Collection

A continuous focal animal sampling method was used to collect durations of individual and social behaviour [22]. We collected data only on adult red pandas of the pairs; Ilosha's cubs were not considered to be focal subjects as they were three months old at the time of the study. To minimize human disturbance to the pandas, the observer collected the data on animal behaviour from the visitor path, after a habituation period of ten days. From the visitors' path, the observer could see the enclosure of both pairs from above and was able to follow the focal animal in almost the whole area. Before the beginning of the study, we planned a preliminary period in which the observer learnt to identify the red pandas through features such as the face mask, body size, hair, and tail characteristics. Red pandas were out of sight only when they were in their dens or in the indoor areas or when they were deeply inside the trees' canopy. Information on enclosure use was also recorded, focusing on whether the red pandas were on the ground, or if they were using elevated areas of the enclosure, such as trees and branches. The ethogram of the study is reported in Tables 1 and 2 (maternal behaviours) and was prepared based on preliminary observation of the subjects and on previous studies of the red panda [14–16,23–27]. For each subject, forty-eight 30 min sessions were carried out, for a total of 1440 min per subject

(96 sessions, 2880 min per pair). Per subject, two sessions per day were done, one in the morning (9.30 AM–12.00 PM) and one in the afternoon (2.00 PM–4.30 PM). Data on all subjects were collected by the same observer (M.A.) over a period of approximately two months. The observer started to collect data in a different enclosure every day, so that subjects of each enclosure were observed uniformly over the study period. Moreover, the order in which each subject was observed was counterbalanced over sessions to ensure that observation of each individual was uniformly distributed over the whole observation time.

Table 1. Ethogram of red pandas of the study. Behaviours of the red pandas have been grouped based on their function in five different classes (Behavioural classes) [14,16,23–27].

Behavioural Class	Behavioural Category	Definition
Exploratory/territorial b. (Expl/terr)	Digging	Digging in ground.
	Human-directed b.	Observing visitors, zookeepers or other humans, following them with attention.
	Licking	Olfactory investigation by licking any object (e.g., branches, leaves, enclosure furnishing) or substrate in the enclosure.
	Individual-play/Manipulation	Tactile investigation of objects in the enclosure, biting or chewing objects, interaction with environmental enrichment devices, carrying objects while moving or rolling on the back.
	Interspecific b.	Watching non-conspecifics, following them with attention (e.g., muntjac, bird).
	Scent-marking	Rubbing genitals on ground, objects, or enclosure furnishing.
	Sniffing	Sniffing any object (e.g., branches, leaves, enclosure furnishing) or substrate in the enclosure.
Locomotive b.	Arboreal locomotion	Climbing on trees, walking, or running on tree branches or in the canopy.
	Ground locomotion	Walking or running on the ground.
	Standing	Standing on back two paws.
Routine b.	Hunt/stalk	Hunting, harassing, following a non-conspecific (e.g., birds).
	Individual resting	Lying sleeping, curled in ball, or flat out.
	Comfort	Self-cleaning of the fur, scratching, and stretching.
	Vigilance	Being watchful, alert, while observing the surroundings. Lying, sitting, or standing with head up and eyes open, head or ears moving.
Consumption b.	Eating	Eating food from the bowls (fruit and vegetables) or eating bamboo from the bamboo feeding point.
	Foraging	Looking for food in the enclosure, searching in the grass, digging on the ground, or eating when browsing.
	Maintenance	Drinking at the water pool of the enclosure; urinating or defecating.

Table 1. *Cont.*

Behavioural Class	Behavioural Category	Definition
Social b.	Aggression	Hitting a conspecific with paws, biting a conspecific. Initiating or receiving an aggression.
	Chasing/display	Agonistic behaviours without physical contact: chase, threat (arching the back and the tail, moving the head up and down) or displacing a conspecific. Receiving an agonistic behaviour without physical contact.
	Grooming	Cleaning the fur by licking a conspecific. Grooming could be mutual, received, or done actively.
	Observing conspecific	Watching a conspecific or being watched by a conspecific.
	Sexual behaviour	Courtship (the female moves, marking heavily while the male follows her, marking over her marks) and mating behaviour.
	Sniffing conspecific	Sniffing a conspecific or being sniffed by a conspecific.
	Social play	Playing with another individual by lunging, wrestling, biting softly.
	Social resting	Resting in contact with a conspecific.
Out of sight	Not observed	The individual is not visible.
Abnormal b.	Abnormal b.	Behaviours such as purposeless locomotion, repetitive route in the enclosure, excessive mouth movements (e.g., tongue flicking).

Table 2. Ethogram of maternal behaviours performed by the female Ilosha [Adapted from 16].

Behavioural Category	Definition
Antagonistic b.	The mother displaces, hits, or bites the cub.
Den	The mother and the cubs are not visible, hiding in the artificial nest.
Grooming cubs	Grooming of the cubs.
Nest building	Building a nest with twigs and grass.
Observing cubs	The mother observes and monitor the cubs.
Play with cubs	The mother plays with the cubs by lunging, wrestling, biting softly.
Rest and sleep with cubs	Lying, sleeping with the cubs.
Transport	The mother carries the cubs in the mouth while moving in the enclosure.

2.3. The Behavioural Variety Index (BVI)

To investigate the behavioural variety in the red pandas of the study, we used the method described by Spiezio and colleagues (2018) [7]. After having examined the existing literature, we prepared a list of species-specific natural behaviours (belonging to the wild behavioural repertoire of the species) collected by previous researchers on the red pandas that included exhaustive ethograms of the species [14–16,23–27]. The list prepared for the current study included 24 behaviours that were grouped into 5 classes according to behavioural function (Table 3). Five specific indices were assigned to the resulting classes: consumption (CO; score: 0–3), exploratory/territorial (EXPL/TERR; score: 0–5), locomotive (LOC; score: 0–3), routine (R; score: 0–4), and social (S; score: 0–8). To investigate the presence of species-specific behaviours related to parental care toward the offspring, a separate class for maternal behaviour (MAT) was prepared. All the indices mentioned above were used to create the Behavioural Variety Index (BVI). When calculating the indices for the red pandas, firstly, each item in Table 3 was scored as 0 or 1, with 1 representing the presence of the behaviour. Then, each index was calculated as the sum of the behavioural items performed by each subject and the indices' score ranged from 0 to the total number of behavioural items found for each class. Each index was calculated for each individual. A BVI was calculated for each subject and resulted from the sum of the five indices. The BVI is calculated based on behaviours recorded in the whole study period, over all sessions.

Table 3. Behaviours of the red panda described in previous research. Behaviours have been grouped in five main groups and the number of behaviours per group were used to develop indices of the red panda behavioural variety. The possible score for each group is calculated starting from the number of behaviours that have been previously collected in the red panda [14,16,23–27].

	Consumption (CO)	Exploratory/Territorial (EXPL/TERR)	Locomotive (LOC)	Routine b. (R)	Social (S)	Maternal b. (MAT)
	Eating	Digging	Arboreal locomotion	Hunting/stalking	Aggression	Antagonistic b.
	Foraging	Licking	Ground locomotion	Resting	Chasing/displacing	Den
	Maintenance	Individual play/manipulation/carrying	Standing bipedal	Comfort	Grooming	Grooming cubs
		Scent-marking		Vigilance	Observe consp	Nest building
		Sniffing			Sexual b.	Observe cubs
		Interspecific b.			Sniffing consp	Play with cubs
					Social play	Rest and sleep with cubs
					Social resting	Transport
Score	0–3	0–6	0–3	0–4	0–8	0–8

2.4. Data Analysis

In the current study, for each red panda, we collected durations (minutes) of different behavioural categories to obtain time budgets, expressed as percentage of time that an animal spends performing different behaviours [28]. As the sample of the study was small, we used descriptive statistics [29], with a single-case research approach [30,31]. For each session, we calculated the total duration of time spent by each subject performing each behavioural class and category. We grouped data collected for each individual per session to obtain the duration of behaviours per pair per session. Both enclosure use (ground vs. arboreal space) and behavioural time budgets were considered to obtain quantitative information on the time spent performing different activities. We used non-parametric statistical tests (Wilcoxon signed-rank tests) to compare behavioural classes between morning and afternoon sessions for each red panda. In the case of the BVI, we collected occurrences (presence or absence) of different behavioural items and used them to assess behavioural variety, a qualitative measure of the behavioural repertoire (the higher the score, the higher the behavioural variety).

3. Results

3.1. Lin and Maituc Pair

On average, the pair without offspring, Lin and Maituc, spent 81.2% of the observation time on elevated areas of the enclosure and 4.2% of the time on the ground, whereas subjects were not visible to the observer (it was not possible to know whether the pandas were hiding in the canopy or in shelters/dens on the ground) for the remaining time.

Regarding activity levels, Lin and Maituc spent, on average, 42.1% of the observation time being inactive (individual and social resting), 40.3% being active, and 17.6% being not visible to the observer. The most performed class of behaviours was routine behaviours (total duration: 1462.4 min), followed by "not observed" (506.3 min), social behaviours (462.4 min), locomotive behaviours (273.9 min), exploratory/territorial behaviours

(112.9 min), and consumption behaviours (62.1 min) (Figure 1). Median (IQR) durations of each behavioural category performed by each red panda are reported in Table 4.

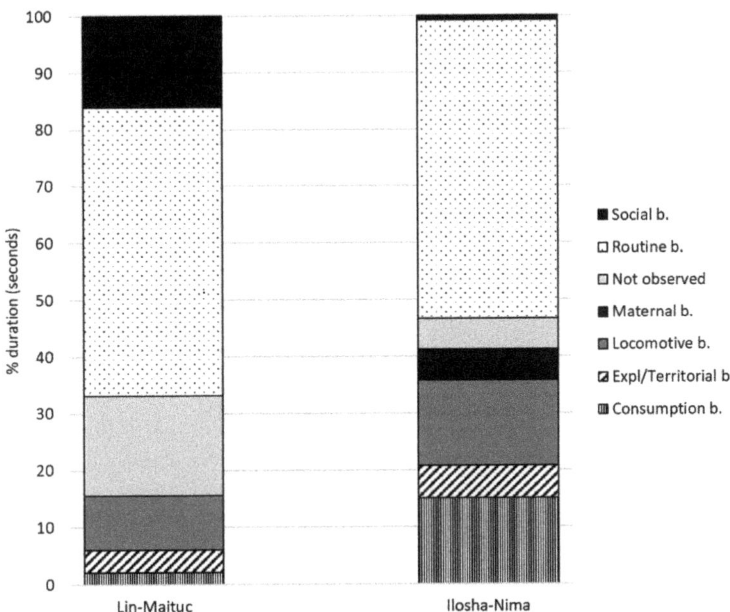

Figure 1. Percentage time spent by the two red pandas' pairs performing different classes of behaviour. For definition of abbreviations and classes see Tables 1 and 2.

Table 4. Median (IQR) duration of behaviours performed by each red panda. Per subject, the table reports the median (interquartile range—IQR) duration of different behavioural categories calculated across session.

		Ilosha	Ny'ma	Lin	Maituc
CONSUMPTION	Eating	568.5 (1161.3)	0 (411.3)	0 (278.5)	0 (7.5)
	Foraging	0 (86.5)	0 (0)	0 (0)	0 (0)
	Maintenance	0 (13)	0 (16.3)	0 (0)	0 (0)
EXPL/TERR	Human-dir behav.	18 (44.3)	49 (117.3)	40.5 (103.8)	32 (91)
	Ind play/manip	0 (0)	0 (0)	0 (0)	0 (0)
	Interspecific b.	0 (0)	0 (2.3)	11.5 (37.3)	0 (13)
	Scent-marking	0 (8.3)	1.5 (38.3)	0 (11.5)	15.5 (76.8)
	Sniff/dig/lick	34.5 (232.5)	0 (14.8)	0 (43.8)	0 (14)
LOCOMOTIVE	Arboreal loc	115 (216.5)	320 (684.8)	184.5 (330)	278.5 (442.3)
	Ground loc	95 (109.5)	81 (525.5)	0 (13.8)	0 (278.8)
	Standing	0 (0)	0 (0)	0 (0)	0 (0)
ROUTINE B.	Comfort	21.5 (226.3)	136.5 (490)	359 (789.8)	253 (717.8)
	Resting ind	811.5 (1783)	1755 (1508.8)	258.5 (1637.3)	1222.5 (2071.5)
	Vigilance	96.5 (270)	191 (253)	197.5 (292.3)	242.5 (440.3)
SOCIAL B.	Aggression	0 (0)	0 (0)	0 (0)	0 (0)
	Chase/displace	0 (0)	0 (0)	0 (0)	0 (0)
	Grooming	0 (0)	0 (0)	0 (0)	0 (0)
	Obs consp	0 (2.3)	19.5 (64.8)	0 (33)	7.5 (35)
	Sexual b.	0 (0)	0 (0)	0 (0)	0 (0)
	Sniffing consp	0 (0)	0 (0)	0 (0)	0 (10.8)
	Social resting	0 (0)	0 (0)	0 (1643.5)	0 (89.3)
OoS	Not observed	73 (350.5)	0 (19)	130.5 (616.5)	36 (631.5)

3.2. Ilosha and Ny'ma Pair

On average, the pair with offspring, Ilosha and Ny'ma, spent 81.7% of the observation time on trees and 15.5% of the time on the ground, whereas they were not visible to the observer (it was not possible to know whether the pandas were hiding in the canopy on in shelters/dens on the ground) for the remaining time.

Regarding activity levels, Ilosha and Ny'ma spent on average 41.2% of the observation time being inactive (individual resting and resting with cubs), 53.4% being active, and 5.4% being not visible to the observer (not observed). The most performed class of behaviours was routine behaviours (total duration: 1516.7 min), followed by locomotive behaviours (434.8 min), consumption behaviours (433.2 min), exploratory/territorial behaviours (162 min), "not observed" (155.6 min), maternal behaviours (155.3 min), and social behaviours (22.4 min) (Figure 1). Median (IQR) durations across sessions of each behavioural category performed by each red panda are reported in Table 4.

3.3. Morning vs. Afternoon Sessions

For Ilosha and Ny'ma, we found no significant differences in duration of behavioural categories performed in the morning and in the afternoon (see Table 5 for median, IQR, and statistical values of Wilcoxon test).

Table 5. Median (IQR) duration of behavioural classes performed by each red panda in the morning and in the afternoon. Per subject, the table reports median and interquartile range (IQR) duration of different behavioural classes (OOS: Out of Sight), calculated across morning (AM) and afternoon sessions (PM) in each study subject. Below, medians and IQR in the table report *V* and *p* values of the Wilcoxon tests between morning and afternoon sessions.

		CO	EXPL/TERR	LOC	OOS	R	S	MAT
Ilosha	AM	100.5 (705)	1 (83.8)	68.5 (255.8)	0 (46.3)	158 (1442.3)	0 (10.5)	24 (498.8)
	PM	176 (1053.8)	31 (113)	104 (232.8)	45 (152.5)	676.5 (1557)	0 (10)	0 (19.5)
		$V = 96$; $p = 0.322$	$V = 87$; $p = 0.948$	$V = 84$; $p = 0.274$	$V = 64.5$; $p = 0.570$	$V = 88$; $p = 0.339$	$V = 40$; $p = 0.937$	$V = 102$; $p = 0.079$
Ny'ma	AM	0 (195.3)	36 (144.5)	191.5 (1131)	0 (0)	1301.5 (1579.5)	0 (21.5)	
	PM	0 (139.5)	40.5 (85.8)	75 (285)	0 (0)	1471.5 (819.5)	5.5 (27)	
		$V = 55$; $p = 0.776$	$V = 126$; $p = 0.715$	$V = 183.5$; $p = 0.064$	$V = 13.5$; $p = 0.933$	$V = 83$; $p = 0.158$	$V = 66.5$; $p = 0.636$	
Lin	AM	0 (0)	8.5 (106)	9 (194.8)	0 (351.8)	524.5 (1433.3)	28.5 (1652)	
	PM	0 (258.3)	30 (146.8)	22.5 (236)	0 (192.8)	767.5 (1232.8)	0 (388.8)	
		$V = 12$; $p = 0.062$	$V = 88.5$; $p = 0.538$	$V = 103$; $p = 0.940$	$V = 38$; $p = 0.657$	$V = 88$; $p = 0.211$	$V = 80$; $p = 0.084$	
Maituc	AM	0 (0)	13.5 (241)	6 (581.8)	27.5 (412)	437.5 (1170)	10.5 (226.8)	
	PM	0 (0)	0 (41.8)	0 (219.5)	0 (0)	1662 (1085.5)	0 (43.3)	
		$V = 10$; $p = 0.499$	$V = 136$; $p = 0.028$ *	$V = 107$; $p = 0.044$ *	$V = 59$; $p = 0.117$	$V = 30$; $p = 0.005$ *	$V = 92$; $p = 0.069$	

* Significant difference between morning and afternoon.

For Lin and Maituc, we found no significant differences for the female Lin. The male Maituc performed significantly more exploratory/territorial and locomotory behaviours in the morning than in the afternoon, whereas the opposite pattern was found for routine behaviours (see Table 5 for median, IQR, and statistical values of Wilcoxon test).

3.4. Behavioural Variety

The distribution of the behavioural items reported in red pandas of the study and the five corresponding indices are presented in Figure 2. For consumption behaviours (C = 3) and exploratory/territorial behaviours (EXPL/TERR = 6), all items were performed by the subjects, with 25 and 75% of the pandas having a score of 3 and 4, respectively. For locomotive behaviours, 50% of the subjects had a score of 2 (L = 2) and 50% had a score of 3, performing all the behavioural items of the class (L = 3) (Figure 2). For routine behaviours (R = 5), all behavioural items apart from hunting/stalking were found. All subjects performed four items and had a score of 4. For social behaviours, all behavioural items apart from social play were performed by the red pandas, with 50% of the subjects performing 5 behavioural items and 50% performing 6 of them. In summary, the BVI scores revealed that all pandas had a score $\geq$ 17 and the mean BVI was 17.5 $\pm$ 0.6. In particular, two subjects (the females, Lin and Ilosha) had a BVI of 17 and two subjects (the males, Ny'ma and Maituc) had a BVI of 18.

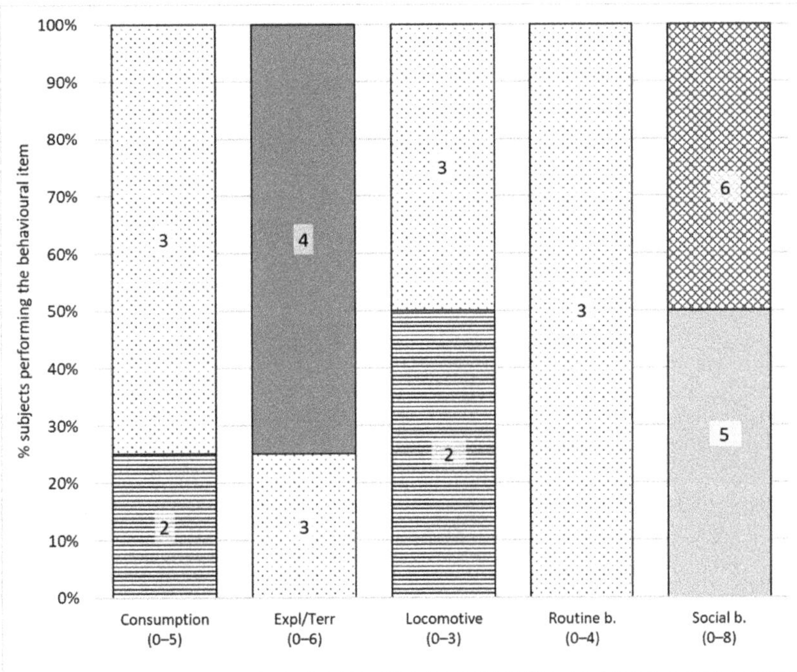

Figure 2. Distribution of scores for the five behavioural indices. Different bar boxes report the number of behavioural items that have been performed per class of behaviours (consumption behaviours, expl/terr: exploratory/territorial behaviours, locomotive behaviours, routine behaviours, social behaviours). Below each bar is reported the name of the class and the score (number of items per class). The y-axis indicates the percentage of subjects that performed the number of behavioural items reported within boxes (and thus had the same BVI score).

3.5. Maternal Behaviour

We investigated the interaction of the female Ilosha with her offspring, describing the average time spent performing different parental care behaviours (Figure 3). The most performed category was "den", meaning that the female was in the nest box with the cubs (91.2 min, 3.2% of the observation time, median [IQR]: 0 [175.3]). "Den" was followed by rest and sleep with the cubs (26.3 min, 0.9%, median [IQR]: 0 [0]), grooming (23 min, 0.8%, median [IQR]: 0 [33.8]), observe cubs (8.8 min, 0.3%, median [IQR]: 12.5 [38.5]), antagonistic behaviours (3.7 min, 0.1%, median [IQR]: 0 [0]), transport (1.8 min, 0.1%, median [IQR]: 0 [0]), and social play (0.5 min, $\simeq$0%, median [IQR]: 0 [0]). Based on the BVI analysis, all the behavioural items described for maternal behaviours (Table 2, Table 3) have been reported in the female except for nest building (MAT = 7).

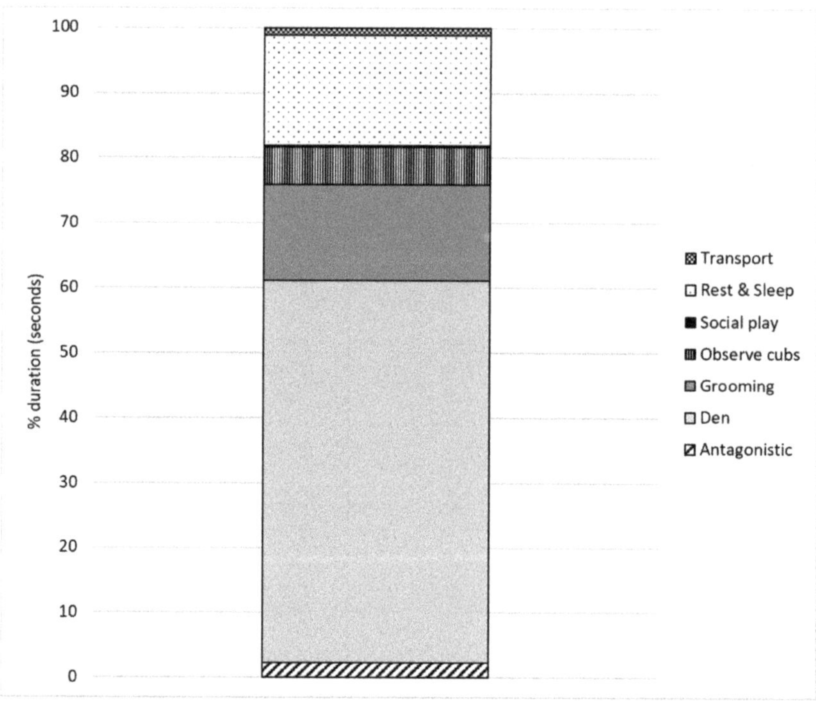

Figure 3. Percentage duration of time spent by Ilosha performing different maternal behaviours.

4. Discussion

First, we observed no abnormal behaviour in any of the pandas of this study. On average, the red pandas' activity time ranged from 40.3% (Lin and Maituc) to 53.4% (Ilosha and Ny'ma) of the total observation time. On the other hand, inactivity was approximately 40% in both pairs (41.2 and 42.1%). These findings agree with previous studies focusing on time budgets of red pandas in zoos, with inactivity ranging between approximately 40 and 50% [26], suggesting that zoo-housed red pandas show daily activity patterns [32]. In the wild, red pandas have been found to be active for 45 to 60% of the day, depending on temperature and food availability, with conspicuous periods of inactivity and long rests, especially in winter. As data collection took place in the late autumn when temperature started to become low, these results seem to agree with data collected in wild red pandas [14,15], suggesting good activity levels of the subjects, together with factors such as temperature and climate, daily husbandry, enclosure design, and environmental enrichment program which appeared to also have influenced the activity level of red pandas of this study.

Red pandas in the wild are arboreal, and in a controlled environment, they have been found to prefer the vertical space, off the ground [14,16,23,33]. The red pandas of this study spent, on average, approximately 80% of the time on elevated areas of the enclosure, specifically on trees, as reported for their wild counterparts. However, Ilosha and Ny'ma spent more time on the ground than Lin and Maituc (approximately 15 and 4%, respectively). This could be due to the presence of the offspring, leading to an increased use of the ground to look for food and to control the territory, especially in the female Ilosha. Indeed, energetic costs of lactation have been found to be particularly high in red pandas: during the lactation period, females might increase food consumption up to 200% above the rate observed during non-lactation [25,34]. This could also partially explain the higher amount of time spent by the pair with Ilosha on consumption behaviour (15%

vs. 2% reported for Lin and Maituc), exploratory/territorial behaviours (6% vs. 4%), and locomotive behaviour (15% vs. 9%), especially ground locomotion and use of the enclosure ground. In particular, the pair with offspring performed more "eating" and "foraging" than the other pair, and these behaviours were frequently performed on the ground rather than on trees.

In both pairs of the study, the most performed behavioural category was routine behaviour (more than 50% of the observation time), including resting, comfort behaviours (grooming, scratching, and stretching), and vigilance, intended as watchful observation of surroundings. This finding is in line with previous study of this species in zoos [11] and in the wild, as the red panda spends several hours resting, as mentioned above, and comfort behaviours are well-represented and may take up to 16% of daily activity [16]. Self-grooming is particularly relevant in this species and takes place mainly after awakening or eating [16,25,35]. When grooming themselves, red pandas spend a lot of time licking their body and limbs, washing their muzzles, stretching, or rubbing their back [16,25,35]. The second most performed behavioural classes were "not observed" and social behaviours for Lin and Maituc (the pair without offspring); consumption and locomotive behaviour for Ilosha and Ny'ma. In the latter pair, these behaviours might be common due to offspring presence, as previously described. In the case of Lin and Maituc, they spent more time in the canopy and in the nest boxes, and were not visible, presumably due to the enclosure design and to different needs in the absence of offspring. The behavioural class "not observed" is particularly relevant in zoos as animals in zoological institutions must have the opportunity to hide or escape from stressors or negative stimuli, such as the presence of visitors [5]. Moreover, they performed more social behaviour, as Lin and Maituc spent time observing each other as well as interacting (grooming, sniffing) and sleeping (social resting) together.

We compared activity budgets of each red panda between morning and afternoon sessions. Overall, we found no significant differences in durations of behavioural classes in different parts of the day, even if the male Maituc performed more exploratory/territorial behaviours and locomotion in the morning than in the afternoon, whereas routine behaviours such as resting and comfort were shown more in the afternoon. This finding agrees with previous literature on wild red pandas, with peaks of activity in the morning (700–1000h) and in the evening (1700–1800h) [27].

To assess the welfare of red pandas, we investigated behavioural variety of the two pairs using the BVI. This tool allows to compare the behavioural repertoire of our subjects with that reported in the species, both in in the wild and in controlled environment. Based on our results, the mean BVI was 17.5, meaning that red pandas performed 73% of the behavioural items described in the species. In particular, males of the two pairs performed 71% of the behavioural items whereas females performed 67% of all items. Based on previous literature on red pandas, males scent-mark and patrol their territory more than females [12,13].

In general, subjects of the study performed all behavioural items described in previous literature except for hunt/stalk (routine behaviour) and social play (social behaviour). Regarding hunt/stalk, red pandas are housed in a naturalistic enclosure, and they would have the possibility to prey on small reptiles or even small vertebrates. Yet, red pandas of the study are fed daily with bamboo and fruits and are also regularly provided with meat (quails). Therefore, it is possible that Ilosha, Ny'ma, Lin, and Maituc did not perform hunting and stalking in the data collection period as they did not need to rely on predation. However, red pandas of this study watched carefully and showed interest for birds (e.g., parakeets, corvids) and individuals of other species in their enclosure (interspecific behaviour) and were not indifferent to their presence as well as to the presence of humans, specifically zookeepers (human-directed behaviour). Regarding social play, even in the wild, this behaviour is common among cubs and juvenile subjects or between mothers and offspring, whereas in adults, it can be performed by males and females during courtship and mating seasons [16]. This may be the reason for failing to report this behaviour in our study. However, Lin and Maituc, the pair without offspring, performed some sexual behaviour, consisting in the male following the female and marking over her scent tracks as well as other social behaviours such as allo-grooming, social resting, and sniffing each other [16]. The percentage of social behaviours showed by Lin and Maituc seems to suggest that housing solitary species in pairs or small groups might promote the performance of a wider array of behaviours, such as affiliative behaviours of Lin and Maituc [5,36]. Positive effects of social housing on solitary species were reported in snow leopards, tigers, and lowland tapirs, showing that housing these animals in pairs or small groups might promote exploratory behaviour and/or reduce the performance of abnormal behaviour [36–40].

Red pandas of the study performed almost all behavioural items in the class exploratory/territorial behaviour, especially scent-marking. These behaviours are typical of the species and represent important indicators of behavioural variability as well as appropriate intraspecific and interspecific communication [12,41]. In addition, these behaviours have been considered as positive welfare indicators in different species, including the red panda, as their presence highlights good health and imply that several needs of the animals are met [10,11]. Measures of variability such as behavioural diversity [42] can be useful to assess overall welfare through the analysis of behavioural changes, allowing to implement positive welfare changes deriving from training, enrichment, or other husbandry practice [28,42,43].

Regarding maternal behaviour, we focused on the interaction of Ilosha with her cubs. Based on our results, the most performed behavioural category was "den" (59% of the observation time), intended as the mother being in the nest box with cubs. Even in the wild, early maternal care takes place mainly in the den and cubs start to leave the den around two months of age. At this stage, they spend approximately four hours a day in the den, although this period can vary based on weather, external temperature, and predator/human disturbance [16]. The second most performed behavioural category was "resting and sleeping" (17%). The mother with her cubs has been found to sleep together in contact, until new-born red pandas are eleven months old [16]. The third most performed behaviour was grooming, intended as the mother licking her cubs. This is one of the most common maternal care behaviours during denning period as the female cleans cubs' fur and stimulates urination and defecation by licking the ano-genital region of her offspring [16]. The female Ilosha paid particular attention to her offspring, observing cubs for 6% of the observation time, and intervening in the case of threats or cubs vocalizing to ask for consideration. Other behavioural categories such as antagonistic behaviours, cub transport, and social play were performed less frequently (<2% of the observation time), presumably due the age of cubs. Indeed, in the current study, the three-month-old cubs left the den, moved autonomously in the enclosure, and played with each other, with decreased need to be transported, and reduced direct interaction with their mother [16]. Regarding the behavioural variety in maternal behaviour, Ilosha performed all behavioural items described previously as maternal behaviours of red pandas except for nest building. Usually, female red pandas start building the nest several weeks before parturition and continue through the denning period [16]. However, cubs of Ilosha were autonomous and started to spend more time out of the nest, therefore, it needed less or no upkeeping, explaining the lack of nest-building activity reported in the current study. Based on our results, the female Ilosha showed almost all maternal behaviours typical of the species, suggesting good reproductive skills and competent parental care behaviour, which need to be preserved in the ex-situ population. Ilosha successfully raised other offspring in years preceding the current study. Thus, behavioural variety and competence reported in Ilosha seems to suggest a positive welfare of that female, leading to good reproductive success and cubs' survival.

We acknowledge that the current study has limitations since we only observed four red pandas during daytime and in only one season of the year. However, our study provides further information on the behaviour of red pandas in zoos, especially maternal behaviour, and tests the validity of the BVI to measure behavioural variety in animals under human care. Variability measures such as behavioural diversity and enclosure use variability [42] can be useful to assess welfare as they allow within-subject comparisons of behavioural changes. Thus, future studies could use the BVI as a measure of behavioural variation aside from traditional ethograms, allowing to implement positive welfare changes deriving from training, enrichment, or other husbandry practice [28,42,43].

5. Conclusions

In conclusion, findings of this study highlighted that the red pandas showed no abnormal behaviours, whereas we found different positive behaviours that have been described both in zoological gardens and in the wild. These behaviours need to be maintained in ex situ contexts, to obtain physically and psychologically healthy subjects as well as viable populations. Studies like this are important to improve the knowledge on endangered species biology and needs, enhancing their husbandry standards (e.g., keeping pairs or single animals/stopping reproduction) and breeding success as well as the in-situ conservation efforts. Finally, monitoring the behaviours of pairs housed in different conditions (e.g., female with offspring vs. female with contraceptive implant) might be a valuable tool to make informed decisions about husbandry and management of animals under human care, such as breeding control and social housing of solitary species.

Author Contributions: Conceptualization, C.S., B.R. and M.A.; methodology, C.S.; investigation, M.A.; data curation, M.A. and B.R.; writing—original draft preparation, B.R.; writing—review and editing, C.S. and J.W.; visualization, J.W.; supervision, C.S. and J.W. All authors have read and agreed to the published version of the manuscript.

Funding: This research received no external funding.

Institutional Review Board Statement: The study was carried out through the live observation of the animals, using non-invasive techniques. The procedure of the study was in accordance with the EU Directive 2010/63/EU and the Italian legislative decree 26/2014 for Animal Research. All procedures performed in the study were in accordance with the ethical standards of Parco Natura Viva, as the research was approved by Parco Natura Viva ethical committee and by the local veterinary authority.

Data Availability Statement: The data presented in this study are available on request from the corresponding author.

Acknowledgments: We would like to thank Cesare Avesani Zaborra and Camillo Sandri for allowing this study to take place in Parco Natura Viva. Furthermore, we acknowledge Nedeljka Stevanovic and Sara Castellazzi, Parco Natura Viva zookeepers, for their hard work and for their precious help with the red panda monitoring. Finally, very special thanks should be given to Angela Glatston for her helpful comments and revisions of the manuscript's first draft.

Conflicts of Interest: Caterina Spiezio is employed by Parco Natura Viva as head of the Research and Conservation Department. The category of potential conflict of interest is "Employment". Barbara Regaiolli is employed by Parco Natura Viva as researcher in the Research and Conservation Department. The category of potential conflict of interest is "Employment". All these authors have disclosed those interests fully to the journal, and they have in place an approved plan for managing any potential conflicts arising from the involvement. Janno Weerman and Mariangela Altamura have no conflict of interest.

References

1. Kappelhof, J.; Weerman, J. The development of the red panda Ailurus fulgens EEP: From a failing captive population to a stable population that provides effective support to in situ conservation. *Int. Zoo Yearb.* **2020**, *54*, 102–112. [CrossRef]
2. Zhang, Z.; Hu, J.; Yang, J.; Li, M.; Wei, F. Food habits and space-use of red pandas Ailurus fulgens in the Fengtongzhai Nature Reserve, China: Food effects and behavioural responses. *Acta Theriol. (Warsz.)* **2009**, *54*, 225–234. [CrossRef]
3. Hill, S.P.; Broom, D.M. Measuring zoo animal welfare: Theory and practice. *Zoo Biol.* **2009**, *28*, 531–544. [CrossRef] [PubMed]
4. Bacon, H. Behaviour-based husbandry—A holistic approach to the management of abnormal repetitive behaviors. *Animals* **2018**, *8*, 103. [CrossRef]
5. Hosey, G.; Pankhurst, S.; Melfi, V. *Zoo Animals: Behaviour, Management, and Welfare*; Oxford University Press: New York, NY, USA, 2013.

6. McCormick, W. Recognizing and Assessing Positive Welfare: Developing Positive Indicators for Use in Welfare Assessment. In Proceedings of the Measuring Behavior, Utrecht, The Netherlands, 28–31 August 2012.
7. Spiezio, C.; Valsecchi, V.; Sandri, C.; Regaiolli, B. Investigating individual and social behaviour of the Northern bald ibis (Geronticus eremita): Behavioural variety and welfare. *PeerJ* **2018**, *6*, e5436. [CrossRef]
8. Miller, L.J.; Pisacane, C.B.; Vicino, G.A. Relationship between behavioural diversity and faecal glucocorticoid metabolites: A case study with cheetahs (Acinonyx jubatus). *Anim. Welf.* **2016**, *25*, 325–329. [CrossRef]
9. Mason, G.J.; Latham, N.R. Can't stop, won't stop: Is stereotypy a reliable animal welfare indicator? *Anim. Welf.* **2004**, *13*, 57–69.
10. Duncan, I.J. Behavior and behavioral needs. *Poult. Sci.* **1998**, *77*, 1766–1772. [CrossRef] [PubMed]
11. Andres-Bray, T.; Moller, P.; Powell, D.M. Preliminary model of personality structure in captive red pandas (Ailurus fulgens). *JZAR* **2020**, *8*, 29–36. [CrossRef]
12. Conover, G.K.; Gittleman, J.L. Scent-marking in captive red pandas (Ailurus fulgens). *Zoo Biol.* **1989**, *8*, 193–205. [CrossRef]
13. EAZA Small Carnivor Tag. EAZA Best Practice Guidelines Red Panda (Ailurus fulgens). *EAZA* **2013**, *1*, 1–42.
14. Reid, D.G.; Jinchu, H.; Yan, H. Ecology of the red panda Ailurus fulgens in the Wolong Reserve, China. *J. Zool.* **1991**, *225*, 347–364. [CrossRef]
15. Wei, F.; Zhang, Z. Red Panda Ecology. In *Red Panda: Biology and Conservation of the First Panda*; Glatston, A.R., Ed.; Academic Press: Cambridge, MA, USA, 2021; pp. 193–212.
16. Glatston, A.R. *Red Panda: Biology and Conservation of the First Panda*; Academic Press: London, UK, 2021.
17. Gittleman, J.L. Behavioral Energetics of Lactation in an Herbivorous Carnivore, the Red Panda (Ailurus fulgens). *Ethology* **1988**, *79*, 13–24. [CrossRef]
18. Delfour, F.; Vaicekauskaite, R.; García-Párraga, D.; Pilenga, C.; Serres, A.; Brasseur, I.; Pascaud, A.; Perlado-Campos, E.; Sánchez-Contreras, G.J.; Baumgartner, K.; et al. Behavioural diversity study in bottlenose dolphin (Tursiops truncatus) groups and its implications for welfare assessments. *Animals* **2021**, *11*, 1715. [CrossRef]
19. Princée, F.P.G.; Glatston, A.R. Influence of climate on the survivorship of neonatal red pandas in captivity. *Zoo Biol.* **2016**, *35*, 104–110. [CrossRef]
20. Goswami, S.; Tyagi, P.C.; Malik, P.K.; Pandit, S.J.; Kadivar, R.F.; Fitzpatrick, M.; Mondol, S. Effects of personality and rearing-history on the welfare of captive Asiatic lions (Panthera leo persica). *PeerJ* **2020**, *8*, e8425. [CrossRef]
21. Spiezio, C.; Vaglio, S.; Vandelle, C.; Sandri, C.; Regaiolli, B. Effects of rearing on the behaviour of zoo-housed chimpanzees (Pan troglodytes). *Folia Primatol.* **2021**, *92*, 91–102. [CrossRef]
22. Altmann, J. Observational study of behaviour: Sampling methods. *Behaviour* **1974**, *49*, 227–266. [CrossRef]
23. Jule, K.R. Effects of Captivity and Implications of Ex Situ Conservation: With Special Reference the Red Panda (Ailurus fulgens). Ph.D. Dissertation, University of Exeter, Exeter, UK, 2008.
24. Bray, T.C. Assessing Behavioral Syndromes in Captive Red Pandas (Ailurus fulgens) Using an Ethological Approach. *CUNY Acad. Works* **2017**, *180*, 1–46. Available online: https://academicworks.cuny.edu/hc_sas_etds/ (accessed on 30 October 2021).
25. Roberts, B.M.S.; Gittleman, J.L. Ailurus fulgens. *Mamm. Species* **1984**, *222*, 1–8. [CrossRef]
26. van de Bunte, W.; Weerman, J.; Hof, A.R. Potential effects of GPS collars on the behaviour of two red pandas (Ailurus fulgens) in Rotterdam Zoo. *PLoS ONE* **2021**, *16*, e0252456. [CrossRef]
27. Zhang, Z.; Hu, J.; Han, Z.; Wei, F. Activity patterns of wild red pandas in Fengtongzhai Nature Reserve, China. *Ital. J. Zool.* **2011**, *78*, 398–404. [CrossRef]
28. Sanders, K.; Fernandez, E.J. Behavioral implications of enrichment for golden lion tamarins: A tool for ex situ conservation. *J. Appl. Anim. Welf. Sci.* **2020**, *25*, 1–10. [CrossRef]
29. Bateson, M.; Martin, P. *Measuring Behaviour: An Introductory Guide*, 4th ed.; Cambridge University Press: Cambridge, UK, 2021. [CrossRef]
30. Rose, P.E.; Scales, J.S.; Brereton, J.E. Why the "Visitor Effect" is complicated. unravelling individual animal, visitor number, and climatic influences on behavior, space use and interactions with keepers—A Case Study on Captive Hornbills. *Front. Vet. Sci.* **2020**, *7*, 236. [CrossRef]
31. Regaiolli, B.; Sandri, C.; Rose, P.E.; Vallarin, V.; Spiezio, C. Investigating parental care behaviour in same-sex pairing of zoo greater flamingo (Phoenicopterus roseus). *PeerJ* **2018**, *6*, e5227. [CrossRef] [PubMed]
32. Khan, A.S.; Rai, U.; Chand, P.; Baskaran, N. Summer Activity and Feeding pattern of Captive Red panda (Ailurus fulgens fulgens) at Padmaja Naidu Himalayan Zoological Park, Darjeeling, India. *Lab-2-Land Mag.* **2019**, *2*, 8–14.
33. Wei, F.; Feng, Z.; Wang, Z.; Hu, J. Habitat use and separation between the giant panda and the red panda. *J. Mammal.* **2000**, *81*, 448–455. [CrossRef]
34. Wei, F.; Feng, Z.; Wang, Z.; Zhou, A.; Hu, J. Nutrient and energy requirements of red panda (Ailurus fulgens) during lactation. *Mammalogy* **1999**, *63*, 3–10. [CrossRef]
35. Lawlor, T.E.; Kowalski, K. Mammals, an outline of theriology. (English translation of original 1971 Polish edition by M. Bibrich for the Smithsonian In-stitute and National Science Foundation). *J. Mammal.* **1979**, *60*, 240. [CrossRef]
36. Spiezio, C.; Manicardi, S.; Vallisneri, M.; Regaiolli, B. Don't leave me alone! Housing solitary species in pairs: The case of the lowland tapir (Tapirus terrestris). *Int. Zoo News* **2016**, *63*, 244–254.
37. Clauss, M.; Müller, D.; Steinmetz, H.; Hatt, J.M. The more the merrier or happy when alone? Hypothesis on stress susceptibility in captive individuals of solitary species. *Second Int. Conf. Dis. Zoo Wild Anim.* **2010**, *12*, 92–95.

38. Macri, A.M.; Patterson-Kane, E. Behavioural analysis of solitary versus socially housed snow leopards (Panthera uncia), with the provision of simulated social contact. *Appl. Anim. Behav. Sci.* **2011**, *130*, 115–123. [CrossRef]
39. Pitsko, L.E. Wild Tigers in Captivity: A Study of the Effects of the Captive Environment on Tiger Behavior. Master's Thesis, Virginia Polytechnic Institute and State University, Blacksburg, VA, USA, 2003.
40. Swaisgood, R.R.; Shepherdson, D.J. Scientific approaches to enrichment and stereotypies in zoo animals: What's been done and where should we go next? *Zoo Biol.* **2005**, *24*, 499–518. [CrossRef]
41. Mason, R.T.; LeMaster, M.P.; Müller-Schwarze, D. *Chemical Signals in Vertebrates*; Springer Press: New York, NY, USA, 2005; Volume 10.
42. Brereton, J.E.; Fernandez, E.J. Which index should I use? A comparison of indices for enclosure use studies. *Anim. Behav. Cogn.* **2022**, *9*, 119–132. [CrossRef]
43. Fernandez, E.J. Training as enrichment: A critical review. *Anim. Welf.* **2022**, *31*, 1–12. [CrossRef]

Journal of
Zoological and Botanical Gardens

MDPI

Article

Baseline Behavioral Data and Behavioral Correlates of Disturbance for the Lake Oku Clawed Frog (*Xenopus longipes*)

Jemma E. Dias [1], Charlotte Ellis [2], Tessa E. Smith [1], Charlotte A. Hosie [1], Benjamin Tapley [2] and Christopher J. Michaels [2,*]

1 Department of Biological Sciences, University of Chester, Parkgate Road, Chester CH1 4BJ, UK; jemmaedias@gmail.com (J.E.D.); tessa.smith@chester.ac.uk (T.E.S.); l.hosie@chester.ac.uk (C.A.H.)
2 Zoological Society of London Outer Circle, Regent's Park, London NW1 4RY, UK; charlotte.ellis@zsl.org (C.E.); ben.tapley@zsl.org (B.T.)
* Correspondence: christopher.michaels@zsl.org; Tel.: +44-2074496430

Abstract: Animal behavior and welfare science can form the basis of zoo animal management. However, even basic behavioral data are lacking for the majority of amphibian species, and species-specific research is required to inform management. Our goal was to develop the first ethogram for the critically endangered frog *Xenopus longipes* through observation of a captive population of 24 frogs. The ethogram was applied to produce a diurnal activity budget and to measure the behavioral impact of a routine health check where frogs were restrained. In the activity budget, frogs spent the vast majority of time swimming, resting in small amounts of time devoted to feeding, foraging, breathing, and (in males) amplexus. Using linear mixed models, we found no effect of time of day or sex on baseline behavior, other than for breathing, which had a greater duration in females. Linear mixed models indicated significant effects of the health check on duration of swimming, resting, foraging, feeding, and breathing behaviors for all frogs. This indicates a welfare trade-off associated with veterinary monitoring and highlights the importance of non-invasive monitoring where possible, as well as providing candidates for behavioral monitoring of acute stress. This investigation has provided the first behavioral data for this species which can be applied to future research regarding husbandry and management practices.

Keywords: amphibian; behavior; welfare; zoo research

Citation: Dias, J.E.; Ellis, C.; Smith, T.E.; Hosie, C.A.; Tapley, B.; Michaels, C.J. Baseline Behavioral Data and Behavioral Correlates of Disturbance for the Lake Oku Clawed Frog (*Xenopus longipes*). *J. Zool. Bot. Gard.* **2022**, *3*, 184–197. https://doi.org/10.3390/jzbg3020016

Academic Editors: Kris Descovich, Caralyn Kemp and Jessica Rendle

Received: 10 March 2022
Accepted: 14 April 2022
Published: 19 April 2022

1. Introduction

Animal welfare is a central component of the management of animals in captivity, yet the basic tools to properly assess it are absent for many species [1]. A holistic understanding of behavior [2,3] alongside a scientific framework [4] can facilitate welfare management, but this requires species-specific data. Although welfare may be partly measured through the use of stress hormone analyses [5], behavior correlates of welfare are important non-invasive tools for routine management of captive animals, such as the use of quantitative observations of the spatial distribution of animals and of behavioral 'indicators' [3]. Behavior is the result of numerous extrinsic and intrinsic processes, both physical and mental, and so is sensitive to welfare state [6,7]. Additionally, behavior can be readily and non-invasively monitored and measured, often in real time, by husbandry staff with minimal resource requirements. Validated behavioral measures are powerful tools for managing and improving welfare, but one reliant on an understanding of activity patterns in a focal species and of which behaviors are effective indicators of welfare.

Amphibians are highly threatened as a group [8]. They are widely maintained in captivity for the purposes of research [9], conservation [10–12], education [13], and as pets [14], and yet suffer from negative bias in welfare science [15]. Moreover, amphibians are a diverse group with high degrees of species-level specialization [16], making it important to understand behavioral repertoires, activity budgets, and measures of welfare for individual species [17,18]. In the handful of species where these data have been collected under captive conditions, welfare impacts of basic husbandry conditions have been identified through behavioral measures [19–24], highlighting the importance of the development of such tools. Furthermore, for animals involved in ex situ conservation, welfare may have impacts on conservation success [25].

The Lake Oku clawed frog (*Xenopus longipes*) is a small, fully aquatic anuran species occurring in Lake Oku, Mount Oku, North West Region Cameroon [26]. The species was assessed as critically endangered on the IUCN Red List of Threatened Species [27] and was subject to a mass mortality event between 2006 and 2010, the cause of which remains unknown [28,29]. A population is maintained for captive husbandry research at ZSL London Zoo [30]. This population, one of only two populations in zoos globally, has been used to develop husbandry guidelines for the species [31], to document reproductive and life history data [30,32], and for research into foraging behavior [33] and individual identification systems for the species [34]. However, basic behavioral data, including an ethogram and activity budget and identification of behavioral indicators of welfare, are still lacking for the species.

In this study, we developed an ethogram for *X. longipes*, and used this to document the diurnal activity budget for the species and to identify behavioral correlates of welfare through validation in association with handling events.

2. Materials and Methods

2.1. Study Subjects

The study sample included 24 adult wild-caught adult *X. longipes*, consisting of seven males and seventeen females, housed at ZSL London Zoo since collection from the wild in 2008. The subjects were housed in a large unit [30] containing five occupied tanks. Tank dimensions are 45 × 45 × 45 cm, with water depth 35 cm. Each tank contains several terracotta tubes, some large stones, two plants (*Microsorium pteropus*), and 5 cm aperture plastic mesh for animals to rest on. Subjects were distributed across the five tanks in mixed sex groups with at least one male per tank. There were three tanks of five individuals, one tank of six, and one tank of three. The legal acquisition, provenance, and husbandry of the animals is provided by Michaels et al. [30].

2.2. Ethical Approval

Ethical approval of these methods was provided by the Faculty Research and Ethics Committee at the University of Chester (1708/20/JD/BS) and full ethical review was deemed unnecessary by the Ethics Committee at ZSL as all methods fell within normal husbandry practice (ZDR435).

2.3. Ethogram

Observations to establish an ethogram for the study species lasted a total of three hours and were conducted live between 1030 h and 1530 h on 22 February 2021. During the continuous observation period, the observer noted all behaviors witnessed ad-lib across the subject group including males and females, and frogs in different tanks. Descriptions were provided for each behavior. All event and state behaviors were noted and adapted from previous work with closely related species, such as *Xenopus laevis* [35].

2.4. Baseline Behavioral Data

In order to generate an activity budget for the species, each occupied tank was recorded for a total of three hours in February 2021 using Samsung S10 HMX-H200BP, Canon Legria HFR706, and SONY DCR-SX30 camcorders. Data were collected at three times per day: a morning session (10:30 to 11:30 a.m.), a noon session (12:30 to 13:30 p.m.), and an afternoon session (14:30 to 15:30 p.m.), hereafter Time of Day, on the 22 and 26 February 2021. Nocturnal observations were not possible due to coronavirus-related limitations on staffing and protocol, and consequential concerns regarding health and safety, despite evidence of circadian rhythms and nocturnal activity in captivity in the close relative *X. laevis* [36] and in the wild for *X. longipes* [37]. Importantly, all fundamental behaviors including reproduction are routinely observed in *X. longipes* during the day [30,31]. Cameras were positioned directly in front of the tank to be recorded to ensure maximum visibility and that the whole of each aquarium was visible. Any husbandry, including cleaning and feeding animals, was conducted after the final recording in order to avoid affecting behavior. However, small invertebrate organisms on which frogs preyed were resident in the aquarium throughout the study.

Following the completion of this filming, the BORIS software was used to record durations of swimming, resting, foraging, feeding, breathing, and amplexus behavior using the ethogram (Table 1). The use of this software allowed for continuous recording and focal sampling, as the video was repeated with the observer watching and only scoring for a different focal frog each time. This method allowed for the computation of durations of behaviors, as well as for matching of experimental data for individual frogs. Individuals were identified using belly markings and by following individuals manually through footage. An activity budget was produced to illustrate the proportion of time spent by each individual performing each behavior.

Table 1. Ethogram of state behaviors for captive *X. longipes*, adapted from work on *X. laevis* [35].

Behavior	Definition
Swimming	Subject is moving from one location to another through the water, exercising front limbs, back limbs or both to travel. This may be horizontally or vertically.
Resting	Subject is stationary. None of the subject's limbs are being exercised to actively travel in any direction. This may be in the water or resting on a substrate.
Foraging	Subject is actively searching for food through a substrate using the forelimbs. This may be followed by feeding, for which a separate event is recorded.
Feeding	Subject is consuming a food item, rapidly wafting the item towards the face and mouth with forelimbs and often tilting body forwards following.
Breathing	Subject is breathing at the surface of the water with the nares breaching the surface.
Sloughing	Subject forces out limbs in order to removed shed skin. Swimming will likely become rapid and uncontrolled. The slough is often consumed.
Amplex	A male frog grips a female around her inguinal region as part of reproductive behaviour.

The data were collected by only one observer who was trained in the software by a member of staff at the University of Chester. At the time of study, the observer was an MRes Biological Sciences student (graduated October 2022 with Merit); the observer has experience of behavioral study in a range of taxon, gained through a BSc Animal Behavior and Welfare and experience in the zoological industry. The observer has experience recording the behavior of other *Xenopus* species and was trained in doing so by members of the Amphibian Behaviour and Endocrinology Group at the University of Chester. The observer received additional training relevant to *X. longipes* on section prior to the study with staff who work with the species professionally.

Statistical Analysis

Total durations of behaviors were calculated separately for each frog for each time of day (AM, noon, PM). These data were analyzed using linear mixed models via the Lmer and lme4 packages [38,39] in R version 4.1.1 using RStudio Version 1.4.17. Model choice was informed by the Akaike Information Criterion (AIC); interactions were not included as these models resulted in an increased AIC value. Models used each behavior as a response variable, with sex, individual ID, tank number, and time of day being explanatory variables. Individual ID, nested within Tank, was a random factor to control for repeated measures and nested aspects of the design. The anova (model) function was used to test for effects of explanatory variables, and the emmeans package [40] was used for pairwise comparison when a significant effect of session (the only explanatory variable of interest with more than two levels) was detected. Swimming, Resting, Foraging, Feeding, and Breathing were included in analysis. No Other or Sloughing behavior were recorded and too few observations of male-only behavior Amplexus were recorded for meaningful analysis, so these categories were not analyzed. We confirmed that model assumptions were met through visual inspection of residuals via the ggResidpanel function in R [41].

2.5. Behavioral Response to Stressor

The frogs underwent a routine health check, with one tank, chosen at random, being subject to the routine procedure each day for a five-day period, until each tank had been subjected to the health check once. These health checks involved removing all of the frogs in a tank at once from the water by hand, placing them in a separate container of approximately 2.5 L of water taken from their aquarium, selecting a frog at random, catching it in a gloved hand, and visually inspecting it for 30 s. Frogs were also handled on their backs in order to be swabbed on the underbelly for routine chytrid fungus surveillance using a sterile swab. Frogs were then placed in a second identical container until all individuals in the group had been checked and the group could be returned at once to the main tank simultaneously. Health checks commenced at 10:15 a.m. so that frogs were returned to the home tank and observed at a similar time to the behavioral observation sessions (time of day (AM)). All frogs were returned to the tank at the same time. Observations began immediately upon return to the tank. Observations lasted an hour, and the video cameras were set up in the same manner as for experiment two. Humans were not present for the duration of the recording sessions.

The footage was analyzed in the same way using the BORIS software, ethogram, and individual IDs to allow for pairing of data in the control and in the health check. The use of this software facilitated continuous recording and focal sampling. The data was collected by the same observer.

Statistical Analysis

Behavioral data matched for time of day (i.e., time of day (AM) data) were used to test for effects of health check on behavior. Data were analyzed for baseline data using linear mixed models via the lmer and lme4 packages in R [38,39]. Model choice was informed by the Akaike Information Criterion (AIC); interactions were not included as these models resulted in an increased AIC value. Models used each behavior as a response variable, with health check status (yes or no), sex, and individual ID nested within a tank number being explanatory variables. Individual ID, nested within a tank, was a random factor to control for repeated measures and nested aspects of the design. The anova(model) function was used to test for effects of explanatory variables, and the emmeans package [40] was used for pairwise comparison when a significant effect of session (the only explanatory variable of interest with more than two levels) was detected. Swimming, Resting, Foraging, Feeding, and Breathing were included in analysis, but No Other or Sloughing behavior were recorded and too few observations of male-only behavior Amplexus were recorded for meaningful analysis, so these categories were not analyzed. We confirmed that model assumptions were met through visual inspection of residuals via the ggResidpanel function in R [41] for baseline data.

3. Results

3.1. Ethogram

An ethogram was produced to identify the state behaviors exhibited by *X. longipes* in captivity (Table 1).

3.2. Baseline Behavioral Data

Sloughing was so rarely observed that although it was recorded once during the ethogram construction, it was not observed at all during subsequent observations. Linear mixed models, with sex, individual ID, tank number, and time of day being explanatory variables, showed that there was no effect of frog sex or time of observation on any behavior other than an effect of sex on duration of breathing (Table 2). An activity budget pooled across all sessions and both sexes is presented in Figure 1. Parameter estimates of the models are presented in Table 4. Sloughing and Amplexus behavior were almost never recorded, and data were not analyzed. Amplexus accounted for 1.3% of total budget and was only observed three times across all frogs and all observations; it could also only be exhibited by males.

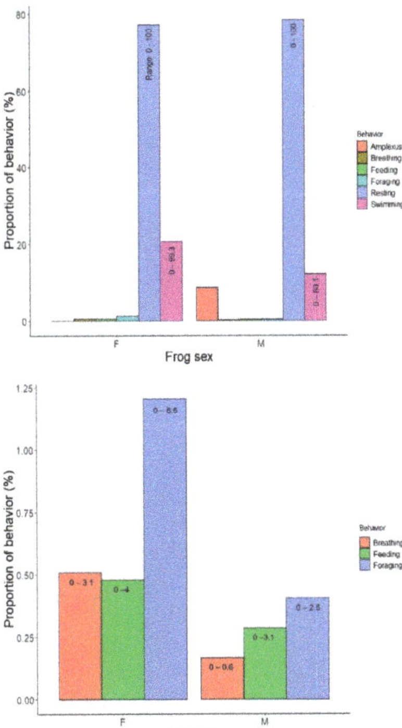

Figure 1. Diurnal activity budget of male and female *X. longipes*; data are expressed as percentages of total behavior duration pre-disturbance across AM, noon, and PM observations. Sloughing is not shown as it was not recorded during the observation periods. The lower pane shows the three behaviors that are not visible in the top pane at a different scale for clarity. The range of percentages has been displayed for each behavior by subtracting the smallest percentage of total behavior pre-disturbance from the greatest percentage of total behavior pre-disturbance.

Table 2. Results of linear mixed models with sex and time of observation as explanatory variables. Significant p values are in bold.

Behavior	Effect of Sex	Effect of Session
Swimming	$F_{1,22} = 2.54$, $p = 0.13$	$F_{2,46} = 3.14$, $p = 0.053$
Resting	$F_{1,22} = 0.16$, $p = 0.69$	$F_{2,46} = 0.73$, $p = 0.49$
Foraging	$F_{1,22} = 4.0$, $p = 0.06$	$F_{2,46} = 0.02$, $p = 0.98$
Feeding	$F_{1,22} = 0.79$, $p = 0.38$	$F_{2,46} = 1.5$, $p = 0.2286$
Breathing	$F_{1,22.003} = 5.36$, $\boldsymbol{p = 0.03}$	$F_{2,46.003} = 0.53$, $p = 0.59$

3.3. Behavioral Response to Stressor

All behaviors measured were significantly affected by the health check (Table 3; Figure 2), but no effect of sex was detected (Table 3). The proportion of time spent exhibiting Swimming and Feeding behaviors increased, whilst Resting, Foraging, and Breathing behaviors decreased. Parameter estimates of the models are presented in Table 3. Sloughing and Amplexus behavior data were not analyzed as the former was not recorded in main observation sessions and the latter was too rarely detected to yield data for meaningful analysis.

Table 3. Results of linear mixed models with health check and sex as explanatory variables. Significant p values are in bold.

Behavior	Effect of Health Check	Effect of Sex
Swimming	$F_{1,22.58} = 171.5$, p < **0.001**	$F_{1,34.3} = 1.755$, $p = 0.19$
Resting	$F_{1,22.58} = 171.5$, p < **0.001**	$F_{1,34.3} = 1.756$, $p = 0.19$
Foraging	$F_{1,45} = 6.2$, p = **0.016**	$F_{1,45} = 2.38$, $p = 0.13$
Feeding	$F_{1,45} = 7.46$, p = **0.009**	$F_{1,45} = 1.40$, $p = 0.24$
Breathing	$F_{1,22.9} = 12.62$, p = **0.002**	$F_{1,28.3} = 3.06$ $p = 0.09$

Table 4. Effect parameters from linear mixed models of baseline behaviors, as a factor of sex and time of day, and of behaviors as a factor of sex and health check.

Model	Response Variable	Parameter	Estimate (SD for Random Effect)	Standard Error of Estimate	t Value	Lower 95% CI of Estimate	Upper 95% CI of Estimate
Behavior = sex + time of day + frog (tank)	Swimming	Intercept	656.43	159.51	4.115	348.306	964.553
		Sex (M)	−399.70	250.75	−1.594	−889.705	90.303
		Time (noon)	174.00	145.98	1.192	−111.794	459.794
		Time (pm)	365.70	145.98	2.505	79.906	651.494
		R^2 Marginal	0.104	-	-	-	-
		R^2 Conditional	0.525	-	-	-	-
		Random effect	475.9	-	-	-	-
	Resting	Intercept	2781.37	200.78	13.853	2392.367	3170.374
		Sex (M)	136.04	335.39	0.406	−519.360	791.448
		Time (noon)	−99.15	150.05	−0.661	−392.906	194.606
		Time (pm)	−180.75	150.05	−1.205	−474.506	113.006
		R^2 Marginal	0.013	-	-	-	-
		R^2 Conditional	0.638	-	-	-	-
		Random effect	683.9	-	-	-	-

Table 4. *Cont.*

Model	Response Variable	Parameter	Estimate (SD for Random Effect)	Standard Error of Estimate	*t* Value	Lower 95% CI of Estimate	Upper 95% CI of Estimate
	Foraging	Intercept	39.59	10.96	3.611	18.463	60.725
		Sex (M)	−26.72	13.36	−2.000	−52.830	0.6149
		Time (noon)	3.00	14.30	0.210	−24.934	30.934
		Time (pm)	2.10	14.30	0.147	−25.834	30.034
		R^2 Marginal	0.057	-	-	-	-
		R^2 Conditional	0.082	-	-	-	-
		Random effect	8.188	-	-	-	-
	Feeding	Intercept	23.609	5.718	4.129	12.592	34.625
		Sex (M)	−6.373	7.184	−0.887	−20.412	7.666
		Time (noon)	−8.400	7.275	−1.155	−22.642	5.842
		Time (pm)	−12.450	7.275	−1.711	−26.692	1.792
		R^2 Marginal	0.050	-	-	-	-
		R^2 Conditional	0.112	-	-	-	-
		Random effect	6.649	-	-	-	-
	Breathing	Intercept	$1.930 \times e^{01}$	4.205	4.589	11.186	27.408
		Sex (M)	$-1.165 \times e^{01}$	5.031	−2.315	−21.308	1.986
		Time (noon)	$-4.174 \times e^{-14}$	5.558	0.00	−10.757	10.757
		Time (pm)	−4.950	5.558	−0.891	15.707	5.807
		R^2 Marginal	0.083	-	-	-	-
		R^2 Conditional	0.088	-	-	-	-
		Random effect	1.395	-	-	-	-
Behavior = health check status + sex + frog (tank)	Swimming	Intercept	608.284	114.36	5.319	384.294	829.647
		Healthcheck (yes)	1397.250	106.70	13.095	1184.948	1609.553
		Sex (M)	−234.632	177.11	−1.325	−578.974	119.898
		R^2 Marginal	0.671	-	-	-	-
		R^2 Conditional	0.820	-	-	-	-
		Random effect	336.5	-	-	-	-
	Resting	Intercept	608.284	160.25	17.547	384.294	829.647
		Health check (yes)	1397.250	183.11	−6.839	1184.948	1609.553
		Sex (M)	−234.632	239.44	0.132	−578.974	119.898
		R^2 Marginal	0.671	-	-	-	-
		R^2 Conditional	0.820	-	-	-	-
		Random effect	336.5	-	-	-	-
	Foraging	Intercept	37.187	8.444	4.404	20.836	53.537
		Health check (yes)	−27.150	10.875	−2.497	−48.206	−6.093
		Sex (M)	−18.469	11.962	−1.544	−41.632	4.694
		R^2 Marginal	0.155	-	-	-	-
		R^2 Conditional	0.155	-	-	-	-
		Random effect	0.00	-	-	-	-

Table 4. *Cont.*

Model	Response Variable	Parameter	Estimate (SD for Random Effect)	Standard Error of Estimate	*t* Value	Lower 95% CI of Estimate	Upper 95% CI of Estimate
	Feeding	Intercept	30.251	17.36	1.743	−3.589	63.861
		Health check (yes)	61.050	22.35	2.731	17.765	104.334
		Sex (M)	−29.148	24.59	−1.185	−76.762	18.467
		R^2 Marginal	0.159	-	-	-	-
		R^2 Conditional	0.159	-	-	-	-
		Random effect	0.00	-	-	-	-
	Breathing	Intercept	18.164	3.025	6.004	12.307	24.024
		Health check (yes)	−12.900	4.630	−3.553	−20.137	−5.663
		Sex (M)	−7.762	4.439	−1.749	−16.541	0.826
		R^2 Marginal	0.235	-	-	-	-
		R^2 Conditional	0.326	-	-	-	-
		Random effect	4.612	-	-	-	-

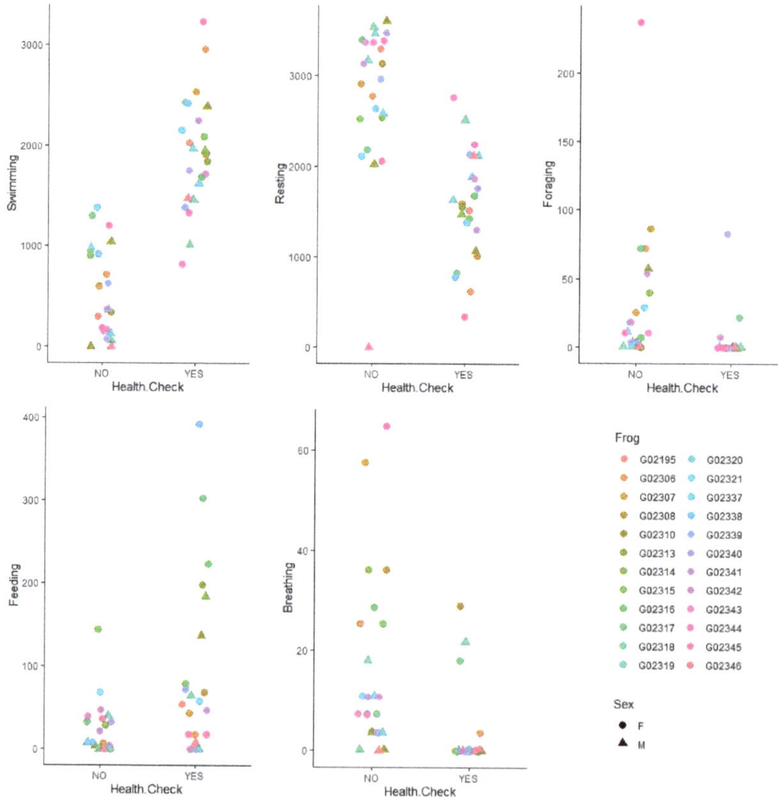

Figure 2. Mean durations of behaviors under baseline and post-health check conditions in *X. longipes*. There was a significant effect of health check on each behavior (see Table 3). Sloughing and Amplexus behavior data were not analyzed as the former was not recorded in main observation sessions and the latter was too rarely detected to yield enough data for analysis.

4. Discussion

We created the first ethogram for *X. longipes*, describing swimming, resting, foraging, feeding, breathing, amplexus, and sloughing behaviors observed by captive *X. longipes*. Although the behavioral repertoire seen in the ethogram is limited in comparison to that of *X. laevis* [35], it is more comprehensive than ethograms available for other amphibian species [42].

Behaviors which have been previously identified as potential stress indicators in other *Xenopus* species were not observed in *X. longipes*. For instance, walling behavior in *X. laevis* was previously described [21] as "Fast swimming back and forwards along a tank wall; rapid rear limb kicks; scrabbling at tank walls with forelimbs; snout against tank wall". Whilst swimming behaviors were detected in this species, the threshold for walling behavior could not be met as "rapid rear limb kicks", "scrabbling at tank walls with forelimbs", and "snout against tank wall" was not present during the observation period (JED, personal observation). Furthermore, although the speed of swimming may have increased in some instances in this study, it was not quantified to identify as "fast swimming".

Little is known about the biology of *X. longipes* [30,43]. The analogue species concept is widely used in the development of amphibian conservation breeding programs [16,44] whereby common relatives of a threatened species are used as models to develop husbandry protocols prior to working with target species [45]. Previously published studies have demonstrated the limitations of the analogue species concept with regard to assumptions made regarding reproductive biology and larval development [30,46]. This study could indicate further limitations of this concept when comparing behaviors of congeneric species. The production of this ethogram highlights the importance of species-specific behavior and welfare research and the caution which should be taken when comparing the behavior of species, even within the same genus. Therefore, husbandry and care practices should be reflective of species-specific natural behavioral biology.

There was no significant difference in any behavior across the sessions at three different times of the day. However, there is evidence of increased nocturnal locomotor activity in *X. laevis* [36], which could also be the case for *X. longipes*. Therefore, a comparison of diurnal and nocturnal behavior may yield significant differences; this was outside the scope of this study. As a result, future applications of this work may not have to control for the time of day. Although nocturnal observation would be useful to inform baseline activity budgets in this study, this was impossible within resource constraints as similar video cameras equipped with infrared night-vision simply created a glare from the glass that prevent ed observation. *X. longipes* is relatively diurnal compared with *X. laevis*; although greater shoreline activity is noted at night in the field [37], captive animals routinely exhibit all fundamental behaviors including locomotion, feeding, and reproduction during the day [30,31], and the data presented here demonstrate that a range of behaviors was detected. From the perspective of practical application, husbandry interventions causing stress, and keeper observations to quantify welfare, all take place during the day, so diurnal behavioral patterns are most relevant. Future work should include nocturnal data collection.

A significant difference exists in breathing duration between the sexes in *X. longipes* where females spend more time breathing than males. Given the much smaller size of males than females [43], this may be the result of differing volume:surface area ratios and implications thereof on the proportion of gas exchange requirements that can be met through cutaneous routes. However, our data do not allow for a clear reason to be identified and other mechanisms may exist. Consequently, differences in behavior between the sexes should still be considered in future work regarding this species.

Our data show that these frogs spend the vast majority of their time swimming and resting, with little of their activity budget allocated to other behaviors. The proportion of time spent swimming was broadly similar (between 10 and 20% of total time) to that reported for *X. laevis* previously [20 (under the condition where refuge was present in this experiment), 21]. Comparisons for other behaviors are not available in the literature. This species does engage in complex feeding behavior when food is present [33], and it is important to note that the present behavioral budget is specifically for frogs outside of when food is delivered to systems; a predominance of foraging and feeding behavior would be expected at these points. Although the breeding season for this species has not yet been identified and in captivity it appears to be sporadic and linked to favorable environmental parameters [31], Amplexus was relatively rarely observed, both in terms of duration (Figure 1), but especially in terms of number of bouts (only three across all observations). We recognize that during breeding periods this may increase substantially. Additionally, our data derive from groups of frogs, which will inevitably perturb individual behavior through interactions between conspecifics. However, given that this species is routinely kept in groups in captivity [30] and observed in groups in close proximity to one another in the field [44], we believe that our data are a good representation of the norm for this species.

The models used for the baseline data have relatively low marginal and (other than for Swimming and Resting) conditional R^2 values and relatively broad confidence intervals around parameter estimates, indicating a large amount of variation in behavior durations, and supportive of sex and time of day explaining little variation. For Swimming, Resting, and Feeding, the conditional R^2 is much higher than the marginal, and for the former two in this list, these values are close to one. Standard deviations of the random effect are also reasonably high. This suggests that in these models, frog identity (nested within tank) explained a substantial amount of variation, and that there may be consistency between individuals in the durations of these behaviors that is not linked to their sex.

Swimming, resting, foraging, feeding, and breathing behaviors were all significantly affected by the health check (Figure 2). A change in behavior was seen in *X. laevis* when subjected to unnatural environmental conditions and was linked with an increase in corticosterone [21]. One explanation for the increase in swimming following the health check in *X. longipes* could be the presence of an escape response which likely mirrors the increase in walling behavior in *X. laevis* during the stress response. Although we did not identify 'walling' behavior [21] as a qualitatively separate behavior from Swimming in our study, increased Swimming could be compared to the increase in walling seen in stressed *X. laevis*; the relatively small physical size of *X. longipes* individuals relative to tank size may have reduced boundary interaction, which is part of the definition of walling. Whilst walling may have been observed over a longer observation period, the 1 h period after a health check was selected as previous work on *X. laevis* has recorded walling behavior within half an hour of experiencing a stressor [21]. Furthermore, anecdotally, walling has not been reported by keeping staff in this species. It seems likely that Swimming behavior induced by stress in *X. laevis* becomes walling once animals interact with a transparent barrier, while *X. longipes* follows the barrier but does not react by swimming up the barrier.

The increase seen in Feeding behavior is likely the result of frogs encountering potential food items in the aquarium more frequently due to the increase in Swimming behavior. The models used for the health check data have moderate to high marginal and conditional R^2 values, indicating that these models are a good fit, and that there is a relatively strong effect of health check despite substantial variation in the data (Figure 2. For two behaviors (Foraging and Feeding), the random effect standard deviation is zero (the *lme4* package reports outcomes of zero when the value is very close to zero), indicating that differences between individuals that cannot be explained by the rest of the model are negligible in this case.

One notable difference in the experimental design of this study and investigations applying welfare assessment tools to *X. laevis* [21,47,48] is that frogs in previous works have been separated into individual tanks for the observation periods. As the subjects are usually kept in groups [31], separation was deemed to be an unnecessary cause of stress in this study. Nonetheless, although data analysis controlled for tank, it is possible that behavior was influenced by interactions between individuals within a tank. This interaction is relevant to the practical application of the data, however, as this species is usually kept in a group.

Although the behavioral changes we detected, given the context, are strongly suggestive of a stress response and align with research in congeners [20,21], validation of this would require that behavior be correlated with corticosterone levels [43]. However, the methods used to quantify corticosterone release rates for *X. laevis* have not yet been validated for use in *X. longipes*. In order to do so, rigorous validation experiments would be required to undergo technical validation, to confirm the sensitivity, specificity, accuracy, and precision of the assay, and biological validation before application to this species [44]; this was outside the scope of this study. In the interim, we suggest that behavioral changes shown here may be used as an indicator of probable stress response to at least short-term disturbance, which may be used to inform husbandry practices. Our results indicate a behavioral impact of the capture of frogs for veterinary monitoring, consistent with a stress response in this genus [20,21], with respect to duration of swimming behavior and repetitive swim patterns. These data highlight the importance of tempering the need to monitor the health of captive animals with the impact of doing so on their welfare and emphasize the need to use non-invasive methods to monitor animals where possible.

There are minimal studies regarding the impact of health checks on amphibian species. Capture, restraint, and handling has been used in the biological validation of corticosterone detection methods for *Ambystoma andersoni* [49] which elicited an increase in corticosterone release, inactivity and gill beat rates following a health check. The contrasting increase in inactivity in *A. andersoni* and increase in activity in *X. longipes* further highlights the need for species-specific research in this area.

This study provides a strong foundation for further research on *X. longipes*, following models used for other pipid taxa to optimize husbandry [20,21,50]. Using the behaviors identified in the ethogram and as potential indicators of stress, investigations can begin assessing husbandry and housing conditions for the species in captivity in order to enhance conservation goals. Investigations into husbandry practices, and other welfare related questions, could be confirmed with the use of corticosterone analysis. If the methods used to quantify corticosterone release rates available for *X. laevis* can be validated for *X. longipes*, further investigation could confirm the potential for increased swimming as an indicator of stress. If a significant rise in the behavior correlates with greater corticosterone release rates, the potential for this behavior as a non-invasive welfare assessment tool can be established [21].

5. Conclusions

This investigation has produced a detailed ethogram for *X. longipes*, establishing six recognizable and observable behaviors. These behaviors can now be applied to further research into husbandry and management practices for the species. Application of these findings may enhance conservation and animal welfare goals by providing evidence needed to better evaluate captive husbandry protocols.

Comparison of behavior in the control and following the health check revealed a significant difference in many behaviors, including increases in Swimming and Feeding alongside decreases in Resting, Foraging, and Breathing. An increase in swimming was linked to walling behavior, although not all aspects of walling were observed. Swimming also became more repetitive, which was illustrated by the decrease in Breathing and Foraging. Increased swimming duration and repetitiveness could be a potential stress-indicator behavior for the species, although this should be confirmed with corticosterone analysis.

Corticosterone analysis could be used to further investigate the duration of the stress response if methods used to quantify corticosterone release rates in *X. laevis* can be applied to *X. longipes*. This would confirm the potential for these behaviors as non-invasive indicators of welfare for this species in captivity.

Author Contributions: Conceptualization, J.E.D., C.J.M., C.E., T.E.S. and C.A.H.; methodology, J.E.D. and C.J.M.; software, J.E.D. and C.J.M.; validation, J.E.D. and C.J.M.; formal analysis, J.E.D. and C.J.M.; investigation, J.E.D. and C.J.M.; resources, J.E.D., C.J.M., C.E. and B.T.; data curation, J.E.D., C.J.M. and C.E.; writing—original draft preparation, J.E.D. and C.J.M.; writing—review and editing, J.E.D., C.J.M., B.T., T.E.S. and C.A.H.; visualization, J.E.D. and C.J.M.; supervision, C.J.M., T.E.S. and C.A.H.; project administration, J.E.D. and C.J.M.; funding acquisition, J.E.D., T.E.S. and C.A.H. All authors have read and agreed to the published version of the manuscript.

Funding: This research was funded by a small student bursary grant from the University of Chester, covering transport and equipment.

Institutional Review Board Statement: Ethical review and approval were waived for this study as it did not deviate from normal husbandry practice. Please see Methods for details of institutional review and approval.

Informed Consent Statement: Not applicable.

Data Availability Statement: Data are available https://github.com/CJMichaels/Dias-et-al-Xenopus-longipes-health-check-welfare (accessed on 15 April 2022).

Acknowledgments: The authors would like to thank Francesca Servini, Daniel Kane and Unnar Aevarsson at ZSL London Zoo for supporting the practical elements of the study, carrying out husbandry on the animals and Daniel Kane for comments on an earlier version of the manuscript. The authors would also like to thank staff within the Biological Sciences department at the University of Chester, including Michal Zatrak who helped to fine tune the experimental design. The captive colony of *X. longipes* was exported in 2008 and 2012 under permit from the Cameroon Ministry of Forestry & Wildlife (0928/PRBS/MINFOF/SG/DFAP/SDVEF/SC and 0193/CO/MINFOF/SG/DFAP/SDVEF/SC), following prior consultation with the Oku community.

Conflicts of Interest: The authors declare no conflict of interest.

References

1. Melfi, V.A. There Are Big Gaps in Our Knowledge, and Thus Approach, to Zoo Animal Welfare: A Case for Evidence-Based Zoo Animal Management. *Zoo Biol.* **2009**, *28*, 574–588. [CrossRef] [PubMed]
2. Dawkins, M.S. Behavior as a tool in the assessment of animal welfare. *Zoology* **2003**, *106*, 383–387. [CrossRef] [PubMed]
3. Dawkins, M.S. Using behavior to assess animal welfare. *Anim. Welf.* **2004**, *13*, 3–7.
4. Maple, T.L. Toward a Science of Welfare for Animals in the Zoo. *J. Appl. Anim. Welf. Sci.* **2007**, *10*, 63–70. [CrossRef] [PubMed]
5. Tillbrook, A.J.; Ralph, C.R. Hormones, stress and the welfare of animals. *Anim. Prod. Sci.* **2017**, *58*, 408–415. [CrossRef]
6. Mench, J.A.; Mason, G.J. Chapter 9: Behavior. In *Animal Welfare;* Appleby, M.C., Hughes, B.O., Eds.; CAB International: Cambridge, MA, USA, 1997; pp. 127–141.
7. Moberg, G.P.; Mench, J.A. *The Biology of Animal Stress: Basic Principles and Implications for Animal Welfare;* CABI Publishing: New York, NY, USA, 2000.
8. Beebee, T.J.C.; Griffiths, R.A. The amphibian decline crisis: A watershed for conservation biology? *Biol. Conserv.* **2005**, *125*, 271–285. [CrossRef]
9. Home Office. *Annual Statistics of Scientific Procedures on Living Animals Great Britain 2018;* The Stationery Office: London, UK, 2018; ISBN 978-1-5286-1336-1.
10. Griffiths, R.A.; Pavajeau, L. Captive Breeding, Reintroduction, and the Conservation of amphibians. *Conserv. Biol.* **2008**, *22*, 852–861. [CrossRef]
11. Browne, R.K.; Wolfram, K.; Garcia, G.; Bagaturov, M.F.; Pereboom, Z.J.J.M. Zoo based amphibian research and Conservation Breeding Programs (CBPs). *Amphib. Rept. Conserv.* **2011**, *5*, 1–14.
12. Zippel, K.; Johnson, K.; Gargliardo, R.; Gibson, R.; McFadden, M.; Browne, R.; Martinez, C.; Townsend, E. The Amphibian Ark: A global community for ex situ conservation of amphibians. *Herpetol. Conserv. Biol.* **2011**, *6*, 340–352.
13. Pavajeau, L.; Zippel, K.C.; Gibson, R.; Johnson, K. Amphibian Ark and the 2008 Year of the Frog Campaign. *Int. Zoo Yearb.* **2008**, *42*, 24–29. [CrossRef]
14. Mohanty, N.P.; Measey, J. The global pet trade in amphibians: Species traits, taxonomic bias, and future directions. *Biodiv. Conserv.* **2019**, *28*, 3915–3923. [CrossRef]
15. Brod, S.; Brookes, L.; Garner, T.W.J. Discussing the future of amphibians in research. *Lab. Anim.* **2019**, *48*, 16–18. [CrossRef] [PubMed]
16. Michaels, C.J.; Gini, B.F.; Preziosi, R.F. The importance of natural history and species-specific approaches in amphibian ex-situ conservation. *Herpetol. J.* **2014**, *24*, 135–145.
17. Harding, G.; Griffiths, R.A.; Pavajeau, L. Developments in amphibian captive breeding and reintroduction programs. *Conserv. Biol.* **2015**, *30*, 340–349. [CrossRef]

18. Tapley, B.; Bradfield, K.S.; Michaels, C.; Bungard, M. Amphibians and conservation breeding programmes: Do all threatened amphibians belong on the ark? *Biodiv. Conserv.* **2015**, *24*, 2625–2646. [CrossRef]
19. Hurme, K.; Gonzalez, K.; Halvorsen, M.; Foster, B.; Moore, D.; Chepko-Sade, B.D. Environmental enrichment for dendrobatid frogs. *J. Appl. Anim. Welf. Sci.* **2003**, *6*, 285–299. [CrossRef]
20. Archard, G.A. Refuge use affects daily activity patterns in female *Xenopus laevis*. *Appl. Anim. Behav. Sci.* **2013**, *145*, 123–128. [CrossRef]
21. Holmes, A.M.; Emmans, C.J.; Jones, N.; Coleman, R.; Smith, T.E.; Hosie, C.A. Impact of tank background on the welfare of the African clawed frog, *Xenopus laevis* (Daudin). *Appl. Anim. Behav. Sci.* **2016**, *185*, 131–136. [CrossRef]
22. Passos, L.F.; Garcia, G.; Young, R. Neglecting the call of the wild: Captive frogs like the sound of their own voice. *PLoS ONE* **2017**, *12*, e0181931. [CrossRef]
23. Passos, L.F.; Garcia, G.; Young, R. Do captive golden mantella frogs recognise wild conspecifics calls? Responses to the playback of captive and wild calls. *J. Zoo Aquar. Res.* **2020**, *9*, 49–54. [CrossRef]
24. Boultwood, J.; O'Brien, M.; Rose, P. Bold Frogs or Shy Toads? How Did the COVID-19 Closure of Zoological Organisations Affect Amphibian Activity? *Animals* **2021**, *11*, 1982. [CrossRef]
25. Swaisgood, R.R. Current status and future directions of applied behavioral research for animal welfare and conservation. *Appl. Anim. Behav. Sci.* **2007**, *102*, 139–162. [CrossRef]
26. Browne, R.K.; Blackburn, D.C.; Doherty-Bone, T. Species Profile: Lake Oku Clawed Frog (Xenopus longipes), Amphibian Ark, 2009. Available online: http://www.amphibianark.org/wp-content/uploads/2018/08/Species-Profile-Xenopus-longipes-July-09.pdf (accessed on 25 November 2021).
27. IUCN SSC Amphibian Specialist Group. *Xenopus longipes* (Amended Version of 2017 Assessment). The IUCN Red List of Threatened Species, 2020. Available online: https://doi.org/10.2305/IUCN.UK.2020-3.RLTS.T58176A177346697.en (accessed on 16 April 2022). [CrossRef]
28. Blackburn, D.C.; Evans, B.J.; Pessier, A.P.; Vredenburg, V.T. An enigmatic mortality event in the only population of the Critically Endangered Cameroonian frog *Xenopus longipes*. *Afr. J. Herpetol.* **2010**, *59*, 111–122. [CrossRef]
29. Doherty-Bone, T.M.; Ndifon, R.K.; Nyingchia, O.N.; Landrie, F.E.; Yonghabi, F.T.; Duffus, A.L.J.; Price, S.; Perkins, M.; Bielby, J.; Kome, N.B.; et al. Morbidity and mortality of the Critically Endangered Lake Oku clawed frog (*Xenopus longipes*). *Endanger. Species Res.* **2013**, *21*, 115–128. [CrossRef]
30. Michaels, C.J.; Tapley, B.; Harding, L.; Bryant, Z.; Grant, S.; Gill, I.; Nyingchia, O.; Doherty-Bone, T. Breeding and rearing the Critically Endangered Lake Oku Clawed Frog (*Xenopus longipes* Loumont and Kobel 1991). *Amphib. Rept. Conserv.* **2015**, *9*, 100–110.
31. Tapley, B.; Michaels, C.; Harding, L.; Bryant, Z.; Gill, I.; Grant, S.; Chaney, N.; Dunker, F.; Freiermuth, B.; Willis, J.; et al. *Lake Oku Frog—EAZA Best Practice Guidelines*; EAZA: Amsterdam, The Netherlands, 2016; Available online: https://www.eaza.net/assets/Uploads/CCC/2016-Lake-Oku-frog-EAZA-Best-Practice-Guidelines-Approved.pdf (accessed on 25 November 2021).
32. Tapley, B.; Michaels, C.J.; Doherty-Bone, T.M. The tadpole of the Lake Oku clawed frog *Xenopus longipes* (Anura; Pipidae). *Zootaxa* **2015**, *3981*, 597–600. [CrossRef]
33. Michaels, C.J.; Das, S.; Chang, Y.M.; Tapley, B. Modulation of foraging strategy in response to distinct prey items and their scents in the aquatic frog *Xenopus longipes* (Anura: Pipidae). *Herpetol. Bull.* **2018**, *143*, 1–6.
34. Aevarsson, U.A.E.; Graves, A.; Carter, K.C.; Doherty-Bone, T.M.; Kane, D.; Servini, F.; Tapley, B.; Michaels, C.J. Individual identification of the Lake Oku clawed frog (*Xenopus longipes*) using a photographic identification technique. *Herpetol. Conserv. Biol.* **2022**, in press.
35. Holmes, A.M.; Emmans, C.J.; Smith, T.; Hosie, L. Unpublished data. Available from the authors (contact LH).
36. Oishi, T.; Nagai, K.; Harada, Y.; Narus, M.; Ohtani, M.; Kawano, E.; Tamotsu, S. Circadian Rhythms in Amphibians and Reptiles: Ecological Implications. *Biol. Rhythm Res.* **2004**, *35*, 105–120. [CrossRef]
37. Doherty-Bone, T.M.; Nyingchia, O.N.; Tapley, B. Cannibalism in the Critically Endangered Lake Oku Clawed Frog: A possible cause of morbidities and mortalities? *Herpetol. Notes* **2018**, *11*, 667–669.
38. Bates, D. Fitting linear mixed models in R. *R News* **2005**, *5*, 27–30.
39. De Boeck, P.; Bakker, M.; Zwister, R.; Nivard, M.; Hofman, A.; Tuerlinckx, F.; Partchev, I. The Estimation of Item Response Models with the lmer Function from the lme4 Package in R. *J. Stat. Softw.* **2011**, *39*, 1–28. [CrossRef]
40. Lenth, R. Emmeans: Estimated Marginal Means, Aka Least-Squares Means. R Package Version 1.4.7. 2020. Available online: https://CRAN.R-project.org/package=emmeans (accessed on 7 January 2022).
41. Goode, K.; Rey, K. ggResidpanel Tutorial and User Manual. 2019. Available online: https://goodekat.github.io/ggResidpanel-tutorial/tutorial.html (accessed on 2 October 2021).
42. Rose, P.; Evans, C.; Coffin, R.; Miller, R.; Nash, S. Using student-centred research to evidence-base exhibition of reptiles and amphibians: Three species-specific case studies. *J. Zoo Aquar. Res.* **2014**, *2*, 25–32. [CrossRef]
43. Loumont, C.; Kobel, H.R. *Xenopus longipes* sp. nov., a new polyploid pipid from western Cameroon. *Rev. Suisse Zool.* **1991**, *98*, 731–738. [CrossRef]
44. Preece, D.J. The captive management and breeding of poison-dart frogs, family Dendrobatidae, at Jersey Wildlife Preservation Trust. *Dodo* **1998**, *34*, 103–114.

45. Lisboa, C.S.; Vaz, R.I.; Brasileiro, C.A. Captive breeding program for *Scinax alcatraz* (Anura: Hylidae): Introducing amphibian ex situ conservation in Brazil. *Amphib. Rept. Conserv.* **2021**, *15*, 279–288.
46. Michaels, C.J.; Fahrbach, M.; Harding, L.; Bryant, Z.; Capon-Doyle, J.S.; Grant, S.; Gill, I.; Tapley, B. Relating natural climate and phenology to captive husbandry in two midwife toads (*Alytes obstetricans* and *A. cisternasii*) from different climatic zones. *Alytes* **2016**, *33*, 2–11.
47. Hosie, C.A.; Smith, T.E. Behavioral Biology of Amphibians. In *Behavioral Biology of Laboratory Animals*; Coleman, K., Schaprio, S.J., Eds.; Routledge: Oxfordshire, UK, 2021; pp. 345–359.
48. Holmes, A.M.; Emmans, C.J.; Coleman, R.; Smith, T.E.; Hosie, C.A. Effects of transportation, transport medium and re-housing on *Xenopus laevis* (Daudin). *Gen. Comp. Endocrinol.* **2018**, *266*, 21–28. [CrossRef]
49. Emmans, C. Using Behaviour and Glucocorticoids as Non-Invasive Measures of Stress and Welfare in Captive *Ambystoma andersoni*, a Critically Endangered Species of Salamander Found in Lake Zacapu, Mexico. Master's Thesis, University of Chester, Chester, UK, 2015.
50. Cooke, G.M. Use of a translucent refuge for *Xenopus tropicalis* with the aim of improving welfare. *Lab. Anim.* **2018**, *52*, 304–307. [CrossRef]

Journal of
Zoological and
Botanical Gardens

MDPI

Communication

Impact of Broad-Spectrum Lighting on Recall Behaviour in a Pair of Captive Blue-Throated Macaws (*Ara glaucogularis*)

Zoe Bryant *, Eva Konczol and Christopher J. Michaels

Zoological Society of London, London NW1 4RY, UK; eva.konczol@zsl.org (E.K.); christopher.michaels@zsl.org (C.J.M.)
* Correspondence: zoe.bryant@zsl.org

Abstract: Many birds, including macaws, are highly visual animals able to detect a wide band of light wavelengths ranging into ultraviolet A, but in captivity, full-spectrum lighting is not universally employed. Where purpose-made bird lighting is used, this is typically made with the provision of ultraviolet B radiation and vitamin D_3 synthesis in mind. Limited research in this field suggests behavioural and physiological benefits of broad-spectrum lighting provision, but more work is needed to broaden the taxonomic scope and to investigate its impacts on understudied areas of husbandry, including behavioural management. We compared the duration of time a bonded pair of blue-throated macaws at ZSL London Zoo opted to remain in an inside den after being recalled from an outdoors flight aviary, with and without the presence of artificial lighting in the form of High Output T5 Fluorescent lamps, which are rich in UVA and UVB wavelengths as well as those visible to humans. We hypothesized that the birds would remain inside for longer when T5 lighting was on, as they would be more visually comfortable. Using randomization analyses, we show that, over 54 trials split between winter and spring, the mean duration spent inside after recall increased from 81.04 to 515.13 s with the presence of the lighting unit, which was highly statistically significant. Our results are likely to be explained by much higher visibility of indoor surroundings creating a more hospitable indoor environment for the birds and will have implications for captive macaw management.

Keywords: Psitaccidae; husbandry; ultraviolet light; artificial lighting; T5 lamps; vision; behaviour

Citation: Bryant, Z.; Konczol, E.; Michaels, C.J. Impact of Broad-Spectrum Lighting on Recall Behaviour in a Pair of Captive Blue-Throated Macaws (*Ara glaucogularis*). *J. Zool. Bot. Gard.* **2022**, 3, 177–183. https://doi.org/10.3390/jzbg3020015

Academic Editors: Kris Descovich, Caralyn Kemp and Jessica Rendle

Received: 7 March 2022
Accepted: 7 April 2022
Published: 13 April 2022

1. Introduction

The Blue-throated macaw *Ara glaucogularis* (Dabbene 1921) is endemic to the Beni savanna in north central Bolivia [1] and is assessed as Critically Endangered by the IUCN [2], with the remaining wild population size said to be fewer than 500 individuals [3]. Threats include illegal poaching, habitat loss, and a lack of breeding sites [2,3]. The species is 85 cm in length with a wingspan of 90 cm, and the feather colouration of adult birds is turquoise on the back, crown, and dorsal sides, with yellow on the chest and a turquoise feather patch on the throat, which gives them their name. They have yellow eyes, a black to greyish beak, turquoise tail feathers, a white-feathered face with dark blue feather stripes, and a bare cere [1,4,5]. Their habitat is hot and humid, with average diurnal ambient temperatures ranging from 24 to 32 °C across the year [6].

Like most birds, macaws possess excellent eyesight, with proportionately large eyes on the sides of the head providing a wide field of vision but a small field of stereovision. The lenses are held in place by the ciliary body that contains striated muscles, and these muscles allow birds to have conscious control over their iris and pupil size; macaws use this to rapidly dilate and constrict pupils in a behaviour termed 'pinning' [7]. Macaws have a thinner cornea and a larger posterior chamber when compared with mammals, as well as a disc-like flat posterior chamber which enables their retinas to have broad visual acuity. Avian retinas are avascular and thick, and the larger number of cones ensures they have good eyesight clarity. Birds also have four types of single cones, which act as photoreceptor cells in their retinas, and these give them their tetrachromatic colour vision, allowing them to discern around 100 million colours, including longer wavelengths in the ultraviolet spectrum (UVA radiation). Macaws rely on this heightened primary sensory system to inform them about their environment and conspecifics, and the function of ultraviolet vision in birds is also thought to be related to orientation, foraging, as well as signalling [8,9]. Like reptiles (see [10]), in addition to wavelengths of light important for vision, birds are also reliant on UVB wavelengths for cutaneous synthesis of vitamin D_3 and, therefore, for calcium homeostasis [11,12]. However, unlike in the case of captive reptiles, where artificial lighting provision is relatively advanced [9], broad-spectrum lighting is not universally used for indoors-housed captive birds, and there is a need for research to underpin changes in practice in this area. Limited published research, which does not focus on psitaccids, showed that the absence of UVA light created the suboptimal condition for birds in general [13] and proposed that environments where UVA is provided are enriching for birds as their environmental perception is enhanced [14]. Ross et al. [15] investigated lighting preferences between supplemental UVA and standard artificial light in 18 non-psittacine bird species housed at the Lincoln Park Zoo, finding that they spent more time in the area of their exhibit that was lit by the UVA light with birds from high light native ecological niches showing increased preference. The more comprehensive work on the broiler and laying hens (see [16] for a review) demonstrates the importance of UVA and UVB lighting for the welfare and health of this species. In addition to the need to broaden and deepen research in this field generally, the impact of broad-spectrum lighting on behavioural management success has not been previously evaluated, and psittacids have also been poorly studied in this field in general.

Macaws must have access to an indoor area when kept in captivity in regions with temperate climates, as they dislike wind and are susceptible to frostbite [17]. Behavioural management in the form of recall training should ideally be implemented with captive macaws, particularly with those housed in exhibits with an indoor and outdoor area. The ability to recall birds is important not only for day-to-day husbandry but also to ensure effective non-invasive health management of the birds. The success of behavioural management, such as recall training, is partly dependent on the environment in which it occurs, as negative stimuli from the environment will cancel out some or all of the impact of positive reinforcement for behaviour [18]. It is therefore important for birds to feel comfortable when called indoors to remove any negative stimuli that could hinder the effectiveness of training. Moreover, enclosure usage is at least partly determined by the suitability of enclosure zones to the species in question, and areas of lower suitability may result in animals not fully using available space and resources [19].

Given the sensitivity of the avian eye to light, we hypothesized that light levels in the indoor part of an aviary might impact the willingness of blue-throated macaws to remain in this part of their environment. We tested the impact of full-spectrum lighting installed in the indoor part of a macaw aviary on the latency to leave this area after being recalled in order to inform enclosure design and bird management moving forward.

2. Materials and Methods

We collected data on one bonded pair of blue-throated macaws, Zoological Information Management System Global Accession Numbers PYR12-00063 and ZRS15-08246, housed in our 'MAC 1' enclosure. This comprised of an indoor den area measuring approximately 327 cm length × 315 cm width × 243 cm height and an outdoor area measuring approximately 530 cm length × 320 cm width × 530 cm height—both with appropriate perching, food, and water provisions and linked with a remotely operated door and shutter. Prior to this study, the indoor light provision for these animals was comprised of one circular wall light approximately 35 cm in diameter containing one GE 2D butterfly CFL square double D 4 pin 16-watt lamp behind frosted glass, and a glass-panelled window measuring 78 cm width × 154 cm length, with glass approximately 9 cm thick, allowing some natural ambient daylight to penetrate the den. None of these light sources emitted any UVB or UVA light. Birds were recall trained through positive reinforcement prior to the study, such that they would enter the indoor den from outside in response to a bell, after which they were rewarded with food. Birds always had access to both in and outdoor parts of their aviary.

On the MAC1 indoor den ceiling, we installed a hydroponic light housing unit (Light Wave 4-tube 54W, Growth Technology, Taunton, UK) fitted with two High Output (HO) T5 fluorescent hydroponic lamps (Blue Growth Lights, Growth Technology, UK) and two 54W HOT5 fluorescent UVB-emitting lamps (D3+ 12% HOT%, Arcadia Reptile, UK). The lighting unit was meshed in with a cage of galvanized steel welded mesh with 2.5 cm apertures to prevent birds from tampering with the lighting. The lighting was calibrated to provide an ultraviolet index (UVI; Baines et al., 2016) of a maximum 2.0 at bird head height on the perch closest to the lamps. UVI was measured with a Solarmeter 6.5 UVI meter (Solartech, Greensboro, NC, USA). The lighting unit was installed but not used for several weeks prior to the beginning of the study, so it was present throughout.

We then conducted two series of experimental recall sessions, one in winter (November 2020) and one in the subsequent spring (March 2021). Both series consisted of twelve consecutive sessions with T5 lamps off and fifteen consecutive sessions with T5 lamps on, for a total of 54 sessions. Each treatment series was preceded by one week of unmonitored exposure to allow birds to habituate.

All sessions took place between 14:00 and 14:30 when the birds were not already inside the den. At each session, one keeper took a UVI reading using a Model 6.5 UV Index Solarmeter at the perching located directly under where the broad-spectrum light unit was mounted on the ceiling, at macaw head height, and a lux intensity reading at perch level using a RoHs Digital LUX meter MT30 model as well as a perching surface temperature using an infrared thermometer (Ketotek, Fujian, China). We also took a UVI and temperature reading on one of their outdoor perches (the same one each time) as well as an outdoor LUX reading immediately before or after the indoor reading.

After these parameters were collected, the keeper would conduct necessary husbandry indoors so that the birds were not disturbed after their recall. Both animals were then recalled into the den by the keeper. After the training concluded at each session, the keeper left their indoor enclosure via the keeper access door, and using two stopwatches, one for each animal, we viewed the birds through a gap in the keeper access door where we were hidden from view of the birds and recorded the time it took each individual bird to leave their indoor den to go back outside. The timing was capped at 900 s; if a bird had not left the den at this point, 900 was recorded as the duration.

Data were analysed using RStudio (Version 4.1.1). Data from both seasons were combined. To confirm the environmental effects of the lighting on the environment in the enclosure, Mann–Whitney tests (using the *stats* package in RStudio) were used to compare the effects of treatments on indoor and outside lux, UVi, and temperature. In order to determine whether the two birds could be treated as separate entities in analysis, we tested for autocorrelation between the birds using a Spearman's Rank (again using the *stats* package in RStudio); this showed strong autocorrelation ($Rs_{53} = 0.87$, p (2-tailed) < 0.001). We, therefore, used mean data from the two animals in the analysis (i.e., for each trial, the mean duration spent inside by the birds was used).

We used the *shuffle* function within the *Mosaic* package [20] to run a randomization analysis with 10,000 iterations in order to test for the effect of lighting on the duration spent inside after recall. Randomization is a valid strategy for analysing small- and single-n samples and is useful when working with small sample sizes in zoo contexts [21,22]. The residual between the means of each treatment is used as a test statistic, and the data are then shuffled randomly 10,000 times and a new test statistic calculated; the p value is derived from the overlap of simulated test statistics with the observed test statistic.

3. Results

3.1. Statistical Analyses

No effects were found on outside temperature (Z = 0.57, $p = 0.57$), indoors temperature (Z = 0.74, $p = 0.61$), outside lux (Z = 0.91, $p = 0.36$) or outside UVi (Z = 0.31, $p = 0.76$). The lighting unit significantly increased indoors lux (Z = 6.26, $p < 0.0001$) and Uvi (Z = 6.27, $p < 0.0001$). Environmental parameters are given in Table 1. Indoor temperature was significantly higher than outdoor temperature (Z = −6.4056, $p < 0.0001$), while outdoors lux was significantly higher than indoor lux, even with the T5 lighting turned on (Z = 3.52871, $p = 0.00042$).

Table 1. Environmental parameter means and associated standard deviations in and outdoors, under T5 on and off treatments, along with Z scores and *p* values from Mann–Whitney U Tests comparing treatments for each parameter. Significant *p* values are in bold.

Parameter	Treatment	Mean (s)	SD	Z	P (T5 Array on vs. off)
Lux—inside	On	595.67	112.75	6.26	**<0.0001**
	Off	50.65	18.98		
Lux—outside	On	816.23	264.79	0.91	0.36
	Off	846.08	505.28		
UVi—inside	On	1.9	0.06	6.27	**<0.0001**
	Off	0	0		
UVi—outside	On	1.12	1.41	0.31	0.76
	Off	0.63	0.77		
Temperature—inside (°C)	On	17.96	1.98	0.74	0.61
	Off	16.90	1.88		
Temperature—outside (°C)	On	10.32	6.95	0.57	0.57
	Off	9.33	5.13		

Mean (SD) time spent inside with T5 lighting off was 81.04 (45.57) s, with a range 18–215 s, and with T5 lighting on, it was 515.13 (278.38) s with a range 33–900 s (900 being the cut-off maximum). Fourteen of thirty observations with T5 lighting exceeded the 900 s cut-off. There was a highly significant effect of lighting on time spent inside after recall ($p < 0.0001$; Figure 1).

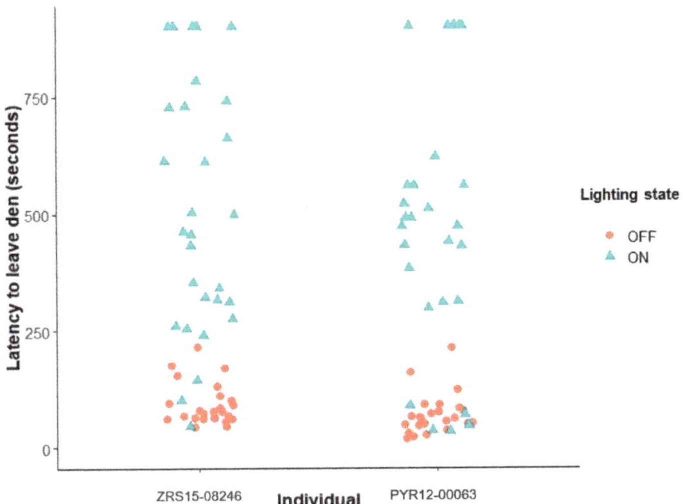

Figure 1. Latency for both individual macaws, ZRS15-08246 and PYR12-00063, to leave the den after recall under different T5 lighting statuses—on or off. Mean data between individuals were used for analysis.

3.2. *Qualitative Observations*

Behaviour changes were observed immediately after the UV lights were installed. The birds were perceived by keepers to move around in the inside den more confidently and use more space than before. As well as the measured increased time spent inside, birds were perceived to respond to recall more quickly. The male, ZRS15-08246, was especially affected, having been particularly reluctant to enter the dens before the installation of lighting and refusing to enter without the presence of the female, PYR12-00063, but afterwards becoming very confident and routinely entered the den on his own. Additionally, since the lighting was installed, keepers observed that the birds spent time inside even when not recalled and remained inside to feed, whereas previous observation showed that although the birds went inside for food, they only collected the items they wanted and consumed them outside.

4. Discussion

Our data show that the presence of the broad-spectrum lighting increased the time both animals spent in their inside den following each recall session. Improved vision due to the presence of UVA and overall higher colour rendering index (CRI) and lux with the additional lighting (all other measured environmental factors did not differ between treatments) may allow the birds to feel more secure or more visually comfortable in their environment, and therefore, more willing to remain inside after their training concluded. It is important to note that these birds are both parent-reared and a bonded pair and, therefore, exhibit less attachment to keepers compared with hand-reared birds.

These data are congruent with findings in other taxa [13–16] and could have wide applicability to other psittacid birds. The provision of broad-spectrum lighting in the captive environment could increase enclosure usage and improve behavioural management, as indoor training is more likely to be a success if the animals can see their environment more clearly and feel more confident navigating their indoor spaces. The birds' overall welfare is likely improved with the additional broad-spectrum lighting inside because there is no longer a trade-off for the birds using the indoor space [19]. Anecdotally, the macaws historically avoided spending time inside, even during cold weather, but after the installation of the broad-spectrum T5 lamps, this is no longer the case. The small overlap (Figure 1) in latency to leave the den between with and without broad-spectrum lighting conditions reinforces the fact that birds did sometimes still choose to leave the den quickly in order to use outside resources and indicates that the birds were able to express a wider range of behavioural choice.

We do not know which parts of the UV spectrum were specifically important to the birds, so other types of lamps may prove to be just as effective. We had planned to conduct a UVB and UVA blocking study as part of our data gathering, which would have helped to inform us of this; however, all data collection had to cease as the birds started nesting and consequently reared three chicks. Further work could investigate this, as well as expand the data collection to other bird species, and include measures of enclosure and resource use. The inclusion of blood sampling or radiography could also indicate the impact of UV lighting on calcium metabolism and health [11,12]. However, based on the present small dataset and existing literature, we suggest that broad-spectrum lighting can facilitate the use of behavioural management in macaws, as well as impact other areas of health and welfare, which supports its use in indoor space for captive birds.

Author Contributions: Conceptualization, Z.B. and E.K.; methodology, Z.B. and E.K.; formal analysis, C.J.M.; investigation, Z.B. and E.K.; data curation, C.J.M.; writing—original draft preparation, Z.B., C.J.M. and E.K.; writing—review and editing, Z.B., C.J.M. and E.K.; supervision, C.J.M.; project administration, Z.B. All authors have read and agreed to the published version of the manuscript.

Funding: This research received no external funding.

Institutional Review Board Statement: Ethical review and approval were waived for this study due to all aspects of the work being within best practice husbandry; the project was registered and internally reviewed as ZDZ153 on the ZSL Projects Database.

Informed Consent Statement: Not applicable.

Data Availability Statement: Data are available at https://github.com/CJMichaels/Macaw-Recall-Data (accessed 12 April 2022).

Acknowledgments: The authors would like to thank Lewis Rowden and Matthew Rendle RVN for their input into the project at an earlier stage, Zuzana Boumrah and Suzi Hyde, who were involved with the husbandry of these animals throughout the project, and Robert Harland and Gary Ward for their support in completing the work.

Conflicts of Interest: The authors declare no conflict of interest.

References

1. Yamashita, C.; Machado de Barros, Y. The Blue-throated Macaw Ara glaucogularis: Characterization of its distinctive habitats in savannahs of the Beni, Bolivia. *Ararajuba* **1997**, *52*, 141–150.
2. BirdLife International. Ara glaucogularis. The IUCN Red List of Threatened Species 2021: E.T22685542A196624397. Available online: https://doi.org/10.2305/IUCN.UK.2021-3.RLTS.T22685542A196624397.en (accessed on 30 September 2021).
3. Herzog, S.K.; Maillard, O.; Boorsma, T.; Sanchez-Avila, G.; Garcia-Soliz, V.H.; Paca-Condori, A.C.; De Abajo, M.V.; Soria-Auza, R.W. First systematic sampling approach to estimating the global population size of the critically endangered Blue-throated Macaw Ara glaucogularis. *Bird Cons. Int.* **2021**, *31*, 293–311. [CrossRef]
4. Duffield, G.E.; Hesse, A.J. Ecology and conservation of the Blue-throated Macaw. *PsittaScene* **1997**, *9*, 10–11.
5. World Parrot Trust. Available online: https://www.parrots.org/encyclopedia/blue-throated-macaw (accessed on 30 September 2021).
6. Climates to Travel world Travel Guide. Available online: https://www.climatestotravel.com/climate/bolivia#forest (accessed on 27 January 2022).
7. Sandmeier, P. Anatomy and physiology. In *BSAVA Manual of Avian Practice*, 1st ed.; Chitty, J., Monks, D., Eds.; British Small Animal Veterinary Association: Gloucester, UK, 2018; pp. 31–32.
8. Bennett, A.T.D.; Cuthill, I. Ultraviolet vision in birds: What is its function? *Vis. Res.* **1994**, *34*, 1471–1478. [CrossRef]
9. Olsson, P.; Lind, O.; Mitkus, M.; Delhey, K.; Kelber, A. Lens and cornea limit UV vision of birds—A phylogenetic perspective. *J. Exp. Biol.* **2021**, *224*, jeb243129. [CrossRef]
10. Baines, F.M.; Chattell, J.; Dale, J.; Garrick, D.; Gill, I.; Goetz, M.; Skelton, T.; Swatman, M. How much UVB does my reptile need? The UV-Tool, a guide to the selection of UV lighting for reptiles and amphibians in captivity. *JZAR* **2016**, *4*, 42–63.
11. Watson, M.K.; Mitchell, M.A. Vitamin D and ultraviolet B radiation considerations for exotic pets. *J. Exot. Pet Med.* **2014**, *23*, 369–379. [CrossRef]

12. Stanford, M. Effects of UVB radiation on calcium metabolism in psittacine birds. *Vet. Rec.* **2006**, *159*, 236–241. [CrossRef] [PubMed]
13. Maddocks, S.A.; Church, S.C.; Cuthill, I. The effect of the light environment on prey choice by zebra finches. *J. Exp. Biol.* **2001**, *204*, 2509–2515. [CrossRef] [PubMed]
14. Moinard, C.; Sherwin, C.M. Turkeys prefer fluorescent light with supplementary ultraviolet radiation. *Appl. Anim. Behav. Sci.* **1999**, *64*, 261–267. [CrossRef]
15. Ross, M.R.; Gillespie, K.L.; Hopper, L.M.; Bloomsmith, M.A.; Maple, T.L. Differential preference for ultraviolet light among captive birds from three ecological habitats. *Appl. Anim. Behav. Sci.* **2013**, *147*, 278–285. [CrossRef]
16. Lewis, P.D. A review of lighting for broiler breeders. *Br. Poult. Sci.* **2006**, *47*, 393–404. [CrossRef] [PubMed]
17. Low, R. *Parrots: Their Care and Breeding*, 2nd ed.; Blandford Press Ltd.: Dorset, UK, 1986; p. 23.
18. Brando, S.I. Animal learning and training: Implications for animal welfare. *Vet. Clin. Exot. Anim. Pract.* **2012**, *15*, 387–398. [CrossRef] [PubMed]
19. Hill, S.P.; Broom, D.M. Measuring zoo animal welfare: Theory and practice. *Zoo Biol.* **2009**, *28*, 531–544. [CrossRef] [PubMed]
20. Pruim, R.; Kaplan, D.T.; Horton, N.J. The mosaic package: Helping students to think with data using R. *R J.* **2017**, *9*, 77–102. [CrossRef]
21. Dugard, P.; File, P.; Todman, J. *Single-Case and Small-N Experimental Designs: A Practical Guide to Randomization Tests*; Routledge: Abingdon-on-Thames, UK, 2012.
22. Michaels, C.J.; Gini, B.F.; Clifforde, L. A persistent abnormal repetitive behaviour in a false water cobra (Hydrodynastes gigas). *Anim. Welf.* **2020**, *29*, 371–378. [CrossRef]

Journal of
Zoological and Botanical Gardens

MDPI

Article

Social Behavior Deficiencies in Captive American Alligators (*Alligator mississippiensis*)

Zane Cullinane Walsh *, Hannah Olson, Miranda Clendening and Athena Rycyk

Division of Natural Sciences, New College of Florida, Sarasota, FL 34243, USA; hannah.olson19@ncf.edu (H.O.); miranda.clendening19@ncf.edu (M.C.); arycyk@ncf.edu (A.R.)
* Correspondence: zane.cullinanewals19@ncf.edu

Abstract: Understanding how the behavior of captive American alligator (*Alligator mississippiensis*) congregations compares to wild congregations is essential to assessing the welfare of alligators in captivity. Wild alligator congregations perform complex social behaviors, but it is unknown if such behaviors occur in captive congregations as frequently. We observed the behaviors of a captive and wild congregation of American alligators in Florida, USA in January 2021. Social behaviors were, on average, 827% more frequent in the wild congregation than the captive, and the wild congregation had a richer repertoire of social behaviors, with growling and HOTA (head oblique tail arched) behaviors being particularly common. High walking, a nonsocial behavior, dominated the behavioral repertoire of the captive congregation (94% of behaviors, excluding feeding) and may be a stereotypy that can be used as an indicator of welfare. Both congregations experienced human disturbance and displayed flushing as a species-specific defense reaction. Captive environments differ from the wild with respect to size, structure, stocking density, resource availability, and human presence. These differences translate into behavioral differences between wild and captive congregations. We identified important behavioral differences between wild and captive alligator congregations that can serve as a platform for more detailed investigations of alligator welfare in captivity.

Keywords: alligator; animal welfare; behavioral observation; comparative; social behavior

Citation: Walsh, Z.C.; Olson, H.; Clendening, M.; Rycyk, A. Social Behavior Deficiencies in Captive American Alligators (*Alligator mississippiensis*). *J. Zool. Bot. Gard.* **2022**, 3, 131–146. https://doi.org/10.3390/jzbg3010011

Academic Editors: Kris Descovich, Caralyn Kemp and Jessica Rendle

Received: 15 February 2022
Accepted: 17 March 2022
Published: 21 March 2022

1. Introduction

The American alligator (*Alligator mississippiensis*) is a predatory species found throughout the Southeastern region of the United States [1]. Within Southern Florida, American alligators are both a keystone predator and an ecosystem engineer [2]. As of 2019, the International Union for Conservation of Nature lists the American alligator as a species "of least concern" [3]. Nonetheless, American alligators are categorized as a threatened species under the Endangered Species Act [4] due to their similarity in appearance to the threatened American crocodile (*Crocodylus acutus*). Thus, regulations in place for the protection of American alligators (hereafter referred to as alligators) also serve to protect American crocodiles [5].

Alligators perform complex social behaviors within congregate settings, both in the wild and in captivity. However, their social behavior between settings has not been compared in a singular study. Alligators can be successfully kept in a captive environment [6], but the behaviors of animals can differ in types and frequency performed when in captivity versus a natural environment [7].

Although there is a lack of published comparative studies in captive and wild alligators, there are extensive comparative behavioral studies that have been previously conducted on mammalian species. For example, chimpanzee (*Pan troglodyte*) groups in captivity increase their time spent grooming, but decrease their time spent foraging relative to wild groups [8]. Additionally, enclosure design and the number of human visitors that captive cotton-top tamarins (*Saguinus oedipus oedipus*) receive correlate with significant differences in social interactions within the group [9]. These findings demonstrate that social behavioral discrepancies can occur between captive and wild animal congregations. As wild alligators commonly display social behavior, it is prudent to evaluate differences in social behavior between captive and wild alligator congregations [10].

While alligator social behavior has not been compared between captivity and natural settings, nesting behaviors have been compared between these settings. Rates of hatching and nesting successes in wild and captive individuals were closely correlated, denoting an insignificant difference between these groups [6]. However, congregation density and stocking rates were significantly different [6].

Interest in reptile welfare in captive settings has been growing and the similarity in reproductive success between captive and wild settings is positive but is only one dimension of welfare. Behavioral expression, both general and social, is also an indicator of animal health and wellbeing [7]. Alligator behaviors have been studied in a captive environment, but they have not been directly compared between wild and captive settings in a singular comprehensive study [6,10,11].

The objective of the current study was to assess social and general behavioral differences between alligator congregations in a captive and wild environment.

2. Materials and Methods

2.1. Wild Observation Site and Congregation

A congregation of wild alligators was observed at Myakka River State Park, Sarasota, Florida. The study site, known as the Deep Hole, is an area containing a large freshwater sinkhole located within the Park's protected Wilderness Preserve [12]. Human visitation is limited in the Wilderness Preserve to 30 permits per day to minimize interference. The area is open and free from tree cover, with the exception of a small cluster of trees on the Western basking area of the sinkhole (Figure 1). It provides alligators with a large prey supply and many suitable basking areas around the sinkhole [12]. The sinkhole connects the lower Myakka Lake and lower Myakka River [12]. During the dry season, when the alligators were observed, the sinkhole was approximately 41 m deep and approximately 90 m in diameter [12].

Figure 1. (**a**) The Deep Hole site in the Myakka Wilderness Preserve where wild alligators were observed. The body of water pictured is the sinkhole. (**b**) The captive enclosure at Croc Encounters, a wildlife sanctuary, where captive alligators were observed. The land area is utilized as basking space.

The Deep Hole site is occupied by a large congregation of alligators [12]. We estimated the congregation density by dividing the average number of individuals observed with binoculars each hour by the estimated area of the sinkhole (6362 m^2). On average, 0.0150 alligators/m^2 are found within the sinkhole. However, the alligators were not evenly distributed throughout the space, rather they were often clustered on or near the banks. Additionally, the number of alligators varied over time during our study, averaging 95 $\pm$ 4 individuals.

2.2. Captive Observation Site and Congregation

A congregation of captive alligators was observed at Croc Encounters, a wildlife sanctuary located in Tampa, Florida. This facility accepts and maintains various reptilian species which have been surrendered to the institution or taken in after being deemed nuisance wildlife. The alligator congregation was fed a dry crocodilian diet mix by passing groups on guided tours and Croc Encounters employees. Guided tours typically occur one to four times per day. The diameter of the captive enclosure was approximately 15.2 m, providing an estimated 182.4 m^2 area.

The captive congregation was composed of 84 adult alligators that varied in size, sex, and age; despite these differences, all were considered adults. The congregation density of 0.461 alligators/m^2 was higher at this site.

2.3. Behavioral and Environmental Data Collection

Two ethograms were created and used to quantify behavior of both the wild and captive alligator congregations. Both ethograms include behaviors previously described in the literature and some behaviors observed during pilot observations. The first ethogram consisted of "social" behaviors including bellowing, bellowing with water dancing, chumpfing, fighting, food theft, growling, hissing, head colliding, headslapping, raised posture, raised snout, roaring, HOTA (head oblique tail arched), and tail wagging (detailed descriptions of the behaviors and associated sources can be found in Table S1) [10,11]. HOTA posturing was included in the social behaviors ethogram due to its known function as a precursor to other social behaviors in a congregation and performance in conjunction with other social displays [11]. In a social context, this behavior can serve as a signifier of alertness [11]. The second ethogram consisted of "general" behaviors, including basking, death rolling, group feeding, group basking, high walking, individual feeding, and low walking (detailed descriptions of the behaviors and associated sources can be found in Table S2) [10,11]. Behavioral data were collected at each site for 5.5 h a day between 9:00 and 14:30 for seven continuous days at each site (38.5 total hours per site). The wild alligators were observed 5–11 January 2021 and the captive alligators 14–20 January 2021. During behavioral observations, the occurrence of behaviors from both the general and social ethogram were recorded with a time and date stamp. When the same behavioral event occurred multiple times in succession, the time interval was recorded and the number of times this behavior was performed. The identification of individual members of each congregation was not feasible. Therefore, behavioral states and events were recorded without the identification of the performing individual or individuals. However, the number of alligators performing these behaviors was recorded.

Additionally, at the wild site, human presence and activity were documented throughout the observational period. Human presences included the presence of park visitors in the vicinity, activities performed by the visitors, and any observed reaction by the alligators. At the captive site, we recorded every time a caretaker entered the interior area of the enclosure, any time a tour group approached the exterior area of the enclosure, and any observed reaction by the alligators. There was constant anthropogenic noise at this site that originated outside the sanctuary, particularly from a nearby interstate with heavy vehicle traffic.

Ambient air temperature and humidity were recorded every hour during the observation periods using a digital thermometer. At no time during the study was any contact made with an alligator in either congregation. Precautions were taken to minimize disturbance, including maintaining a minimum of approximately 30 m from all wild alligators, maintaining a non-disruptive tone when speaking, and minimizing speaking when possible [13]. At the captive location, there was glass separating the enclosure and viewing area. The observers were within 2 m of the glass and remained as still and quiet as possible. As alligators moved around the enclosure, their proximity to the observers varied.

2.4. Data Analysis

The mean ± SE number of social behaviors performed per individual per hour for each congregation was calculated. Data from the 14:00–14:30 period were not included in this analysis because it was not a full hour. We then calculated the difference between the two means and conducted a permutation test using the statistical software R with the *perm* package [14,15]. The test shuffles the data between conditions ("Captive" and "Wild") and calculates the difference in means for a set number of permutations (n = 100,000). The *p*-value, then, is the proportion of those 100,000 values that are greater than our observed value.

The difference in the frequency of the three most commonly observed social (growls, HOTA, and fighting) and general (high walk, flushing, and feeding) behaviors between congregations was compared using Wilcoxon Rank Sum tests. Because multiple comparisons were performed, a Bonferroni correction was used to adjust the alpha level to 0.0083.

To determine the influence of air temperature and human interactions on alligator behavior, all behavior types (social and general) were combined, and behavior/alligator/hour rates were calculated for each congregation. Data from the 14:00–14:30 period were not included in these analyses because it was not a full hour. Linear models were fitted to evaluate hourly trends in the frequency of alligator behaviors relative to air temperature and human interactions. Human interactions at the wild site included the occurrence of humans near the water and humans kayaking on the water. Human interactions at the captive site were quantified as the number of tour groups (there were no human visitors outside tour groups). Analyses and visualizations were performed in MATLAB unless otherwise stated [16].

3. Results

3.1. Social Behaviors

The mean number of social behaviors performed per individual per hour was 0.0942 ± 0.0245 (mean ± SE) for the wild site and 0.0112 ± 0.0027 for the captive site. Of the 100,000 differences in means calculated by the permutation test, none were equal to or greater than the difference in means observed in the current study (0.083). Therefore, the mean number of social behaviors was significantly higher for the wild congregation than the captive congregation ($p < 0.001$) (Figure 2).

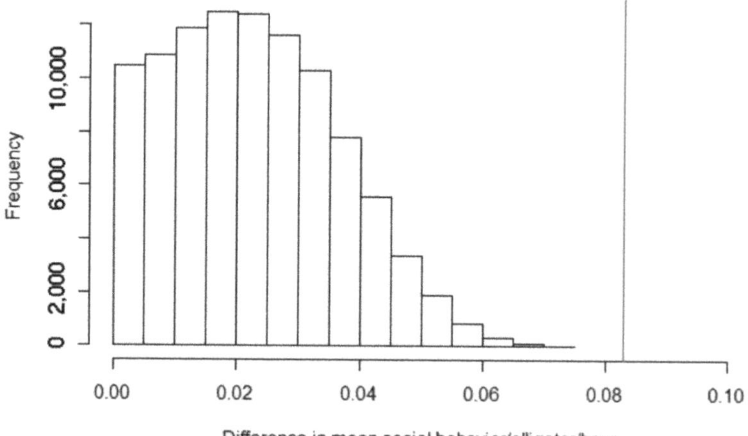

Figure 2. Frequency histogram of the 100,000 permutations of difference in mean social behavior/alligator/hour. This parameter was calculated by subtracting the mean number of social behaviors

per alligator per hour in the captive congregation from the same value in the wild congregation. The permutations are created by randomly resampling the observed data to create hypothetical mean social behavior/alligator/hour for each congregation and the difference in means between congregations is calculated. All 100,000 permutations were higher than the observed difference in means (red vertical line at 0.083).

In total, 320 social behaviors were observed at the wild site and 42 at the captive site. The mean number of social behaviors performed by each congregation per hour was higher at the wild site (8.71 ± 2.41 social behaviors/hour) than at the captive site (0.94 ± 0.23 social behaviors/hour). This increase in behaviors performed across sites indicates that, on average, the wild alligators performed 827% more social behaviors than the captive alligators during the observational periods of this study (Figure 3).

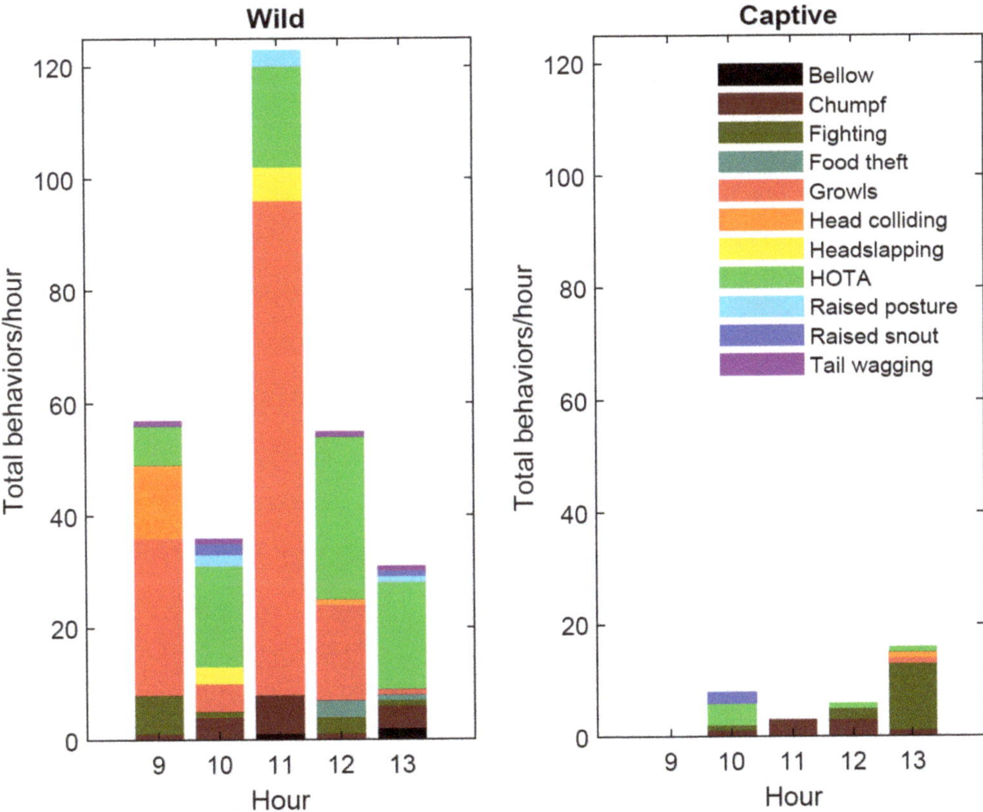

Figure 3. Total number of specific social behaviors (color-coded) during each hour of the day summed across days. The social behavior of the wild congregation is represented on the left and the social behavior of the captive congregation on the right. The y-axis limits are the same for both plots to facilitate comparison of the frequency of social behaviors between the wild and captive congregations. Proportions of general behaviors performed within the wild and captive congregations can be found in Table 1.

The most frequently observed social behavior performed within the wild congregation was growls (46.6% of social behaviors; n = 149) and the second most frequently observed social behavior was HOTA posturing (30.0% of social behaviors; n = 95) (Table 1). Growls (Z

= 3.54, $p < 0.0001$) and HOTA ($Z = 5.32$, $p < 0.0001$) were observed significantly more often in the wild congregation compared to the captive congregation (Figure 4). Growl choruses, in which multiple individuals growl in close succession, occurred in the wild congregation but not the captive. Within the captive congregation, fighting (50% of social behaviors; n = 21) was the most frequently observed social behavior followed by chumpfing (21.4% of social behaviors; n = 9) (Table 1). Fighting was the only social behavior observed more often at the captive site than the wild site (n = 21 at the captive site and n = 10 at the wild site), however the difference was not statistically significant. Roar and water dance were the only social behaviors not observed at either site.

Table 1. The proportion of each social behavior at each site, presented as the percentage of each type of social behavior divided by the total number of social behaviors performed at that site with behavior counts in parentheses (n = 320 for the wild site and n = 42 for the captive site; data collected 09:00–14:30).

Behavior Performed	Wild Site	Captive Site
Bellow	0.94% (n = 3)	0.00% (n = 0)
Chumpf	5.31% (n = 17)	21.43% (n = 9)
Fighting	3.13% (n = 10)	50.00% (n = 21)
Food theft	1.25% (n = 4)	0.00% (n = 0)
Growl	46.56% (n = 149)	2.38% (n = 1)
Head colliding	3.75% (n = 12)	2.38% (n = 1)
Headslapping	2.81% (n = 9)	0.00% (n = 0)
HOTA	29.69% (n = 95)	19.05% (n = 8)
Raised posture	1.88% (n = 6)	0.00% (n = 0)
Raised snout	3.44 % (n = 11)	4.76% (n = 2)
Roar	0.00% (n = 0)	0.00% (n = 0)
Tail wagging	1.25% (n = 4)	0.00% (n = 0)
Water dance	0.00% (n = 0)	0.00% (n = 0)

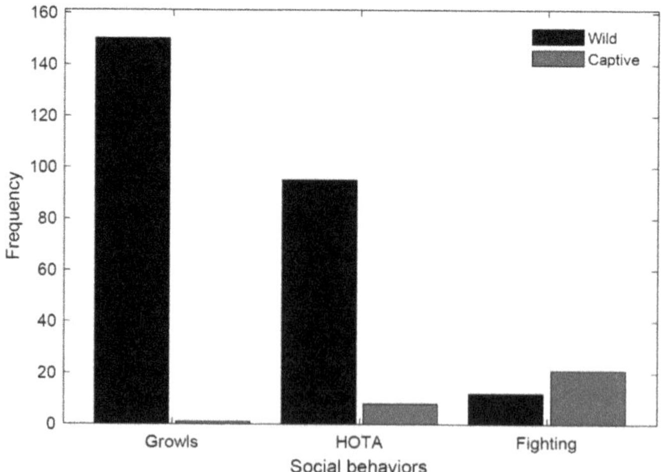

Figure 4. The frequency of the three most commonly observed social behaviors compared between the captive and wild congregations.

3.2. General Behaviors

At the wild site, a total of 422 general behaviors were recorded. The most frequently observed general behavior within the wild congregation was flushing (72.27% of general behaviors; n = 305 at the wild site and 2.70% of general behaviors; n = 74 at the captive site), which most often occurred at the same time as human visitor presence at the Deep Hole site. The death roll behavior was recorded at the wild site while no instances were observed at the captive site (n = 3 at the wild site and n = 0 at the captive site). At the captive site, 2736 general behaviors were recorded. The most frequently observed general behavior within the captive congregation were individuals performing high walks (Figure 5; Table 2). In total, 1973 individual instances were observed of high walking (72.11% of general behaviors; n = 1973).

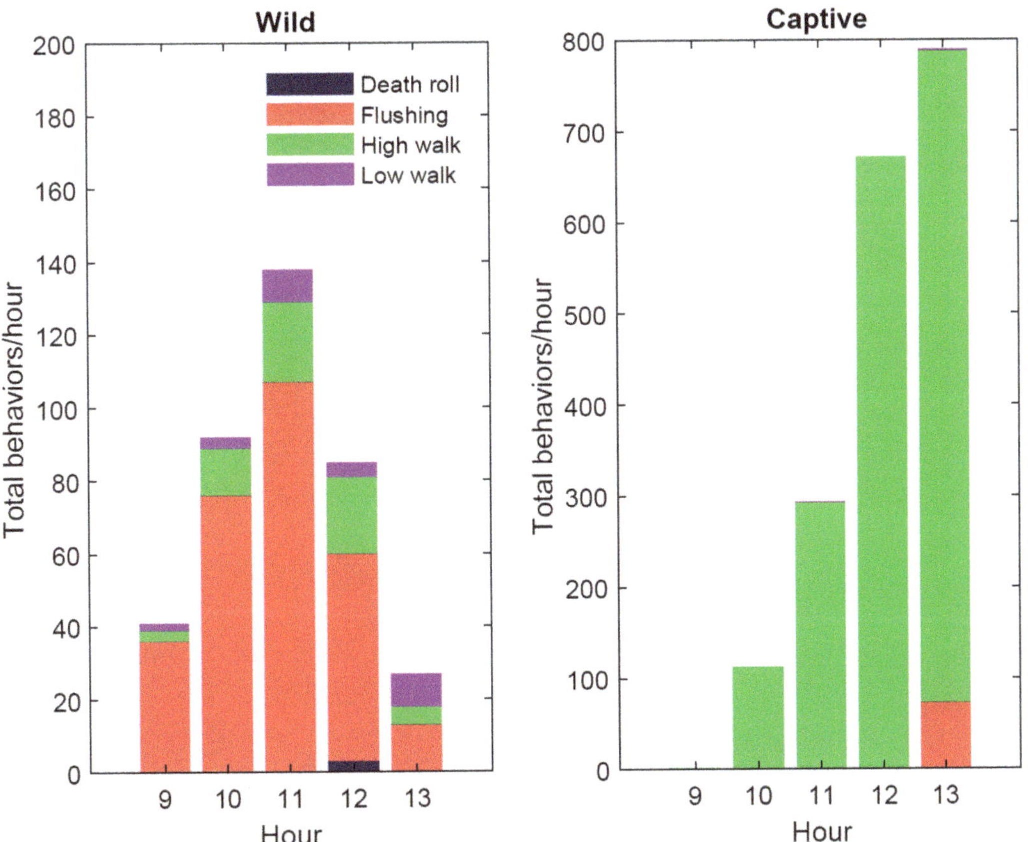

Figure 5. The total number of specific general behaviors (color-coded) during each hour of the day summed across days. The general behavior of the wild congregation is represented in the left plot and the general behavior of the captive congregation in the right plot. Note that the y-axis limits are different between the wild and captive congregation plots. Feeding is excluded because food offerings only occur between 11:30 and 13:27 in the captive congregation, whereas in the wild feeding may occur at any time.

Table 2. The proportion of each general behavior at each site, presented as the percentage of each type of general behavior divided by the total number of general behaviors performed at that site with behavior counts in parentheses (n = 422 for the wild site and n = 2736 for the captive site; data collected 09:00–14:30).

Behavior Performed	Wild Site	Captive Site
Deathroll	0.71% (n = 3)	0.00% (n = 0)
Feeding	2.61% (n = 11)	25.07% (n = 686)
Flushing	72.27% (n = 305)	2.70% (n = 74)
High walk	18.01% (n = 76)	72.11% (n = 1973)
Low walk	6.40% (n = 27)	0.11% (n = 3)

High walking ($Z = -4.21$, $p < 0.0001$) was observed significantly more often in the captive congregation compared to the wild congregation (Figure 6). Flushing ($Z = 4.55$, $p < 0.0001$) was observed significantly more often in the wild congregation compared to the captive congregation (Figure 6). The frequency of feeding events was not statistically compared between congregations, because feeding is regulated by caretakers in the captive setting.

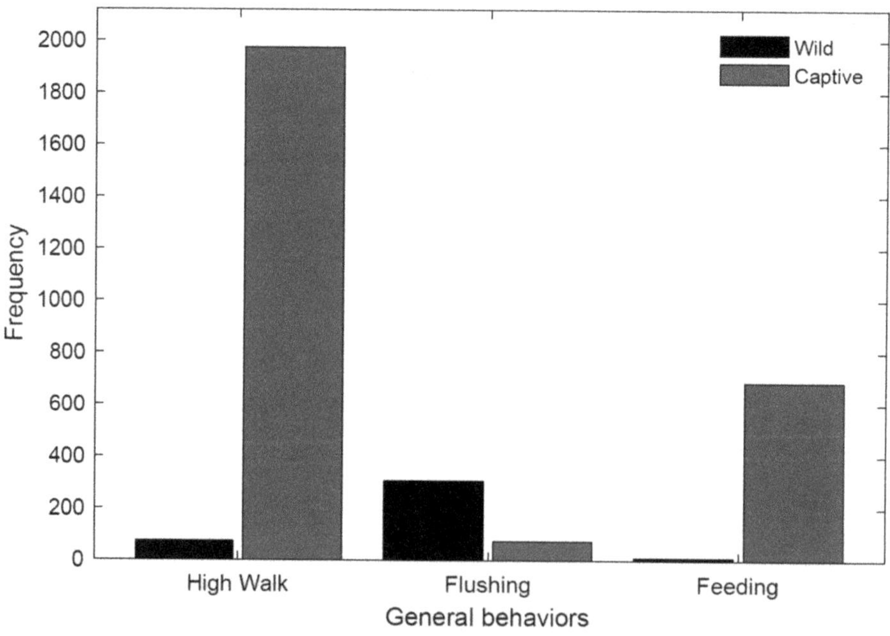

Figure 6. The frequency of the three most commonly observed general behaviors compared between the captive and wild congregations. Feeding opportunities in the captive congregation are regulated by caretakers.

3.3. *Environmental and Human Impacts on Behavior*

The frequency of behavioral events, including both social and general behaviors, generally increased from 9:00 to 12:00 and then decreased in the wild congregation (Figure 7A). In the captive congregation, the frequency of behavioral events increased from 9:00 to 14:00 (Figure 8A).

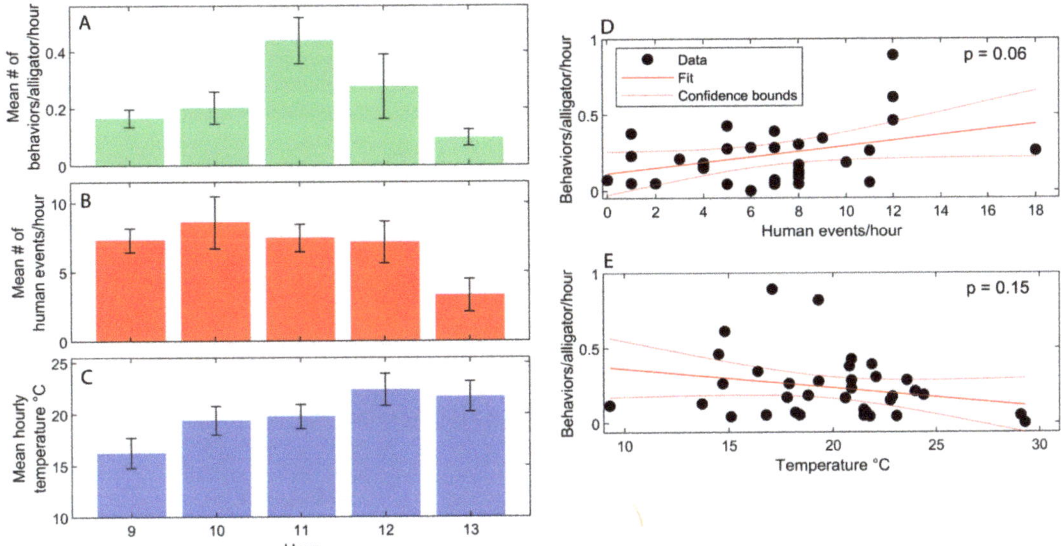

Figure 7. A comparison of trends in behavior, human interactions, and temperature by hour in the wild congregation. (**A**) The mean (±SE) number of behaviors (including social and general) divided by group size per hour by hour of day. (**B**) The mean (±SE) number of human events (including humans walking near the water or kayaking) per hour by hour of day. (**C**) The mean (±SE) hourly temperature (°C) by hour of day. (**D**) Relationship between the frequency of alligator behavior compared to the frequency of human events. The solid line represents a fitted linear model with the dashed lines representing the 95% confidence interval. (**E**) Relationship between the frequency of alligator behavior compared to the air temperature. The solid line represents a fitted linear model with the dashed lines representing the 95% confidence interval.

Human presence at the wild site (from land or water) generally decreased from the morning to the afternoon (Figure 7B). Aggressive social behaviors were not observed in association with human presence at the wild site. Instead, the wild alligators were often recorded flushing due to human presence. Flushing bouts, that involved more than one alligator, in response to human interaction occurred an average of 9.9 times per day at the wild site. A mean of 43.9 individual alligators flushed per day in response to human activity at this location.

At the captive site, tour groups only occurred between 11:00 and 14:00 (Figure 8B). At the captive site there was an increase in aggressive social behaviors between the alligators, such as fighting, when a human caretaker had entered and exited the enclosure. While the caretaker was present inside the enclosure, the alligators showed defensive behavior that included but was not limited to hissing, posturing, and flushing. The mean period of time between the caretaker exiting the enclosure and the first sign of aggressive social interactions between the alligators was 10 min.

The mean hourly temperature at the wild site was 20.3 ± 0.6 °C and the mean hourly humidity level was 46.6%. The mean hourly temperature at the captive site was 17.3 ± 0.5 °C and the mean hourly humidity level was 52.5%. Temperature generally increased from 9:00 to 14:00 at both sites (Figures 7C and 8C).

At the captive site, there was a positive correlation between human presence and the frequency of alligator behavioral events, but no statistically significant correlation at the wild site (Figures 7D and 8D). There was also a positive correlation between the frequency of alligator behavioral events and air temperature at the captive site, but no statistically significant correlation at the wild site (Figures 7E and 8E).

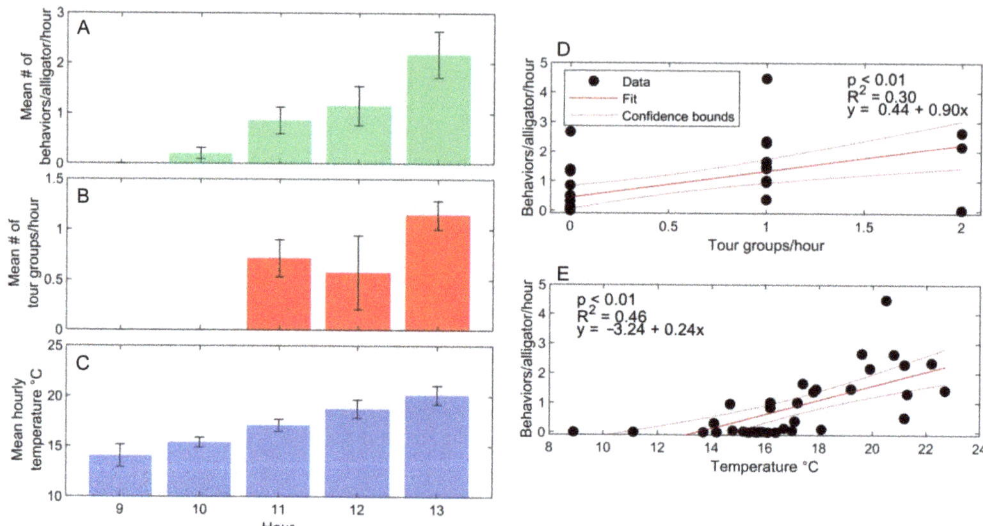

Figure 8. A comparison of trends in behavior, human interactions, and temperature by hour in the captive congregation. (**A**) The mean (±SE) number of behaviors (including social and general) divided by group size per hour by hour of day. (**B**) The mean (±SE) number of tour groups per hour by hour of day. (**C**) The mean (±SE) hourly temperature (°C) by hour of day. (**D**) Relationship between the frequency of alligator behavior compared to the frequency of tour groups. The solid line represents a fitted linear model with the dashed lines representing the 95% confidence interval. (**E**) Relationship between the frequency of alligator behavior compared to the air temperature. The solid line represents a fitted linear model, with the dashed lines representing the 95% confidence interval.

4. Discussion

Social behaviors were consistently observed more often at the wild site than the captive site (Figures 2 and 3; Table 1). In particular, growls and HOTA were observed more often in the wild congregation than the captive (Figure 4). The wild congregation also exhibited a more diverse social behavior repertoire compared to the captive congregation (Figure 3; Table 1). More general behaviors were observed in the captive congregation, but the increase in the performance of general behaviors resulted from the prevalence of a single behavior, high walking, which dominated the behavioral repertoire of the captive congregation (Figures 5 and 6; Table 2). Contrary to high walking, flushing occurred more often in the wild congregation. The frequency of behavioral events of the captive congregation was influenced by human presence (Figure 8).

The difference in population size between congregations is unlikely to account for the observed differences in social behavior. Although there were typically more alligators at the wild site, the average difference in the number of individual alligators (n = 11 which is 12.3%) is small relative to the difference in the frequency of observed social behaviors (7.8/hour which is 827% more social behaviors in the wild congregation). Additionally, the differences in behavior extended far beyond frequency. In the wild congregation, 11 types of social behaviors were observed, while 6 types of social behaviors were observed within the captive congregation. Of particular interest is the prevalence of growling in the wild congregation, but not the captive (Figure 4). Growling is hypothesized to be used for multiple social purposes and may occur in conjunction with or as a component of another social behavior [10,11]. Growling can be used by female individuals as a response to courtship attempts or aggression, although mating behaviors may not have been as prevalent in the mid-January season in which this study was conducted [11]. However, this behavior has been recorded by other researchers as most often being performed by males as a courtship or defensive behavior [10,11]. Growling can also moderate more aggressive behavior like fighting. While the increased fighting in the captive congregation was not statistically significant, it is important to consider the occurrence of growling relative to fighting. In the wild congregation, growling was markedly more prevalent than fighting (14.9 growls:fighting) than in the captive congregation (0.05 growls:fighting) (Table 1). Joanen and McNease (1971) observed three out of five large adult male alligators experienced an increase in the occurrence of aggressive behaviors over the time that they were living in a captive environment. That observed increase in aggressive behavior, along with our findings that fighting was the most commonly performed social behavior by individuals in the captive congregation, suggests that some aspect of the captive environment is influencing the alligators' aggression levels [6]. In the current study, the differences in congregation density, habitat structure, and agitation by the entrance of caretakers into captive enclosures may be the cause of increased aggression in the captive congregation.

The most striking difference in general behavior was the prevalence of high walking in the captive congregation (n = 1973 at the captive site and n = 76 at the wild site). Considering both general and social behaviors (excluding feeding), high walking represents 94% of the observed behaviors in the captive congregation. The elevated high walking behavior at the captive site may be related to overcrowding and space limitation in the captive congregation (the wild site is approximately 35 times the size of the captive site). High walking was observed in the captive congregation on the land area of the enclosure and typically occurred during basking hours. It was necessary for individual alligators to perform the high walking behavior during basking hours in order to maneuver over other alligators on land, though high walking also occurred when not maneuvering over other alligators. The large difference in congregation density for each site ($0.0150/\text{m}^2$ for the wild site and $0.461/\text{m}^2$ for the captive site) resulted in less area available for basking at the captive site than the wild site (Figure 9). High walking in the captive congregation was akin to wandering; this opposes the wild congregation, where alligators typically walked directly from the water to a specific spot for basking. The performance of stereotypic behaviors, like the excessive high walking in the captive congregation, is typically attributed to environmental stress and a lack of adaptation in individual captive animals [17]. Stereotypic behaviors, such as high walking without a clear destination can have negative metabolic consequences beyond the energetic cost of the movement itself, as such activities result in excess post-exercise oxygen consumption (energy continues to be consumed after initial locomotive activities) [18].

Figure 9. (**a**) The basking area of the wild site. (**b**) The basking area of the captive enclosure, showing the typical spacing of individual alligators during basking hours.

The space limitations of the captive enclosure also affected spatial organization of the alligators. At the wild site, alligators commonly arranged themselves based on a size hierarchy, with the largest alligators in the Western and Southern basking areas surrounding the sinkhole whereas the smaller individuals would often be observed basking on the Eastern shoreline. Alternatively, at the captive site, the basking arrangement was in a large pile and there did not appear to be a noticeable size hierarchy or organization in the basking spots. Additionally, direct contact between alligators was commonly observed at the captive site, but not in the wild. Forms of direct contact included alligators forming tightly packed basking piles rather than spaced out groups. Movement is also impacted by the limited area such that alligators commonly crawled over each other, both in basking areas and in the shallow pond (approximately 2 m deep compared to 41 m depth at the wild site) [12]. Stress, as measured through differences in plasma corticosterone levels, has been observed to increase in individual alligators with higher stocking densities in captive environments and decrease in individual alligators in wild environments [19]. Increased stress levels as a result of higher stocking density in the captive site could be a contributing factor to a decrease in social behaviors observed within the captive congregation when compared to the wild congregation.

Differences in the level of confinement may also contribute to behavioral differences observed between congregations. Alligators in an estuary environment exhibit high individual variation in daily movement distance, but travel on average 0.7 to 3.2 km per day [20]. The observed wild site has a larger area than the captive enclosure and has connections with the greater Myakka watershed. The connected open space within the wild site permitted the wild congregation unrestricted movement and migratory opportunities not available to the captive congregation.

While human presence and interactions are generally considered confounding variables in behavioral studies, human presence and interactions with captive animals is commonly intrinsic to housing animals in a captive setting. As such, maintaining a typical pattern of human presence during the study period is important for evaluating behavioral differences between alligators living in wild versus captive conditions. Human presence influenced alligator behavior in both congregations, but the type of influence and magnitude differed. Territorial interactions, which were observed as fighting behaviors between individual alligators within the captive congregation, were observed most often after a caretaker entered the enclosure. A caretaker's entrance always elicited multiple hisses from the alligators, as well as flushing behaviors. Human-induced behavioral changes have been observed in other captive reptiles, such as disruption in social hierarchical perching behaviors of black spiny-tailed iguanas (*Ctenosaura similis*) when humans are close and make eye contact with the iguanas [21]. In the wild congregation, hissing was not observed in response to human presence. The typical response in the wild congregation was flushing and could occur in response to indirect human presence, such as loud speaking voices, humans walking in basking spots, aircrafts traveling overhead, and kayaks traveling through the sinkhole. Conversely, indirect human presence outside of the enclosure was not an observable agitator in the captive site. When humans entered the enclosure there was a more pronounced disturbance to the captive congregation's behavior compared to the wild congregation's reaction to humans. The differences in response to humans could in part be related to proximity between humans and alligators. At the wild site, the distance between human observers and individual alligators was more variable than the captive site. In rare instances, visitors were observed violating park rules to get in close proximity to the wild alligator congregation, at times standing or kneeling within one or two meters of an alligator. In contrast, the captive site had a large barrier that kept tour groups at least three meters from the alligators, maintaining a predictable distance between the guests and the alligators. Both the wild and captive congregations responded to human disturbance when basking by flushing. Alligators are more vulnerable on land, therefore flushing into the water is prudent when disturbed [22]. The consistent presentation of flushing across context when disturbed suggests flushing may be a flight-based species-specific defense reaction to disturbance [23].

Flushing in response to anthropogenic disturbance interrupts basking, which is essential for an alligator to maintain proper homeostatic internal body temperature and the elevated body temperature common to reptiles [24,25]. The maintenance of elevated body temperature during basking hours in reptiles has multiple benefits for the health of individuals, including the performance of typical behaviors and the continuation of normal species-specific growth patterns [22]. The performance of a flushing behavior displaces an individual for a period of time and can therefore negatively impact the individual's energy budget, placing limitations on the performance of other activities [22]. Interruptions to basking behaviors also cause interruptions to the non-rapid eye movement stages of sleep in alligators, [24]. Behaviors like basking are commonly influenced by temperature in crocodilians and indeed we observed a positive correlation between the frequency of behavioral events and temperature at the captive site (Figure 8) [25,26]. Human presence had a more pronounced effect on the frequency of behaviors performed within the captive congregation (Figure 8).

Measuring behavioral differences between wild and captive congregations is only one potential measure of captive animal welfare and should be interpreted with caution [7]. We identified differences between behavior of captive and wild alligators, but welfare requires multiple indices to build a robust picture [27–29]. Ideally, both physiological and behavioral measures should be used, such as cortisol levels, autonomic response, circadian patterns, changes in behavioral repertoire, and stereotypies [30]. We suggest the diminished frequency of social behaviors, impoverished social behavior repertoire, and pronounced occurrence of stereotypies in the captive congregation likely represents compromised welfare. Further study of alligator welfare in captivity is warranted and should consider, in particular, incorporating physiological measurements, non-winter seasons, different congregation densities, and tracking individual differences in behavior.

Concern for reptile welfare in zoos has been growing and includes developing methods to improve welfare [31–33]. One avenue for improving animal welfare is enrichment, such as changes to habitat structure, adding olfactory cues, active food searches, and behavioral training for cognitive stimulation [33–36]. Enrichment methods can decrease stereotypies in captive animals and should be evaluated for improving captive alligator welfare [37,38]. In a study regarding the behavior of captive green sea turtles, researchers discovered that general behaviors impact social behaviors [39]. Food enrichment decreased aggressive behavior displayed and significantly reduced the number of bite wounds inflicted by tank mates [39]. The findings of that study indicate that a change in the general behaviors performed, such as artificially simulated food foraging behavior, by captive reptiles can have a connective relationship with social behaviors performed towards one another [39].

Animal welfare is an important consideration when keeping animals in captivity and care should be taken to assess their welfare and develop species-specific solutions to address deficits [29,36,40]. For example, in the case of American alligators, which display patterns of travel in a natural environment, it is essential to maximize space for the performance of natural locomotive behavior [20].

Despite concerns regarding the welfare of alligators in certain captive environments, there are numerous benefits of captive care of alligators. First, captive environments provide an alternative to euthanizing or relocating nuisance alligators that have come in conflict with humans [41]. Relocation of the alligators is not always successful at resolving the conflict because of the tendency of relocated alligators to travel back to original capture sites [42]. Captive alligators which are kept by private owners and surrendered can also be provided with a living environment if they are unable to be released when habituated to or raised in a captive environment [41]. Second, alligators in captivity provide unique opportunities for scientific research that are not feasible with wild populations. For example, captive research of alligators led to the discovery of a new species of mycoplasma bacteria that causes high mortality rates, and an antibacterial treatment was developed [43]. Third, alligators in captivity offer excellent educational opportunities. Education on wildlife conservation is very commonly taught to the public in zoos and aquariums, with particular programs increasing awareness, avoidance of harmful behaviors towards animal species conservation, and supportive behaviors towards species conservation [44]. Hands-on educational intervention in conservation-related education within zoos and related institutions have been found to retain information learned after an initial educational experience [45,46]. Frequent visitation to zoos increases visitor attitudes, perceptions, and subsequent actions towards the conservation of animal species [47]. Learning experiences from educational intervention on species conservation continue to influence some individuals after the initial lesson [46].

5. Conclusions

Captive alligators, at our study site, displayed fewer social behaviors than wild alligators and wild alligators displayed more variety in social behaviors. These differences may occur because of a range of factors, such as differences in stocking density, enclosure structure and size, human proximity and variation in human behavior, and behavioral variation in individual alligators. Captive alligators also exhibit high-walking stereotypy behaviors that may indicate compromised welfare and warrant consideration of enrichment methods to improve welfare. Building on our behavioral findings in the future by incorporating physiological measures will provide a more complete picture of alligator welfare in captivity.

Supplementary Materials: The following supporting information can be downloaded at: https://www.mdpi.com/article/10.3390/jzbg3010011/s1, Table S1: Social behavior ethogram; Table S2: General behavior ethogram.

Author Contributions: Conceptualization, Z.C.W.; data curation—M.C.; formal analysis—M.C. and A.R.; investigation—Z.C.W. and H.O.; methodology—Z.C.W. and A.R.; project administration—A.R.; software—M.C. and A.R.; supervision—A.R.; visualization—Z.C.W., H.O., M.C. and A.R.; writing—original draft—Z.C.W., H.O. and M.C.; writing—review & editing—Z.C.W., H.O., M.C. and A.R. All authors have read and agreed to the published version of the manuscript.

Funding: This research received no external funding.

Institutional Review Board Statement: This study was conducted under the University of South Florida Institutional Animal Care and Use Committee approval code W IS00008691.

Informed Consent Statement: Not applicable.

Data Availability Statement: The data presented in this study are available on request from the corresponding author.

Acknowledgments: Observations for this study were recorded under Myakka River State Park scientific research permit RP-084, permit 12182014. Thank you to the management of Croc Encounters, John Paner, and the land manager of Myakka River State Park, Stephen Giguere, for granting us access to observe their alligator congregations. Thank you to Tiffany Doan for reviewing the manuscript and providing helpful feedback. Thank you to Helene Gold for assisting our literature review and to Allie Maass for providing feedback on our writing.

Conflicts of Interest: The authors declare no conflict of interest.

References

1. Powell, R.; Conant, R.; Collins, J.T. *Peterson Field Guide to Reptiles and Amphibians of Eastern and Central North America*, 4th ed.; Houghton Mifflin Harcourt: Boston, MA, USA, 2016; ISBN 978-0-544-66249-0.
2. Mazzotti, F.J.; Best, G.R.; Brandt, L.A.; Cherkiss, M.S.; Jeffery, B.M.; Rice, K.G. Alligators and Crocodiles as Indicators for Restoration of Everglades Ecosystems. *Ecol. Indic.* **2009**, *9*, S137–S149. [CrossRef]
3. Elsey, R.; Woodward, A.; Balaguera-Reina, S.A. Alligator Mississippiensis. In *The IUCN Red List of Threatened Species*; IUCN Global Species programme, IUCN SSC, and IUCN Red List Partnership: Gland, Switzerland, 2019.
4. American Alligator. Available online: http://myfwc.com/wildlifehabitats/profiles/reptiles/alligator/ (accessed on 9 March 2021).

5. American Crocodile. Available online: http://myfwc.com/wildlifehabitats/profiles/reptiles/american-crocodile/ (accessed on 22 March 2021).
6. Joanen, T.; Life, L.W.; Grand Chenier, L. Propagation of the American Alligator in Captivity. *J. Wildl. Manag.* **1971**, *30*, 50–56.
7. Veasey, J.S.; Waran, N.K.; Young, R.J. On Comparing the Behaviour of Zoo Housed Animals with Wild Conspecifics as a Welfare Indicator. *Anim. Welf.-Potters Bar* **1996**, *5*, 13–24.
8. Inoue, N.; Shimada, M. Comparisons of Activity Budgets, Interactions, and Social Structures in Captive and Wild Chimpanzees (*Pan troglodytes*). *Animals* **2020**, *10*, 1063. [CrossRef] [PubMed]
9. Glatston, A.R.; Geilvoet-Soeteman, E.; Hora-Pecek, E.; Van Hooff, J.A.R.A.M. The Influence of the Zoo Environment on Social Behavior of Groups of Cotton-Topped Tamarins, *Saguinus oedipus oedipus*. *Zoo Biol.* **1984**, *3*, 241–253. [CrossRef]
10. Vliet, K.A. Social Displays of the American Alligator (*Alligator mississippiensis*). *Integr. Comp. Biol.* **1989**, *29*, 1019–1031. [CrossRef]
11. Garrick, L.D.; Lang, J.W. Social Signals and Behaviors of Adult Alligators and Crocodiles. *Integr. Comp. Biol.* **1977**, *17*, 225–239. [CrossRef]
12. Culter, J.K.; Bowen, C.; Ryan, J.; Perry, J.; Janneman, R.; Lin, W. Exploration of Deep Hole, Myakka River State Park, Florida. In Proceedings of the Joint International Scientific Diving Symposium, Dauphin Island, AL, USA, 24–27 October 2013; pp. 49–60.
13. Blumstein, D.T.; Anthony, L.L.; Harcourt, R.; Ross, G. Testing a Key Assumption of Wildlife Buffer Zones: Is Flight Initiation Distance a Species-Specific Trait? *Biol. Conserv.* **2003**, *110*, 97–100. [CrossRef]
14. R Core Team. *R: A Language and Environment for Statistical Computing*; R Foundation for Statistical Computing: Vienna, Austria, 2019.
15. Fay, M.P.; Shaw, P.A. Exact and Asymptotic Weighted Logrank Tests for Interval Censored Data: The Interval R Package. *J. Stat. Softw.* **2010**, *36*, i02. [CrossRef]
16. *MATLAB, Version 9.10 (R2021a)*; The MathWorks Inc.: Natick, MA, USA, 2021.
17. Vaz, J.; Narayan, E.J.; Dileep Kumar, R.; Thenmozhi, K.; Thiyagesan, K.; Baskaran, N. Prevalence and Determinants of Stereotypic Behaviours and Physiological Stress among Tigers and Leopards in Indian Zoos. *PLoS ONE* **2017**, *12*, e0174711. [CrossRef]
18. Emshwiller, M.G.; Gleeson, T.T. Temperature Effects on Aerobic Metabolism and Terrestrial Locomotion in American Alligators. *J. Herpetol.* **1997**, *31*, 142–147. [CrossRef]
19. Elsey, R.M.; Joanen, T.; McNease, L.; Lance, V. Stress and Plasma Corticosterone Levels in the American Alligator—Relationships with Stocking Density and Nesting Success. *Comp. Biochem. Physiol. Part A Physiol.* **1990**, *95*, 55–63. [CrossRef]
20. Fujisaki, I.; Hart, K.M.; Mazzotti, F.J.; Cherkiss, M.S.; Sartain, A.R.; Jeffery, B.M.; Beauchamp, J.S.; Denton, M. Home Range and Movements of American Alligators (*Alligator mississippiensis*) in an Estuary Habitat. *Anim. Biotelemetry* **2014**, *2*, 8. [CrossRef]
21. Burger, J.; Gochefeld, M.; Murray, B.G., Jr. Role of a Predator's Eye Size in Risk Perception by Basking Black Iguana, *Ctenosaura similis*. *Anim. Behav.* **1991**, *42*, 471–476. [CrossRef]
22. Martín, J. When Hiding from Predators Is Costly: Optimization of Refuge Use in Lizards. *Etologia* **2001**, *9*, 9–13.
23. Bolles, R.C. Species-Specific Defense Reactions and Avoidance Learning. *Psychol. Rev.* **1970**, *77*, 32. [CrossRef]
24. Rial, R.V.; Akaârir, M.; Gamundí, A.; Nicolau, C.; Garau, C.; Aparicio, S.; Tejada, S.; Gené, L.; González, J.; De Vera, L.M. Evolution of Wakefulness, Sleep and Hibernation: From Reptiles to Mammals. *Neurosci. Biobehav. Rev.* **2010**, *34*, 1144–1160. [CrossRef]
25. Smith, E.N. Behavioral and Physiological Thermoregulation of Crocodilians. *Am. Zool.* **1979**, *19*, 239–247. [CrossRef]
26. Boucher, M.; Tellez, M.; Anderson, J.T. Activity Budget and Behavioral Patterns of American Crocodiles (*Crocodylus acutus*) in Belize. *Herpetol. Conserv. Biol.* **2021**, *16*, 86–94.
27. Howell, C.P.; Cheyne, S.M. Complexities of Using Wild versus Captive Activity Budget Comparisons for Assessing Captive Primate Welfare. *J. Appl. Anim. Welf. Sci.* **2019**, *22*, 78–96. [CrossRef]
28. Mendl, M. Assessing the Welfare State. *Nature* **2001**, *410*, 31–32. [CrossRef] [PubMed]
29. Warwick, C.; Arena, P.; Lindley, S.; Jessop, M.; Steedman, C. Assessing Reptile Welfare Using Behavioural Criteria. *Practice* **2013**, *35*, 123–131. [CrossRef]
30. Honess, P.E.; Marin, C.M. Behavioural and Physiological Aspects of Stress and Aggression in Nonhuman Primates. *Neurosci. Biobehav. Rev.* **2006**, *30*, 390–412. [CrossRef] [PubMed]
31. Alligood, C.; Leighty, K. Putting the "E" in SPIDER: Evolving Trends in the Evaluation of Environmental Enrichment Efficacy in Zoological Settings. *Anim. Behav. Cogn.* **2015**, *2*, 200–217. [CrossRef]
32. De Azevedo, C.S.; Cipreste, C.F.; Young, R.J. Environmental Enrichment: A GAP Analysis. *Appl. Anim. Behav. Sci.* **2007**, *102*, 329–343. [CrossRef]
33. Eagan, T. Evaluation of Enrichment for Reptiles in Zoos. *J. Appl. Anim. Welf. Sci.* **2019**, *22*, 69–77. [CrossRef] [PubMed]
34. Manrod, J.D.; Hartdegen, R.; Burghardt, G.M. Rapid Solving of a Problem Apparatus by Juvenile Black-Throated Monitor Lizards (*Varanus albigularis albigularis*). *Anim. Cogn.* **2008**, *11*, 267–273. [CrossRef]
35. Rose, P.; Evans, C.; Coffin, R.; Miller, R.; Nash, S. Using Student-Centred Research to Evidence-Base Exhibition of Reptiles and Amphibians: Three Species-Specific Case Studies. *J. Zoo Aquar. Res.* **2014**, *2*, 25–32. [CrossRef]
36. Warwick, C. Important Ethological and Other Considerations of the Study and Maintenance of Reptiles in Captivity. *Appl. Anim. Behav. Sci.* **1990**, *27*, 363–366. [CrossRef]
37. Fernandez, E.J.; Timberlake, W. Foraging Devices as Enrichment in Captive Walruses (*Odobenus rosmarus*). *Behav. Processes* **2019**, *168*, 103943. [CrossRef]

38. Fernandez, E.J. Appetitive Search Behaviors and Stereotypies in Polar Bears (*Ursus maritimus*). *Behav. Processes* **2021**, *182*, 104299. [CrossRef] [PubMed]
39. Kanghae, H.; Thongprajukaew, K.; Inphrom, S.; Malawa, S.; Sandos, P.; Sotong, P.; Boonsuk, K. Enrichment Devices for Green Turtles (*Chelonia mydas*) Reared in Captivity Programs. *Zoo Biol.* **2021**, *40*, 407–416. [CrossRef] [PubMed]
40. Fraser, D.; Weary, D.M.; Pajor, E.A.; Milligan, B.N. A Scientific Conception of Animal Welfare That Reflects Ethical Concerns. *Ethics Anim.* **1997**, *6*, 187–205.
41. Eversol, C.B.; Henke, S.E.; Ogdee, J.L.; Wester, D.B.; Cooper, A. Nuisance American Alligators: An Investigation into Trends and Public Opinion. *Hum.–Wildl. Interact.* **2014**, *8*, 2. [CrossRef]
42. Janes, D. A Review of Nuisance Alligator Management in the Southeastern United States. In Proceedings of the 4th International Symposium on Urban Wildlife Conservation, Tucson, AZ, USA, 1–5 May 1999; University of Arizona: Tucson, AZ, USA, 1999; pp. 182–185.
43. Clippinger, T.L.; Bennett, R.A.; Johnson, C.M.; Vliet, K.A.; Deem, S.L.; Orós, J.; Jacobson, E.R.; Schumacher, I.M.; Brown, D.R.; Brown, M.B. Morbidity and Mortality Associated with a New Mycoplasma Species from Captive American Alligators (*Alligator mississippiensis*). *J. Zoo Wildl. Med.* **2000**, *31*, 303–314. [CrossRef] [PubMed]
44. Pearson, E.L.; Lowry, R.; Dorrian, J.; Litchfield, C.A. Evaluating the Conservation Impact of an Innovative Zoo-Based Educational Campaign: 'Don't Palm Us Off' for Orangutan Conservation. *Zoo Biol.* **2014**, *33*, 184–196. [CrossRef]
45. Collins, C.; Corkery, I.; McKeown, S.; McSweeney, L.; Flannery, K.; Kennedy, D.; O'Riordan, R. An Educational Intervention Maximizes Children's Learning during a Zoo or Aquarium Visit. *J. Environ. Educ.* **2020**, *51*, 361–380. [CrossRef]
46. Collins, C.; Corkery, I.; McKeown, S.; McSweeney, L.; Flannery, K.; Kennedy, D.; O'Riordan, R. Quantifying the Long-Term Impact of Zoological Education: A Study of Learning in a Zoo and an Aquarium. *Environ. Educ. Res.* **2020**, *26*, 1008–1026. [CrossRef]
47. Kleespies, M.W.; Montes, N.Á.; Bambach, A.M.; Gricar, E.; Wenzel, V.; Dierkes, P.W. Identifying Factors Influencing Attitudes towards Species Conservation—A Transnational Study in the Context of Zoos. *Environ. Educ. Res.* **2021**, *27*, 1421–1439. [CrossRef]

Journal of
Zoological and Botanical Gardens

MDPI

Article

Day Time Activity Budgets, Height Utilization and Husbandry of Two Zoo-Housed Goodfellow's Tree Kangaroos (*Dendrolagus goodfellowi buergersi*)

Katherine Finch * and Amy Humphreys

Chester Zoo, North of England Zoological Society, Upton-by-Chester, Chester CH2 1LH, UK; a.humphreys@chesterzoo.org
* Correspondence: k.finch@chesterzoo.org

Abstract: Goodfellow's tree kangaroos (*Dendrolagus goodfellowi*) are an endangered, arboreal macropod native to the lower, mid-montane rainforests of Papua New Guinea. Despite a number of holders keeping *D. goodfellowi* in zoos across the world, there is a lack of recent published work on this species. Here, we present daytime activity budgets, document height use and provide husbandry information for two Goodfellow's tree kangaroos (*Dendrolagus goodfellowi buergersi*) housed at Chester Zoo, UK. Throughout the observation period, both individuals spent the majority of their time resting within the environment but also spent time engaging in vigilance, travel and feeding behaviour. Additionally, despite the age and sex differences of the study individuals, both animals used the highest height level in the indoor habitat most frequently. We aim to share our information and encourage knowledge transfer with other holders, to both increase understanding and promote evidence-based management of this species.

Keywords: tree kangaroo; zoos; behaviour; activity budget; enclosure use; *Dendrolagus*; understudied

Citation: Finch, K.; Humphreys, A. Day Time Activity Budgets, Height Utilization and Husbandry of Two Zoo-Housed Goodfellow's Tree Kangaroos (*Dendrolagus goodfellowi buergersi*). *J. Zool. Bot. Gard.* **2022**, *3*, 102–112. https://doi.org/10.3390/jzbg3010009

Academic Editors: Kris Descovich, Caralyn Kemp and Jessica Rendle

Received: 26 October 2021
Accepted: 18 February 2022
Published: 4 March 2022

1. Introduction

Tree kangaroos (Genus: *Dendrolagus*) are a unique and fascinating member of the Macropodidae family, which inhabit rainforests across Irian Jaya, Papua New Guinea and Northern Queensland, Australia [1]. As the name suggests, most species of tree kangaroo live an arboreal lifestyle, using trees to rest, forage and travel between different areas [2]. Physiological adaptations such as long claws, rubber-like foot soles, a long tail and specialized limb morphology allow some species, including *D. goodfellowi* and *D. matschiei*, to be specialists in the forest canopy [3,4].

Behavioural research has revealed that many species of tree kangaroo have low activity levels, with studies observing individual engagement in long periods of rest followed by bouts of locomotion [5]. When not resting, tree kangaroos spend time feeding, foraging [6] and performing vigilance behaviour [7]. Patterns of activity are noted to differ across wild *Dendrolagus* species, ranging from crepuscular [8] to nocturnal [9]. However, quantifying activity patterns and time budgets of tree kangaroos has proved challenging, due to the difficulty in tracking wild individuals through the dense forest canopy [10]. As such, the development and evaluation of new non-invasive techniques to monitor tree kangaroos has become increasingly popular [5]. Some low disturbance methods that are currently being trailed include faecal and scratch mark monitoring, utilizing remote cameras and implementing facial recognition software [10].

Identifying new techniques to monitor and observe tree kangaroos has never been more important as wild populations have plummeted in recent decades, leading to 12 of the 14 species of tree kangaroo being listed as threatened by the IUCN [11,12]. Factors which are attributed to this decline include habitat fragmentation [13,14], conflict with human communities and climate change [12]. Dabek and Valentine [12] highlight that the survival of tree kangaroos is most strongly linked to the human communities in their native habitat. However, research suggests that contributions from government, NGO's and zoos are all significant and should not be overlooked [12].

Tree kangaroos have been held in zoos for over a century [15] due to their conservation status and their unique appearance. Extensive population management strategies have been implemented in the form of international studbooks to ensure ex-situ populations of tree kangaroos are as genetically diverse and demographically stable as possible [16]. In addition to this, husbandry practices have been evaluated and vastly improved, with the aim of enhancing the welfare state of zoo-housed individuals [17]. *D. goodfellowi* has received an increasing level of attention from the World Association of Zoos and Aquaria (WAZA), who endorsed a Global Species Management Plan (GSMP) in 2013 for this species. With an ex-situ population of just 55 individuals, low genetic diversity and low number of holders, international cooperation has been suggested to be vital for the maintenance of a healthy, sustainable population of *D. goodfellowi* in zoos [18]. In addition to facilitating ex-situ breeding opportunities for threatened tree kangaroo species, zoos provide an opportunity to study this elusive family of macropods in more detail than would be possible in the wild. As such, they are well placed to contribute to the knowledge base for this understudied species.

Data collection on individual-level behaviour is suggested to be of great importance to the success of keeping tree kangaroos ex-situ [17]. An individual evidence-based management approach has been paramount in providing optimum care [17], allowing keepers to adapt husbandry techniques and resource provision based on the needs of each animal. Throughout the study, we aim to provide husbandry information, outline day-time activity budgets and document height use of two Goodfellow's tree kangaroos housed at Chester Zoo, UK. The aim of this study is two-fold; to provide some information on this understudied species and to use the publication of these data to encourage knowledge transfer between holders of tree kangaroos.

2. Materials and Methods

2.1. Study Subjects

Study subjects were two Goodfellow's tree kangaroo (1M, 1F) (*Dendorolagus goodfellowi buergersi*) housed at Chester Zoo, UK. Goodfellow's tree kangaroo are an endangered species of tree kangaroo native to the mid-montane forests of Papua New Guinea [19]. *D. goodfellowi* are easily distinguishable from other species of tree kangaroo, due to their warm red coloured pelage, golden yellow limbs and long golden and brown non-prehensile tail [20]. Sexual dimorphism is limited in this species, with male specimens only slightly larger than females [3]. *D. goodfellowi* has a life expectancy of approximately 8 years in the wild [21] and over 14 years in captivity [15].

At the time of data collection for this study, the male subject was 2 years of age (date of birth (DOB): 15 May 2017) and the female subject was approximately 18 years of age (DOB: ~16 December 2001) [22]. Due to the age of the female study subject and the species' solitary nature within the wild [19] individuals were housed separately for the duration of the study with alternating access between off-show and on-show facilities. Data were collected when individuals were housed in the 'on-show' facility, so the animals had equal opportunity to utilize the same space, with the same habitat resources but at different data collection periods. Individuals were fed on a species-appropriate diet throughout the data collection period (Appendix A, Figure A1).

2.2. Enclosure Information

Behavioural data collection took place whilst individuals were housed in a custom tree kangaroo exhibit in the 'Islands' expansion of Chester Zoo that opened in 2018. Indoor enclosure dimensions were 3.65 m (Width) $\times$ 0.93 m (Depth) $\times$ 4.43 m (Height), giving a total indoor volume of 15.04 m^3. The open air outdoor enclosure dimensions were 247 m^2, with planting and branching available for individuals to climb to a maximum height of approximately 4 m. All enclosure dimensions were obtained from architectural drawings and not measured by the authors.

The outdoor facility was a mixed-species area, with the study subjects sharing the outdoor environment with dusky pademelons (*Thylogale brunii*). *Thylogale brunii* are a species of forest-dwelling terrestrial macropod endemic to Papua New Guinea, currently listed as 'Vulnerable' to extinction by the IUCN [23]. No interactions between the species were recorded throughout the observation period. To travel between the indoor and outdoor facilities, individuals were provided with branching, allowing access to the indoor facility at a height of 2.6 m. The branching provided differed in texture and width for grip and to encourage a wider range of movement and utilisation of the facility. To quantify height use, both the indoor and outdoor habitats were divided into height 'levels'. Recognisable features were used to distinguish between height levels to ensure accuracy of data collection. Additionally, height levels were divided to include relevant habitat resources (Table 1). Once height levels had been established, each level was measured (by KF) to provide additional information for the reader. The substrate of the indoor environment was coir and the substrate of the outdoor environment was bark chippings. Internal enclosure temperature was kept between 18–22 °C throughout data collection as recommended in the species husbandry guidelines [17]. Food was mainly presented in an elevated bowl (height approximately 1.3 m from the ground), with browse presented at varying heights within the facility to encourage locomotor behaviours.

Table 1. Indoor and outdoor facility height levels were used to quantify height utilisation. Outlined are the name of the height level, height in meters of each level and relevant habitat resources within each area.

Facility	Height Level	Height (m)	Habitat Resources
Indoor	Ground level	0	Coir substrate.
	Level One	0.01–1.22	Fixed wooden structure, fixed and flexible branching.
	Level Two	1.23–2.56	Food bowl, tree stump, fixed and flexible branching.
	Level Three	2.57–4.43	Access to outdoor facility, fixed and flexible branching.
Outdoor	Ground level	0	Bark chipping substrate.
	Level One	0.01–1.32	Flexible branching and ferns.
	Level Two	1.33–2.32	Fixed and flexible branching.
	Level Three	2.33–4.00 (approx)	Access to indoor facility, fixed and flexible branching. Includes higher branches of live trees.

2.3. Husbandry Routine

Throughout data collection, both individuals were managed within the same husbandry routine. Daily keeper routine included the following; provision of fresh food as per species-specific individual diet sheet, cleaning away old food items, removing any soiled substrate, scrubbing and re-filling water drinkers, watering plants, visual health checks of individuals and checking security fencing around the exhibit. If required, the following tasks would also be completed; cleaning of indoor and outdoor windows, replacement of existing furnishings with new ones, e.g., logs and branching, raking substrate and removing old debris such as branches, twigs and stones, trimming back foliage within exhibit, cleaning signage, top-up of substrate within exhibit and public area maintenance.

Procedures and checks are also conducted as part of routine health monitoring. These include but are not limited to; regular weight measurements, nail inspection for nail trims, and the collection of faecal samples for the evaluation of female cyclicity and endo-parasite load. On-site veterinary, endocrinology and specialist animal care staff work holistically to ensure monitoring is timely and with animal well-being as the highest priority.

2.4. Behavioural Data Collection

Behavioural data were collected via live continuous focal sampling, for 60 min observation sessions [24] between the hours of 09:00–17:00 using a pre-determined ethogram (Table 2, Supplementary Materials) at public viewing areas. Data collection schedule was designed to ensure each hour period was observed throughout the study period for each individual. Height utilization data were collected simultaneously using pre-defined enclosure height levels (Table 1, Appendix A, Figures A2 and A3). Behaviour data were recorded using a Microsoft Surface Go (Model 1.824) tablets programmed with an Excel time stamp formula (programmed by KF). Behavioural observations were conducted by trained members of the Behaviour and Welfare team at Chester Zoo. Although inter-observer reliability for this study was not calculated, inter-observer reliability scores [24] were calculated for previous projects with the same observers, with a matched sample score of 85%. The sampling period for each individual was as follows Male: 20/02/2020–12/03/2020 (20/02: 2.8 h, 21/02: 3.3 h, 2/03: 1 h, 3/03: 1 h, 5/03: 2 h, 9/03: 1 h, 10/03: 1 h, 12/03: 1 h). Female: 11/11/2019–15/11/2019 (11/11: 1 h, 12/11: 3 h, 13/11: 2.2 h, 14/11: 1 h, 15/11: 2.1 h). Total observation time was 13.1 h for the male subject and 9.3 h for the female subject. Average dawn and dusk times for the sampling periods were as follows; Male: Dawn 07:04, Dusk 17:31, Female: Dawn 07:31, Dusk 16:18 [25].

Table 2. Ethogram of behaviours, allocated behavioural category used for data visualisation, full description of behaviour and reference for reader to access supplementary files of behaviour. Ethogram adapted from Dabek [26].

Behaviour Category	Behaviour	Description	Supplementary Video File Reference
Vigilance	Vigilance	Sp. is actively observing and aware of surroundings. Eyes are open and individual is alert. If face not visible then head is visibly moving.	S1.1 and S1.2
Feeding	Feeding	Sp. is actively chewing or consuming food items or browse. Includes sp. reaching over to retrieve or manipulate browse or food item. Includes drinking behaviour.	
Grooming	Grooming	Sp. is scratching with fore limb or hind limb at a specific area of the body. Includes sp. rubbing oneself against items within the enclosure.	
Travel	Travel	Sp. is moving from one area of the enclosure to another, in a forward or backward direction. Includes horizontal and vertical climbing, leaping to the ground, descending branches or objects, quadrupedal walking or bipedal hopping.	S1.3 and S1.4
Rest	Rest—awake	Sp. is relaxed with eyes open. Body posture is slightly curled, with tail hanging down in a relaxed manner.	S1.5
Rest	Rest—asleep	Sp. is relaxed with eyes shut. Body curled with face pointing downwards and forelimbs tucked in.	S1.6
Other	Excretion	Sp. is urinating or defecating.	
Other	Yawning	Sp. is opening mouth widely with a deep inhalation of breath seen in diaphragm.	
Other	Sniffing	Sp. nose is extended towards an area or object. Inhalation of air can be seen through flaring of nostrils.	S1.7

The observational data collected here formed part of routine, internal behaviour monitoring, which was commissioned by the collections staff at Chester Zoo and facilitated by Chester Zoo's in-house behaviour and welfare science team to allow for an evidence-based approach to management of individuals of different age and sex classes.

2.5. Statistical Analysis

Due to the small sample size of this study, descriptive statistics were conducted in R (v1.3.2) [27] and presented throughout, with superficial comparisons made between study subjects using raw data. In line with other zoo-based studies, each day of observation was treated as an experimental unit; when more than one observation period occurred each day, an average of these observation sessions were taken [28]. Shapiro–Wilks tests for normality revealed the data to be non-normally distributed, as such the results are presented in the text using median (Mdn) and interquartile ranges (IQR). For results in graphical form, results are presented using Mdn, first quartile (Q1), third quartile (Q3) and range values of the data. For additional information, mean values were also presented graphically. ggplot2 R package [27] was used to construct jitter plot graphs.

3. Results

3.1. Activity Budget

Both study subjects spent the majority of the observation period resting (Male; Mdn: 52.57, IQR: 22.06, Female; Mdn: 79.64, IQR: 16.790, Figure 1), followed by vigilance behaviour (Male; Mdn: 28.70, IQR: 20.76, Female; Mdn 11.06, IQR: 2.29, Figure 1). 'Other' behaviours includes excretion, yawning and sniffing (Table 2). The male spent 5.33% more time feeding and 1.98% more time travelling than the female during the observation period for each individual.

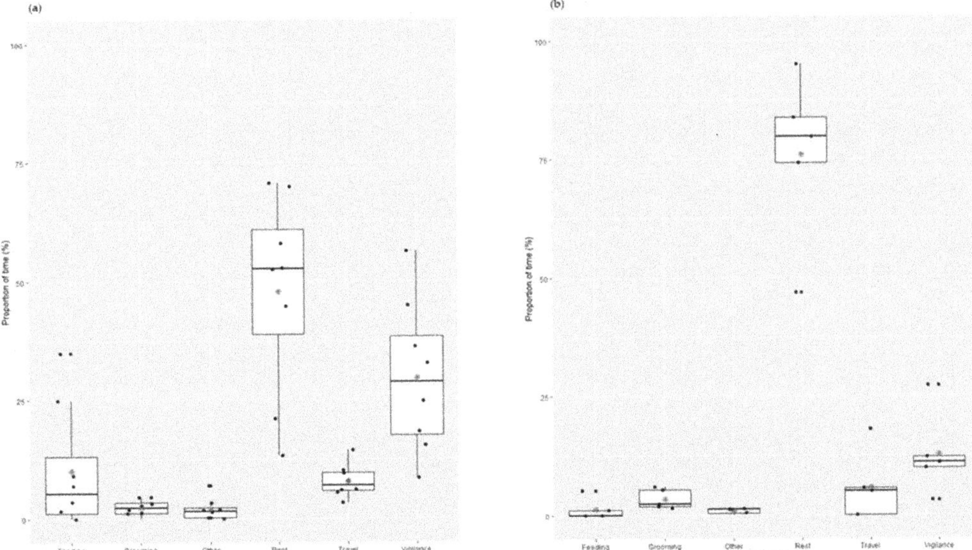

Figure 1. Proportion of time (%) the (**a**) male study subject and (**b**) female study subject engaged in each behavioural category throughout the data collection period. Time budget displayed using boxplots outlining the first quartile, median, third quartile and range of the data. Jitter plots used to visualize each collected data point represented by black circles. Red star denotes the mean value of time spent engaging in each behavioural category.

3.2. Height Utilisation

As the size of the height levels were not consistent between the indoor and outdoor environment, results for individual height utilization during the study period were investigated separately. Average proportion of time spent in the outdoor habitat was low for each individual, with the male and female study subject spending 9.64% and 0.85% of total observation time outdoors, respectively. In consequence, height use data from the outdoor environment were not presented in a graphical format.

When using the indoor habitat, the male subject spent most of his time at height level 3 (Mdn: 63.04, IQR: ± 10.98, Figure 2), followed by height level 2 (Mdn: 34.84, IQR: ± 17.15), ground level (Mdn: 3.91, IQR: ± 4.72) and the least amount of time at height level 1 (Mdn: 3.51, IQR: ± 7.99). When using the indoor habitat, the female study subject also utilised height level 3 most frequently (Mdn: 41.05, IQR: ± 50.19, Figure 2), followed by height level 1 (Mdn: 15.50, IQR: ± 27.24) then height level 2 (Mdn: 12.79, IQR: ± 55.18). The female study subject spent no time at ground level throughout the study period (Mdn: 0, IQR: ± 0).

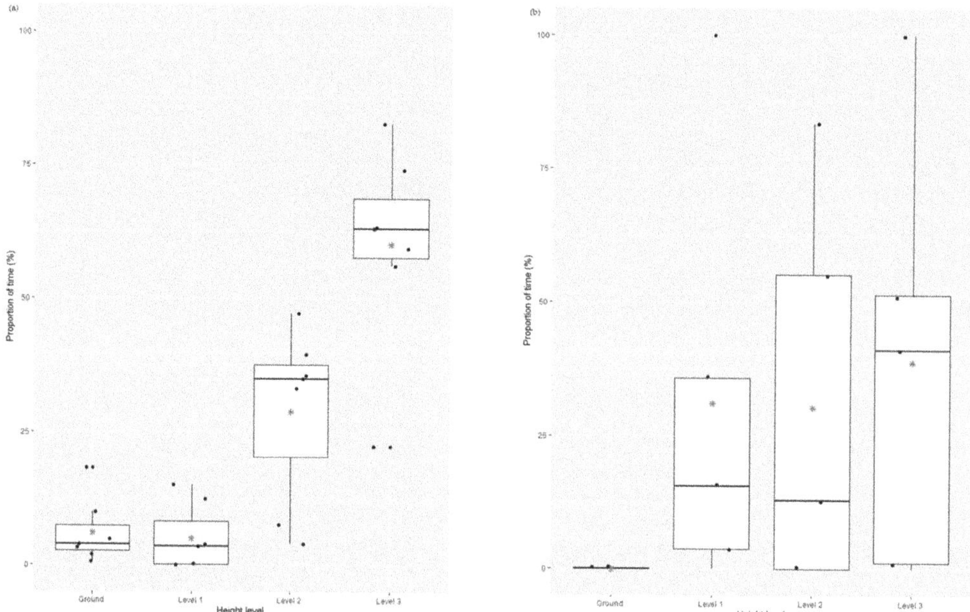

Figure 2. Proportion of time (%) the (**a**) male study subject and (**b**) female study subject spent at different height levels whilst in the indoor habitat. Height use data displayed using boxplots outlining the first quartile, median, third quartile and range of the data. Jitter plots used to visualize each collected data point represented by black circles. Red star denotes the mean value of time spent utilizing each height.

4. Discussion

4.1. Activity Budget

Tree kangaroos are notoriously elusive [29]. Therefore, the observation of these individuals in a zoo environment provides a unique opportunity to add to the knowledge base of this understudied species. Research has highlighted the low activity levels of this species, reporting long periods of resting behaviour interspersed with locomotion and feeding behaviour [5]. A similar trend was found within this study with both individuals spending the majority of their time resting (Figure 1a,b). Additionally, both individuals spent a proportion of time engaging in 'vigilance' behaviour. Although Goodfellow's tree kangaroos do not encounter many predators of a similar body size in their native Papua New Guinea, both arboreal and terrestrial predators, including humans, still pose a risk [7]. Studies on Lumholtz's tree kanagroos (*D. lumholtzi*) highlighted an increase in vigilance behaviour when exposed to odour cues from predatory species [7]. This suggests that as a species vulnerable to predation, vigilance behaviour can form an important part of a tree kangaroos' activity budget. Furthermore, as this data collection took place whilst the zoo was open to the public, there could be the potential for human disturbance to be causing the study animals to display increased vigilance behaviour. Zoo visitors have been known to cause behavioural change in zoo-housed individuals [30], however the visitor effect has yet to be evaluated on this species. Further data collection, over a 24 h period or when individuals are housed off-show, may be useful to consider whether visitor presence may be affecting the activity budget of the study subjects.

When investigating time spent travelling, the female study subject spent 1.98% less time engaging in travel behaviour than the younger male subject. Most species of tree kangaroo are exceptionally well adapted climbers, allowing for individuals to navigate with ease through their native habitat of rainforest canopy [19]. However, to ensure optimum welfare experience, captive facilities may necessarily modify habitat resources to facilitate individual requirements [17]. Additionally, it is important to note that the behaviour data collection only occurred between the hours of 09:00–17:00, thus is not representative of a full 24 h activity budget for these individuals. Tree kangaroo activity budgets have been found to differ amongst species, ranging from crepuscular to nocturnal [8,9]. Thus, natural activity patterns may also be a factor to take into consideration alongside seasonal climate differences, individual differences in behaviour, 'observer error' or the subtle influence of having another species in the outdoor-environment, when interpreting results or using these data as a comparison. Although only descriptive statistical comparisons were made throughout this study, sexual dimorphism should be another factor to consider when interpreting and comparing results between male and female study subjects in other institutions.

4.2. Height Utilization

As a species known to spend a vast majority of their time at height [3], zoos housing tree kangaroos are advised to build environments which both encourage and facilitate an arboreal lifestyle [17]. Despite the differences in age and sex class between the individuals studied, both study subjects utilized the highest height level within the indoor habitat most frequently (Figure 2a,b). During the study, the female did not spend any time at ground level. This is not uncommon for tree kangaroos as it has been observed that some species, including *D. lumholtzi*, are much more vulnerable to predation when on the ground [31]. Further work has stated that Lumholtz's tree kangaroos may come to the ground only as a flight response when startled, but then swiftly return to the forest canopy [32]. These works highlight the importance of understanding wild-type baseline behaviour in order to make accurate and species-appropriate interpretations of data collected in a zoo environment [33]. As such, for an arboreal animal such as a tree kangaroo, no utilization of the ground level of the indoor habitat should not be an area of great concern. However, the suggestion could be made to review the branching and structures within these under-utilised height levels to assess whether this area could be modified to facilitate easier access for the geriatric female, ensuring this individual can exercise a level of choice and control over their environment [34]. Anecdotally, towards the end of the study, observers noted the female subject utilizing a solid box within the exhibit, particularly as an area to rest. Thus, with collections housing older individuals, the implementation of wider more solid structures such as platforms may be of use to provide additional resting opportunities.

4.3. Husbandry Routine

Modern zoos aim to achieve more than simply exhibiting an animal to visitors [35]. Habitats and husbandry routines are continually enhanced to ensure animal welfare is the top priority. In order to create suitable environments and provide optimum care for individuals, a knowledge of the species' natural history is paramount [17]. As discussed, tree kangaroos are arboreal, forest dwelling species [19]. Thus, the habitat at Chester Zoo aims to facilitate an arboreal lifestyle by including a network of branching, plants and structures. Husbandry routines and the facility size allows individuals to be kept separately to replicate their solitary nature in the wild. Provision of a species appropriate diet of vegetables, leafy greens and browse meets the nutritional requirements for a folivore such as the tree kangaroo (Figure A1). The results of this behavioural study highlight that individuals at our facility do not engage in abnormal repetitive behaviour and have a mainly arboreal lifestyle. The provision of this information allows keepers to have an evidence-base from which to make management decisions and highlights areas for further improvement, such as the inclusion of additional structures for resting opportunities.

Overall, we hope the publication of this information will be useful to animal care staff and researchers, promoting individual-based monitoring and evidence-based management of this arboreal macropod. Additionally, the authors hope this work will contribute to the knowledge base surrounding zoo-housed tree kangaroos and will be of particular use to holders housing *D. goodfellowi*, a sub-species of tree kangaroo for which published information on zoo-housed individual behaviour is especially limited.

Supplementary Materials: The following are available online at https://www.mdpi.com/article/10.3390/jzbg3010009/s1, Video: S1.1–S1.7: Full descriptions of behaviour.

Author Contributions: Conceptualization, K.F.; methodology, K.F.; investigation, K.F.; resources, K.F..; data curation, K.F.; writing—original draft preparation, K.F. and A.H.; writing—review and editing, K.F. and A.H.; visualization, K.F. and A.H.; supervision, K.F. and A.H.; project administration, K.F. All authors have read and agreed to the published version of the manuscript.

Funding: This research received no external funding.

Institutional Review Board Statement: Ethical review for this study was waived due to the work conducted being purely observational with no manipulation to the study animal or their environment. Project was commissioned internally by Chester Zoo collections directorate staff.

Informed Consent Statement: Not applicable.

Data Availability Statement: Data available on request.

Acknowledgments: The authors would like to thank the Twilight, Curatorial and Science Teams at Chester Zoo for their support with this project. Particular thanks are extended to the following individuals: Rose Agnew, Megan Leary, Luke Heseltine, Leah Williams, Lisa Holmes, James Waterman, David White, James Andrewes, Kirsten Wicks, Peter Watson, Tim Rowlands, Nick Davis and Andrew McKenzie.

Conflicts of Interest: The authors declare no conflict of interest.

Appendix A

DIET SHEET FOR

Kayjo (C19128) Sangria (C1782)

DIET INGREDIENTS

	QUANTITY	TEMPORARY QUANTITY
Primate Leafeater pellets - Mazuri	50g	
Macropod pellets - Mazuri	50g	
Lucerne pellets - Dengie	50g	
Vegetable mix - select 2 from: Aubergine, butternut squash, broccoli, carrot, parsnip, sweet potato, courgette	350g	
Leafy greens - alternate between: Chicory, pak-choi, spinach, lettuce	350g	
Browse	Min. 1 branch	
Egg, hard-boiled	0.5 item	

Any changes to these quantities must be noted on the daily report

FEEDING SCHEDULE

	M	T	W	T	F	S	S	Other
Primate Leafeater pellets - Mazuri	Daily							
Macropod pellets - Mazuri	Daily							
Lucerne pellets - Dengie			X				X	
Vegetable mix	Daily							
Leafy greens	Daily							
Browse	Daily							
Egg, hard-boiled	X				X			

PREPARATION NOTES	PRESENTATION NOTES
Most produce items should be cut to long "stick"-shapes to make them easier to grab, however some should be cut to be more difficult to grab to increase feeding activity Leafy greens can be offered whole or lightly chopped	

OTHER NOTES

Fruit (apple, pear or banana, max 50g in total per day) can be used as training rewards

Figure A1. Diet sheet of *Dendrolagus goodfellowii* provided to study subjects throughout the observation period. Diet quantities displayed were those provided per individual.

Figure A2. Image outlining pre-determined height levels used throughout data collection within the indoor habitat.

Figure A3. Image outlining pre-determined height levels used throughout data collection within the outdoor habitat.

References

1. Eldridge, M.D.B.; Potter, S.; Helgen, K.M.; Sinaga, M.H.; Aplin, K.P.; Flannery, T.F.; Johnson, R.N. Phylogenetic analysis of the tree kangaroos (*Dendrolagus*) reveals multiple divergent lineages within New Guinea. *Mol. Phylogenet. Evol.* **2018**, *127*, 589–599. [CrossRef] [PubMed]
2. Newell, G.R. Home range and habitat use by Lumholtz's tree-kangaroo (*Dendrolagus lumholtzi*) within a rainforest fragment in north Queensland. *Wildl. Res.* **1999**, *26*, 129–145. [CrossRef]
3. Valentine, P.; Dabek, L.; Schwartz, K.R. What is a tree kangaroo? Evolutionary history, adaptation to life in the trees, taxonomy, genetics, biogeography and conservation status. In *Tree Kangaroos: Science and Conservation*; Dabek, L., Valentine, P., Blessington, J., Schwartz, K., Eds.; Academic Press: Cambridge, MA, USA, 2020; pp. 4–5.
4. Warburton, N.M.; Yakovleff, M.; Malric, A. Anatomical adaptations of the hind limb musculature of tree-kangaroos for arboreal locomotion (Marsupialia: Macropodinae). *Aust. J. Zool.* **2012**, *60*, 246–258. [CrossRef]
5. Whitehead, M. Tree kangaroos. In *The Management of Marsupials in Captivity*; Partridge, J., Ed.; Association of British Wild Animal Keepers: London, UK, 1986.
6. Edwards, M.S.; Ward, A. Nutrition, food preparation and feeding. In *Tree Kangaroo Species Survival Plan: Tree Kangaroo Nutrition Husbandry Manual*; Steenberg, J., Blessington, J., Eds.; American Association of Zoos and Aquariums: Silver Spring, MD, USA, 2013; pp. 3–7.

7. Heise-Pavlov, S.R. Evolutionary aspects of the use of predator odors in antipredator behaviours of Lumholt'z tree-kangaroos (*Dendrolagus lumholtzi*). In *Chemical Signals in Vertebrates 13*; Schulte, B., Goodwin, T., Ferkin, M., Eds.; Springer: Cham, Switzerland, 2016; pp. 261–280.
8. Heise-Pavlov, S.R.; Semper, C.; Burchill, S. Terrestrial activity patterns of the Lumholtz's Tree-Kangaroo (*Dendrolagus lumholtzi*) in a restored riparian habitat—Implications for its conservation. *Ecol. Manag. Restor.* **2021**, *22*, 183–190. [CrossRef]
9. Gillanders, A.; Wilson, C. Tree Kangaroo Tourism as a Conservation Catalyst in Australia. In *Tree Kangaroos: Science and Conservation*; Dabek, L., Valentine, P., Blessington, J., Schwartz, K., Eds.; Academic Press: Cambridge, MA, USA, 2020; pp. 109–125.
10. Byers, J.; Dabek, L.; Gabriel, P.; Stabach, J. Using telemetry and technology to Study the Ecology of Tree Kangaroos. In *Tree Kangaroos: Science and Conservation*; Dabek, L., Valentine, P., Blessington, J., Schwartz, K., Eds.; Academic Press: Cambridge, MA, USA, 2020; pp. 365–378.
11. Leary, T.; Seri, L.; Wright, D.; Hamilton, S.; Helgen, K.; Singadan, R.; Menzies, J.; Allison, A.; James, R.; Dickman, C.; et al. Dendrolagus Goodfellowi. Available online: https://www.iucnredlist.org/species/6429/21957524 (accessed on 17 October 2021).
12. Dabek, L.; Valentine, P. The Future of Tree Kangaroo Conservation and Science. In *Tree Kangaroos: Science and Conservation*; Dabek, L., Valentine, P., Blessington, J., Schwartz, K., Eds.; Academic Press: Cambridge, MA, USA, 2020; pp. 433–437.
13. Newell, G.R. Australia's tree kangaroos: Current issues in their conservation. *Biol. Conserv.* **1999**, *87*, 1–12. [CrossRef]
14. Hutchins, M.; Smith, R. Biology and status of wild tree kangaroos. In Proceedings of the AAZPA Northeastern Regional Conference Proceedings, Washington, DC, USA, 1 May 1990; pp. 500–509.
15. Blessington, J.; Steenberg, J.; Schwartz, K.R.; Schürer, U.; Smith, B.; Richardson MJaffar, R.; Ford, C. Tree kangaoo populations in managed facilities: Global species management plans (GSMPs). In *Tree Kangaroos: Science and Conservation*; Dabek, L., Valentine, P., Blessington, J., Schwartz, K., Eds.; Academic Press: Cambridge, MA, USA, 2020; p. 264.
16. McGreevy, T.J.; Dabek, L.; Husband, T.P. Microsatellite Marker Development and Mendelian Analysis in the Matschie's Tree Kangaroo (*Dendrolagus matschiei*). *J. Hered.* **2010**, *101*, 113–118. [CrossRef] [PubMed]
17. Blessington, J.; Steenberg, J. *Tree Kangaroo (Dendrolagus spp.) Husbandry Manual*, 3rd ed.; Tree Kangaroo Species Survival Plan; Kansas City Zoo: Kansas City, MO, USA, 2007.
18. Schwartz, K.R.; Byers, O.; Miller, P.; Blessington, J.; Smith, B. The Role of Zoos in Tree Kangaroo Conservation: Connecting Ex Situ and In Situ Conservation Action. In *Tree Kangaroos: Science and Conservation*; Dabek, L., Valentine, P., Blessington, J., Schwartz, K., Eds.; Academic Press: Cambridge, MA, USA, 2020; pp. 329–361.
19. Dabek, L. What is a tree kangaroo? Biology, Ecology and Behaviour. In *Tree Kangaroos: Science and Conservation*; Dabek, L., Valentine, P., Blessington, J., Schwartz, K., Eds.; Academic Press: Cambridge, MA, USA, 2020; pp. 17–29.
20. Martin, R. Tree kangaroo taxonomy. In *Tree kangaroos of Australia and New Guinea*; Martin, R., Ed.; CSIRO Publishing: Clayton, Australia, 2005; pp. 15–25.
21. Flannery, T.F.; Martin, R.; Szalay, A.L. *Tree Kangaroos, A Curious Natural History*; Reed Books: Port Melbourne, Australia, 1996.
22. Species360 Zoological Information Management System (ZIMS) (2021). Available online: www.species360.org (accessed on 10 October 2021).
23. Leary, T.; Seri, L.; Flannery, T.; Wright, D.; Hamilton, S.; Helgen, K.; Singadan, R.; Menzies, J.; Allison, A.; James, R. Dusky Pademelon, Thylogale brunii. Available online: https://www.iucnredlist.org/species/21870/21958826 (accessed on 25 January 2022).
24. Martin, P.R.; Bateson, P.P.G. *Measuring Behaviour: An Introductory Guide*, 3rd ed.; Cambridge University Press: Cambridge, UK, 2007.
25. Time and Date Website. Available online: https://www.timeanddate.com/weather/uk/chester/historic (accessed on 1 January 2022).
26. Dabek, L. Mother-Young Relations and the Behavioural Development of the Young in Captive Matschie's Tree Kangaroos (Dendrolagus Matschiei). Master's Thesis, University of Washington, Seattle, WA, USA, 1991.
27. R Core Team. *R: A Language and Environment for Statistical Computing*; R Foundation for Statistical Computing: Vienna, Austria, 2013. Available online: http://www.R-project.org (accessed on 15 October 2021).
28. Bishop, J.; Hosey, G.; Plowman, A. (Eds.) *Handbook of Zoo Research, Guidelines for Conducting Research in Zoos*; BIAZA: London, UK, 2013.
29. Travis, E.K.; Watson, P.; Dabek, L. Health assessment of free-ranging and captive Matschie's tree kangaroos (*Dendrolagus matschiei*) in Papua New Guinea. *J. Zoo Wildl. Med.* **2012**, *43*, 1–9. [CrossRef] [PubMed]
30. Davey, G. Visitors' Effect on the Welfare of Animals in the Zoo: A Review. *J. Appl. Anim. Welf. Sci.* **2007**, *10*, 169–183. [CrossRef] [PubMed]
31. Cianelli, M.; Schmidt, K. Rehabilitation of Lumholtz's Tree Kangaroo Joeys. In *Tree Kangaroos: Science and Conservation*; Dabek, L., Valentine, P., Blessington, J., Schwartz, K., Eds.; Academic Press: Cambridge, MA, USA, 2020; pp. 65–84.
32. Heise-Pavlov, S.; Procter-Gray, E. How an Understanding of Lumholtz's tree kangaroo behavioural ecology can assist conservation. In *Tree Kangaroos: Science and Conservation*; Dabek, L., Valentine, P., Blessington, J., Schwartz, K., Eds.; Academic Press: Cambridge, MA, USA, 2020; pp. 85–107.
33. Blessington, J.; Steenberg, J.; Richardson, M.; Norsworthy, D.; Legge, A.; Sharpe, D.; Highland, M.; Kozlowski, C.; Males, G. Reproductive Biology and Behaviour of Tree Kangaroos in Zoos. In *Tree Kangaroos: Science and Conservation*; Dabek, L., Valentine, P., Blessington, J., Schwartz, K., Eds.; Academic Press: Cambridge, MA, USA, 2020; pp. 309–327.

34. Whitham, J.C.; Wielebnowski, N. New directions for zoo animal welfare science. *Appl. Anim. Behav. Sci.* **2013**, *147*, 247–260. [CrossRef]
35. Roe, K.; McConney, A.; Mansfield, C.F. The Role of Zoos in Modern Society—A Comparisons of Zoo' Reported Priorities and What Visitors Believe They Should Be. *Anthrozoös* **2014**, *27*, 529–541. [CrossRef]

Journal of
Zoological and Botanical Gardens

MDPI

Review

Overlooked and Under-Studied: A Review of Evidence-Based Enrichment in *Varanidae*

Darcy Howard * and Marianne Sarah Freeman

Animal Health and Welfare Research Centre, University Centre Sparsholt, Winchester SO21 2NF, UK; Marianne.freeman@sparsholt.ac.uk
* Correspondence: darcy.b.howard@gmail.com

Abstract: Enrichment has become a key aspect of captive husbandry practices as a means of improving animal welfare by increasing environmental stimuli. However, the enrichment methods that are most effective varies both between and within species, and thus evaluation underpins successful enrichment programs. Enrichment methods are typically based upon previously reported successes and those primarily with mammals, with one of the main goals of enrichment research being to facilitate predictions about which methods may be most effective for a particular species. Yet, despite growing evidence that enrichment is beneficial for reptiles, there is limited research on enrichment for *Varanidae*, a group of lizards known as monitor lizards. As a result, it can be difficult for keepers to implement effective enrichment programs as time is a large limiting factor. In order for appropriate and novel enrichment methods to be created, it is necessary to understand a species' natural ecology, abilities, and how they perceive the world around them. This is more difficult for non-mammalian species as the human-centered lens can be a hinderance, and thus reptile enrichment research is slow and lagging behind that of higher vertebrates. This review discusses the physiological, cognitive, and behavioral abilities of *Varanidae* to suggest enrichment methods that may be most effective.

Keywords: captive; reptile; welfare; cognition; play

Citation: Howard, D.; Freeman, M.S. Overlooked and Under-Studied: A Review of Evidence-Based Enrichment in *Varanidae*. *J. Zool. Bot. Gard.* **2022**, *3*, 32–43. https://doi.org/10.3390/jzbg3010003

Academic Editors: Kris Descovich, Caralyn Kemp and Jessica Rendle

Received: 27 October 2021
Accepted: 11 January 2022
Published: 17 January 2022

1. Introduction

Over the course of the last century, zoos have been transformed from menageries to institutions rooted in science. Long considered the father of zoo biology, Hediger first recognized the inadequacies of the zoo environment in 1950 and emphasized the need to promote the well-being of captive animals [1]. This can be achieved by providing captive animals with opportunities that allow them to display their behavioral capabilities, which was the goal of Markowitz, one that he termed 'behavioral engineering' in the 1970s [2]. Originally based on Skinner's theory of operant conditioning, it has developed to become what we know today as 'environmental enrichment' [3]. Now, over 40 years later, environmental enrichment has become standard practice in the management of captive animals [4]. The focus in enrichment research has moved from the need to provide enhancements in husbandry to how they can be rigorously assessed in order to monitor and improve welfare [5]. Despite the attention this field of zoo science has received, there is still the general opinion that enrichment is a supplementary aspect of care and not integral to the daily husbandry [5,6]. By definition, environmental enrichment is a principle of husbandry that aims to provide stimuli to improve animal care and thus mental and physical wellbeing [7]. Even by this definition it is regarded as extra to standard animal care. What can be agreed upon is that enrichment can be classified into five categories: social, physical, nutritional, occupational, and sensory [8]. Enrichment programs should aim to provide captive animals with enrichment methods from each category, rather than just one [9,10], to improve animal welfare [3] and promote the natural phenotype of their wild counterparts [11]. This is achieved by meeting goals such as increasing activity levels, natural/species-specific behaviors, choice and control, and behavioral diversity [12], as well as reducing the prevalence, or onset, of stereotypic or abnormal behaviors [13]. As has been consistently demonstrated, the extent of zoo research markedly varies across taxa [14,15], and this pattern holds constant for enrichment. A Web of Science search yielded over 1053 and 1256 publications with the terms 'enrichment mammals' and 'enrichment birds', respectively, and yet provided only 143 results for 'enrichment reptiles'. Thus, despite enrichment now being one of the key concepts in captive animal management [16], there are still knowledge gaps, and the lack of progress within non-avian reptiles (reptiles from this point forth) is striking [17]. A growing body of evidence suggests that one group of reptiles, the varanids, have high cognitive abilities and as such understanding how to meet the motivational needs of such a group of species is imperative to improving welfare standards in their care.

2. Reptile Enrichment: What Do We Know?

2.1. *Left in the Cold*

Within the scientific literature, a strong mammal-centric bias is prominent, with a scarcity of studies regarding reptile enrichment [12,14,18–20]. This may be in part due to the long-held misconception that reptiles are stoic, highly adaptable, and tolerant to suboptimal conditions [21], as well as too neurologically simple to suffer [22,23] and thus not requiring enrichment. Where enrichment is utilized, structural or habitat design-based enrichment was the most employed provision for reptiles within U.S. collections, with an average of 86% of holders reporting this provision across all the reptile taxa [17]. This reflects the notion that reptile behavior and cognition tends to be less well-understood compared to that of higher vertebrates [24]. With current understanding being heavily influenced by mammalian-centric paradigms [25], this makes them a low priority for enrichment provision [19]. Enrichment techniques that cater for the animals learning and social functions were reported to be much less utilized by collections [17]. As a result, current reptile husbandry is clearly less than ideal yet deemed acceptable [23].

2.2. *Do Reptiles Benefit?*

Despite the attention bias, this is a developing field (Figure 1), and the studies that have been published in peer-reviewed journals support the notion that reptiles benefit from enrichment [21,26–31] and that it is in fact essential [32]. Evidence for enrichment as a beneficial practice with reptiles is documented through an increase in natural behaviors and relaxed postures under structural enriched environments [33,34]. Additionally, the use of chemosensory enrichment (scent of conspecifics, based upon the species natural ecology) significantly ($p < 0.001$) reduced the occurrence of abnormal behavior (escape attempts) in wild-caught brown wall lizards (*Podarcis liolepis*) by 38% [27]. Similarly, the use of fish-scented enrichment cups resulted in a reduction in escape behaviors of aquarium-housed, freshwater turtles (*Trachemys scripta* and *Pseudemys concinna*), although there was an increase in aggression, which demonstrates the need to assess the efficacy of any new techniques before implementation [35]. When offered multiple forms of enrichment, leopard geckos (*Eublepharis macularius*) interacted with all forms and, in particular, to a feeding puzzle and to structural enrichment placed under a heat source [31]. The alternative forms, sensory (mirror and olfactory) and novel object enrichment, elicited less engagement; however, this highlights that the enrichment should be species-specific and biologically or ecologically relevant. In addition to simply documenting a difference in behavior, when given choice (and thus some level of control), corn snakes (*Pantherophis guttatus)* displayed more time occupying structurally enriched enclosures over standard housing [34]. Clearly, reptiles respond to, and benefit from, the provision of enrichment.

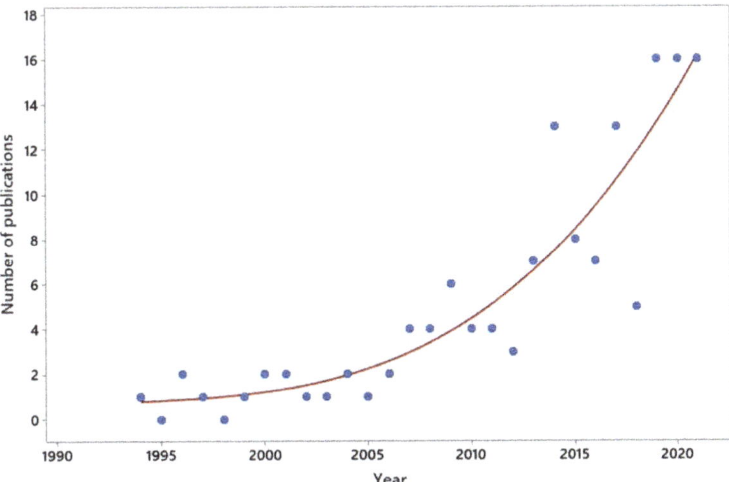

Figure 1. Number of publications using the terms 'reptile enrichment' from Web of Science over the past 30 years, as of October 2021.

Many enrichment studies have been carried out with relatively small sample sizes, often representing case studies of a few individuals. While large sample sizes are important for generalizability to the wider population as a means of predicting the most effective enrichment methods [36], there is also value in studies with small sample sizes [37,38], typically true of zoos [18]. While different complexities in enclosure design [39], confounding factors of sex, age [27], and individuality and the measures used to evaluate the enrichment [40] can lead to issues in generalization of the design, enrichment studies, in many cases, require an individual approach [41,42]. Differing past experiences, temperament, genetics, and coping mechanisms may lead to different preferences, while some individuals may, for example, choose tactile stimulation, others may choose food rewards [43]. The individual approach can also aid enrichment designs for individuals with additional needs [29]. However, for wider application, well designed studies with robust samples sizes are still needed to improve internal validity and ensure any changes are due to the design and not simply chance or impacted by keeper interactions and/or social learning from other individuals when housed together [44]. Thus, both small and large sample sizes are valuable and necessary in advancing reptile enrichment practices.

Furthermore, in addition to improving the welfare of captive individuals, advancements in reptile enrichment may benefit reptile conservation strategies, such as headstarting and translocation, by better preparing animals for wild challenges [45]. This may be particularly true of cognitive enrichment, which may take the form of a challenging puzzle or mentally challenging exploration, or involve training to help the individual cope with challenges and learn new species-specific behaviors [4,46]. Cognitive enrichment has been much less studied in reptiles; however, enrichment itself may help improve the cognitive function of all animals. Thus far, the effects of enrichment on translocation success have had mixed results. DeGregorio [47] found no effect of enrichment on translocation success of ratsnakes (*Pantherophis obsoletus*), and there was a negative correlation of success with time spent in captivity. Conversely, captive common watersnakes (*Nerodia sipedon sipedon*) had an equal rate of survival to that of the wild snakes [48]. This suggests that, while structural enrichment may not improve natural survival traits, enrichment that encourages or helps in learning the processes of natural foraging would be beneficial for maintaining natural behavioral traits. In addition, there are no studies investigating the effects of antipredator training in captive raised and released reptiles (although there are cases of toxin avoidance training in wild reptiles (see later)). Indeed, in painted turtles (*Chrysemys picta*), early life experience is vital to the development of successful navigation [49]. Training to improve navigation, predator recognition, prey location, and foraging training mentally stimulates the animal while providing the tools to improve survival [45], and this should be given more consideration in reptile husbandry going forward. More so, regardless of their level of sociality, no social enrichment studies in the reptilian taxa have been explored. Furthermore, there may exist social stages within the reptilian lifecycle; thus, an understanding of how sociality acts on their learning would be beneficial for optimal captive care [50] and possibly even conservation strategies.

2.3. The Necessity of Evaluation

The value of evidence-based enrichment and the need to broaden the research to a wider range of taxa, such as that of reptiles, is appreciated among zoo professionals [5]. This is particularly the case because keeper perceptions alone may not always be accurate, and personal expectations or lack of time to fully observe the animals can impact judgment. Mehrkam and Dorey [51] found that keepers were least accurate when predicting the preferred enrichment for a reptile (eastern indigo snake (*Drymarchon couperi*)) compared to other taxa. Aside from simply establishing preference, the effectiveness of an enrichment method to meet the required goals must be subjected to empirical evaluation before they can be definitively considered to be 'enriching' [10]. Januszczak et al. [39] found that using an enrichment device that was designed to present 10 live crickets (*Gryllus* spp.) to tree-runner lizards (*Plica plica*) randomly over 40 min was less effective than the commonly used and simpler method of scatter-feeding. Additionally, despite basing enrichment (raised basking platforms) on the natural ecology of eastern fence lizards (*Sceloporus undulatus*), Rosier and Langkilde [52] found that this form of enrichment did not affect activity levels.

Reducing abnormal behaviors is a common goal in animal enrichment, but a lack of undesirable behaviors does not mean the animal is thriving in captivity [14,24]. Behavioral diversity is commonly used to evidence an increase in a wide range of behaviors. However, even this is not without its issues, in that while increasing behavioral diversity can be another aim of enrichment, it is not always a good measure of success, as if the enrichment results in the development of a new abnormal behavior, this will increase behavioral diversity [53]. Appropriate evaluation of enrichment programs is essential so that if the intended goals are not met, alternative strategies can be devised [20,52,54]. Furthermore, within enrichment research, it is vital that researchers report nonsignificant results [23,32], as enrichment that is ineffective does little to improve welfare and is not time- nor cost-effective [55], with time being the largest limiting factor of enrichment provision among keepers [10]. Following frameworks such as the SPIDER framework to Set goals, Plan, Implement, Document, Evaluate, and Readjust if needed ensures that any enrichment offered is maximizing benefits to the individual [20]. The goals and enrichment plans should consider the animal's natural biology and ecology to be species-specific and relevant for the intended recipient.

3. Is Enrichment Vital for Varanids?

3.1. A Brief Background of Varanidae

Native to Africa, Asia, and Australasia [56], there are currently 83 known species of this monotypic family, with the only extant genus *Varanus* [57]. Of these 83, at least 50 are known to have been kept in captivity [58]. Many of these are also popular pet species, and with the global population of exotic reptile pets increasing and considering CITES trade statistics, there are likely to be several thousand individuals in captivity around the world [59,60]. In the wild, varanids (commonly called 'monitor lizards') occupy a diverse range of habitats and niches, in which some are terrestrial, others are aquatic or semi-aquatic, arboreal, or semi-arboreal, and some are saxicolous (rock-dwelling) [61]. Many of these lizards are active predatory species, typically being opportunistic generalists, except for the three known frugivorous species—*V. mabitang*, *V. olivaceus*, and *V. bitatawa*, of the Philippines [62,63]. Despite this variety, monitors are conservative in their morphology, but vary greatly in size [61]. Unlike most reptiles, these lizards are relatively energy-efficient [64], capable of sustaining high metabolic rates and prolonged high-speed movement [65,66]. This is due to their complex lungs that are reminiscent of avian lungs [67] in their unidirectional airflow [68], as well as their gular pump that allows them to breathe while running (unlike other lizards) [65,69], in addition to a high VO_2 max and morphological specializations to the heart and skeletal muscles [69]. Varanids are reputed to be the most intelligent of all lizards and possess a telencephalon (the most highly developed part of the forebrain) that constitutes a larger proportion of their relative brain size in comparison to other lizards [70,71]. These features are in part why the *Varanus* body plan has been so successful [61] and why these lizards would benefit greatly from enrichment, particularly as they are prone to obesity in captivity and require adequate exercise [56]. This is a particular conundrum as varanids are often kept in enclosures that spatially are a fraction of their natural home ranges [72].

3.2. Cognitive Abilities

Cognitive skills are the process that animals acquire, handle, and store information from the environment, and their cognitive abilities refer to the ways that they can act upon this information [73]. The varanid's higher intelligence has long been recognized [74–76], along with their curiosity, perceptiveness, apparent ability to recognize different keepers [56,61], and ability to be successfully target trained [32,77]. Experiments carried out on captive *V. albigularis* even suggest that they may have counting-like skills as they appeared to be able to count to six [78], which is theorized to be attributed to raiding the nests of other reptiles, birds, and mammals, given that the average clutch or litter size would be around six [61]. The 'allostasis concept' argues that animals are adapted to respond to challenge and therefore require cognitive skills to function normally [79]. The wild environment is in stark contrast to the highly predictable environment faced by captive animals [4]. Despite the requirement of wild animals to employ behavioral strategies and cognitive abilities to solve problems (such as the need to access and control limited resources) in ways that minimize threat to self, cognitive challenge is an under-utilized method of enrichment [80]. With any cognitive enrichment, however, the cognitive challenge provided must be appropriate to that animal. If a task fails to challenge an individual animal, either boredom or apathy may result, the state dependent on the cognitive skills of the animal (high or low, respectively) [80]. Alternatively, opportunities that challenge an individual that does not have the appropriate skills to meet the tasks demands can result in anxiety [81]. When the task is well matched to the individuals' skills, it results in the individual becoming absorbed in the task, as well as the experience of pleasure and satisfaction referred to as flow [82]. The flow model is useful in enrichment development [80] but has yet to be applied as these emotional states (boredom, apathy, anxiety, and pleasure) are difficult to measure [4]. Additionally, the animals' cognitive abilities and the way in which they perceive the world must first be understood, a difficult task to perform through a human-centered perspective [22]. As a result, studies investigating animal cognition have typically focused on non-human primates and, to a lesser extent, marine mammals [4].

The cognitive abilities of monitor lizards have been studied a handful of times, concluding that this genus is capable of problem solving and rapid learning [75,83–85], as well as reversal learning [86] and procedural learning [76]. Considering their biology and ecology, varanids, with excellent eyesight and active predatory foraging ability, would be expected to learn and respond to visual stimuli. This was successfully demonstrated with rough-necked monitors (*V. rudicollis)* who were able to discriminate between colors of stimuli as well as showing reverse learning when retrained with different stimuli [86]. Unpalatability of certain prey species (a selective advantage for that prey species as a whole, as predators learn to avoid them in future), means that the taste senses of varanids should also be well established. Indeed, evidence of this is reported through toxic prey avoidance learning in the floodplain monitor (*V. panoptes)* [87]; thus, in addition to visual stimuli, gustatory stimuli should also be able to be discriminated.

Learning is vital to conserve behavioral adaptations and cognitive function that can help individuals to thrive in captivity. The animal's ability to solve problems and retain this skill over time is one way to measure the success of the learning trail. Decreasing latencies to solve a cognitive task suggests that black-throated monitors (*V. albigularis*) can become more efficient at solving food-based puzzles by reducing unnecessary behaviors that do not result in success [85]. Cooper et al. [75] found evidence of problem solving using puzzle feeders in three species of monitor lizards—*V. rudicollis, V. prasinus,* and *V. mertensi.* In a follow-up study, evidence of long-term procedural learning ability was documented in that *Varanus prasinus* and *Varanus mertensi* that were quicker to solve the same puzzle a year later than when they first encounter the task [76].

Given that varanids have the longest incubation period of any lizard, a trait hypothesized to be attributed to their increased brain size [88], the effects of breeding environments should be taken into consideration when assessing cognitive abilities and further, as part of optimal husbandry, to maintain cognitive abilities in the next generation. In particular, the effect of incubation temperature on cognitive abilities should be considered as hatchling velvet geckos *(Amalosia lesueurii)* from 'hot-incubated' eggs (nest temperatures that could be experienced by *A. lesueurii* in the year 2050 [89]: mean = 27 °C, range 14–37 °C) had slower spatial learning abilities than hatchlings from 'cold-incubated' eggs (current nest temperatures: mean = 23.2 °C, range 10–33 °C) [90,91]. Another factor that could be important to consider is social learning abilities. Many lizards are considered 'nonsocial', and even amongst the more social species, this can often be of a temporary nature [50]. However, research is beginning to identify instances of social learning in reptiles [92,93], and while there has been nothing published yet on the varanids, there are many instances of social behaviors (including play behavior) that suggest that social learning is possible.

3.3. Play Behaviour

The tendency of animals to play has been linked with brain size [94] and regarded to be a mode of information acquisition. 'True play' was initially considered a trait only exhibited by mammals and birds; Burghardt [38], however, reports of play in Komodo dragons (*V. komodoensis*) going back over 80 years. More recently, blue-spotted tree monitors (*V. macraei*) and green tree monitors (*V. prasinus*) [77] have been observed to exhibit play behavior that meets the following five criteria of play developed by Burghardt [38]: (1) a behavior that is not fully functional in the context or form in which it is expressed; (2) a behavior that is voluntary, spontaneous, intentional, pleasurable, rewarding, reinforcing, or autotelic; (3) a behavior that differs structurally or temporally from strictly functional behaviors; (4) a behavior that is performed repeatedly in a similar, but not rigidly stereotyped, form; (5) a behavior that is performed when an animal is in a relaxed, unstimulating, or low-stress environment. These tree monitors, housed at two separate collections (*V. macraei* at ZSL London Zoo and *V. prasinus* at Bristol Zoo), were both observed participating in the same object-based play behavior with live plants in the enclosure [77]. This behavior, not observed in wild animals, resembles natural prey-tearing and wiping behavior [95] and may have developed through under-stimulation, as animals may seek to create their own diversions in the absence of extrinsic ones [77]. Alternatively, play may arise when there is minimal stress and sufficient resources to allow time for other behaviors, a theory proposed as the Surplus Resource Theory [96]. In addition to reports of play in captive varanids, a wild yellow monitor (*V. flavescens*) was recently observed vertical swimming in a forward/backward motion in what was perceived to be play [97]. If monitor lizards play, it is an added incentive to provide enrichment [32] that promotes these behaviors, particularly as such behaviors tend to be motivationally robust and do not readily habituate [38]. However, again, adequate enrichment must be facilitated by provision of adequate space [72].

4. Future Directions for Research

Having considered the physiological and cognitive capabilities of monitor lizards, we find that there is a clear need for enhanced provision of enrichment for this family group. They possess a metabolism that is said to 'bridge the gap' between reptiles and mammals [98], they exhibit 'mammal-like' feeding behavior [95], and their cognitive abilities have suggested that they could be regarded as the 'primate of the squamate world' [77,96]. Yet, to the authors knowledge, there have been no empirical studies carried out regarding varanid enrichment or its efficacy, except for one quasi-experimental study by Mendyk and Horn [74]. Two adult black tree monitors (*V. beccarii*), kept as part of Mendyk's private collection, were observed exhibiting skilled forelimb movements to retrieve food through a small gap. Following these observations, a series of four holes narrower than the width of the monitors' heads were drilled into two tree trunks, which were then filled with a variety of prey items. Using coordinated forearm movements, both subjects successfully retrieved all prey types from all four holes located in each tree trunk. Furthermore, despite being housed in separate enclosures, both individuals used identical extraction behaviors. Thus, Mendyk and Horn [74] suspect this behavior to be instinctive. These results were also replicated in an additional female *V. beccarii* kept by another keeper upon request of the authors. Furthermore, all subjects involved continued to show interest in the drilled tree trunk holes. This behavior requires high levels of processing skills, motor coordination, and dexterity, yet again suggesting that varanids share many biological similarities with mammals [74]. Similar extractive behaviors have been observed in other varanid species. The Kimberley rock monitor (*V. glauerti*) will use its claws to widen the diameter of the opening until it is large enough for the head to enter [99], while the sand goanna (*V. gouldii*) will use its tail to flush out prey from rock crevices [100]. These differences may possibly be the result of differences in claw morphometrics [101]. Thus, the efficacy of different types of puzzle feeders as enrichment devices may vary between varanid species, again highlighting the importance of enrichment evaluation. Referring back to the five enrichment categories (which are not mutually exclusive), monitor lizards would likely benefit from puzzle feeders as a means of providing nutritional and occupational enrichment, and these may be made part of the enclosure (such as the drilled tree trunks) as a means of physical enrichment, so long as they are relevant to the species' natural ecology. However, it is important to remember that these can become physical rather than cognitive barriers to food acquisition [80], and with the learning abilities of varanids, it is likely that such devices would need to be modified and updated regularly [80]. The effectiveness of enrichment methods can also be maintained through practices such as the use of partial reinforcement schedules [55].

As demonstrated by Mendyk and Horn [74], case studies that document novel behaviors in captivity, as well as those behaviors observed in wild individuals [99,100], can inform the goals for future captive enrichment programs. For example, recently reported was a rolling prey capture behavior in *V. albigularis*, comparable to the 'death roll' of crocodilians, whereby the animal grips a food item with its jaws and spins on their longitudinal axis with their four limbs pressed to their body [102]. Additionally, social enrichment and cognition research is needed. For varanids, whose social behavior is also highly scent-orientated [56], this would likely require a combination with sensory enrichment via the scent of conspecifics, with scent being their most acute sense, thanks to their deeply forked tongue and Jacobson's organ. These are just some suggestions based upon natural ecology and the findings of this review. New enrichment ideas are typically based upon previously reported successes [10], and with such a large gap in the peer-reviewed literature, anecdotes from monitor lizard keepers are the place to start [32]. While keepers may not accurately perceive the effectiveness of their enrichment provision, it can provide a basis for enrichment creation that can be empirically evaluated. Furthermore, it would provide an insight into how keepers attempt to tackle the constraints of time, cost, and space, and whether they account for the diversity between varanid species when considering enrichment provision.

Given that the ultimate goal of enrichment is typically to improve the welfare of captive animals, we must also focus on the ways in which we assess welfare. Historically, the presence or absence of stereotypical and/or abnormal repetitive behaviors has been a leading indicator of animal welfare; however, the past experience of individual animals can make this an unreliable indicator. If such a behavior undergoes a process of 'establishment', whereby it becomes disassociated from the individual's current welfare [103], then it can appear to be enrichment-resistant, and thus an unreliable indicator of the individual's current welfare [104]. Consequently, further research is needed to develop the use of 'affective states' as a welfare indicator. While an animal's affective state cannot be measured directly, it can be conducted experimentally using cognitive bias testing [4]. This method requires the animal to be trained to discriminate between different stimuli, of which varanids are capable [84,86]. Thus, this would likely prove to be a valuable area of future research. Additionally, future research may also investigate the effectiveness of using the flow model in varanids, beginning with whether the emotional states of flow, boredom, apathy, and anxiety can be accurately measured in this family of lizards. Methods may include measuring levels of motivation, such as willingness to exert high-effort for a high-value reward as a means of measuring apathy [105]; measuring levels of interest, such as the time-oriented to and in contact with multiple stimuli as a means of measuring boredom [106]; measuring absorption in a task, such as how easily and animal is distracted from a task as a means to measure flow [107]; or measuring exploratory behavior, such as in an elevated plus maze as a means of measuring anxiety [108]. However, it must be noted that these measures are based on mammalian studies, and as such, may not translate to use in varanids.

5. Conclusions

It has been over 25 years since Bennett [56] stated that monitor lizards have not been given the attention they deserve, and from the current review, this statement appears to hold true. There is growing evidence that enrichment is beneficial to reptiles and that this should be integral to their care and not just an additional luxury. Their cognition and behavioral flexibility are arguably comparative to that of mammals. Given the impressive cognitive abilities of varanids, as well as their propensity for play, it is likely that they are susceptible to boredom as a result of an unstimulating environment. Research that documents training programs with such species is needed to help inform evidence-based practice and this should include training for reintroductions and translocations. More cognitive studies are needed on varanids to continue to explore the extent of their abilities, including social learning and the extent that social enrichment is needed and to investigate cognitive enrichment that challenges and provides an opportunity to learn new skills that help them cope with the environment. Furthermore, future research is needed to investigate whether the affective states of varanids can be accurately measured in order to provide an additional means of assessing welfare, thus aiding in enrichment evaluation. If these lizards are to be provided with a high quality of life, then they should be provided with appropriate enrichment. This starts with the empirical evaluation of anecdotal methods that have been reported to be successful by varanid keepers. However, it is essential that any subsequent behaviors that are elicited are recognized as those generated within conditions of captivity, and these should never undermine any thorough investigation of species-specific varanid behavior in the wild [72].

Author Contributions: Conceptualization, D.H.; writing—original draft preparation, D.H. and M.S.F.; writing—review and editing, D.H. and M.S.F. All Authors have read and agreed to the published version of the manuscript.

Funding: This research received no external funding.

Institutional Review Board Statement: Not applicable.

Informed Consent Statement: Not applicable.

Data Availability Statement: Data sharing not applicable.

Conflicts of Interest: The authors declare no conflict of interest.

References

1. Rees, P.A. *An Introduction to Zoo Biology and Management*; John Wiley & Sons: Hoboken, NJ, USA, 2011.
2. Markowitz, H. Engineering Environments for Behavioral Opportunities in the Zoo. *Behav. Anal.* **1978**, *1*, 34–47. [CrossRef] [PubMed]
3. Fernandez, E.J.; Martin, A.L. Animal Training, Environmental Enrichment, and Animal Welfare: A History of Behavior Analysis in Zoos. *J. Zool. Bot. Gard.* **2021**, *2*, 531. [CrossRef]
4. Clark, F.E. Cognitive Enrichment and Welfare: Current Approaches and Future Directions. *Anim. Behav. Cognit.* **2017**, *4*, 52–71. [CrossRef]
5. Riley, L.M.; Rose, P.E. Concepts, Applications, Uses and Evaluation of Environmental Enrichment: Perceptions of Zoo Professionals. *J. Zoo Aquar. Res.* **2020**, *8*, 18–28.
6. Banton-Jones, K. Why Enrichment Isn't an Extra. Available online: www.wildenrichment.com (accessed on 11 July 2021).
7. Shepherdson, D.J. Tracing the Path of Environmental Enrichment in Zoos. In *Second Nature: Environmental Enrichment for Captive Animals*; Shepherdson, D., Mellen, J.D., Hutchins, M., Eds.; Smithsonian Institution Press: Washington, DC, USA, 1998; pp. 1–12.
8. Buchanan-Smith, H.M. Environmental Enrichment for Primates in Laboratories. *Adv. Sci. Res.* **2011**, *5*, 41–56. [CrossRef]
9. Bloomsmith, M.A.; Brent, L.Y.; Schapiro, S.J. Guidelines for Developing and Managing an Environmental Enrichment Program for Nonhuman Primates. *Lab. Anim. Sci.* **1991**, *41*, 372–377. [PubMed]
10. Hoy, J.M.; Murray, P.J.; Tribe, A. Thirty Years Later: Enrichment Practices for Captive Mammals. *Zoo Biol.* **2010**, *29*, 303–316. [CrossRef] [PubMed]
11. Skibiel, A.L.; Trevino, H.S.; Naugher, K. Comparison of Several Types of Enrichment for Captive Felids. *Zoo Biol.* **2007**, *26*, 371–381. [CrossRef] [PubMed]
12. Mellen, J.; Sevenich MacPhee, M. Philosophy of Environmental Enrichment: Past, Present, and Future. *Zoo Biol.* **2001**, *20*, 211–226. [CrossRef]
13. Mason, G.; Clubb, R.; Latham, N.; Vickery, S. Why and How Should We Use Environmental Enrichment to Tackle Stereotypic Behaviour? *Appl. Anim. Behav. Sci.* **2007**, *102*, 163–188. [CrossRef]
14. Melfi, V.A. There Are Big Gaps in Our Knowledge, and Thus Approach, to Zoo Animal Welfare: A Case for Evidence-Based Zoo Animal Management. *Zoo Biol.* **2009**, *28*, 574–588. [CrossRef] [PubMed]
15. Rose, P.E.; Brereton, J.E.; Rowden, L.J.; de Figueiredo, R.L.; Riley, L.M. What's New from the Zoo? *An Analysis of Ten Years of Zoo-Themed Research Output. Palgrave Commun* **2019**, *5*, 1–10.
16. Swaisgood, R.R. Current Status and Future Directions of Applied Behavioral Research for Animal Welfare and Conservation. *Appl. Anim. Behav. Sci.* **2007**, *102*, 139–162. [CrossRef]
17. Eagan, T. Evaluation of Enrichment for Reptiles in Zoos. *J. Appl. Anim. Welf. Sci.* **2019**, *22*, 69–77. [CrossRef]
18. De Azevedo, C.S.; Cipreste, C.F.; Young, R.J. Environmental Enrichment: A GAP Analysis. *Appl. Anim. Behav. Sci.* **2007**, *102*, 329–343. [CrossRef]
19. Maple, T.L.; Perdue, B.M. Environmental Enrichment. In *Zoo Animal Welfare*; Springer: Berlin/Heidelberg, Germany, 2013; pp. 95–117.
20. Alligood, C.; Leighty, K. Putting the "E" in SPIDER: Evolving Trends in the Evaluation of Environmental Enrichment Efficacy in Zoological Settings. *Anim. Behav. Cogn.* **2015**, *2*, 200–217. [CrossRef]
21. Case, B.C.; Lewbart, G.A.; Doerr, P.D. The Physiological and Behavioural Impacts of and Preference for an Enriched Environment in the Eastern Box Turtle (Terrapene carolina carolina). *Appl. Anim. Behav. Sci.* **2005**, *92*, 353–365. [CrossRef]
22. Doody, J.S.; Burghardt, G.M.; Dinets, V. Breaking the Social–Non-Social Dichotomy: A Role for Reptiles in Vertebrate Social Behavior Research? *Ethology* **2013**, *119*, 95–103. [CrossRef]
23. Learmonth, M.J. The Matter of Non-Avian Reptile Sentience, and Why It "Matters" to Them: A Conceptual, Ethical and Scientific Review. *Animals* **2020**, *10*, 901. [CrossRef]
24. Warwick, C.; Arena, P.; Lindley, S.; Jessop, M.; Steedman, C. Assessing Reptile Welfare Using Behavioural Criteria. *InPractice* **2013**, *35*, 123–131. [CrossRef]
25. Roth, T.C.; Krochmal, A.R.; LaDage, L.D. Reptilian Cognition: A More Complex Picture via Integration of Neurological Mechanisms, Behavioral Constraints, and Evolutionary Context. *BioEssays* **2019**, *41*, 1900033. [CrossRef]
26. Burghardt, G.M.; Ward, B.; Rosscoe, R. Problem of Reptile Play: Environmental Enrichment and Play Behavior in a Captive Nile Soft-Shelled Turtle, Trionyx Triunguis. *Zoo Biol.* **1996**, *15*, 223–238. [CrossRef]
27. Londoño, C.; Bartolomé, A.; Carazo, P.; Font, E. Chemosensory Enrichment as a Simple and Effective Way to Improve the Welfare of Captive Lizards. *Ethology* **2018**, *124*, 674–683. [CrossRef]
28. Almli, L.M.; Burghardt, G.M. Environmental Enrichment Alters the Behavioral Profile of Ratsnakes (Elaphe). *J. Appl. Anim. Welf. Sci.* **2006**, *9*, 85–109. [CrossRef]
29. Therrien, C.L.; Gaster, L.; Cunningham-Smith, P.; Manire, C.A. Experimental Evaluation of Environmental Enrichment of Sea Turtles. *Zoo Biol.* **2007**, *26*, 407–416. [CrossRef]

30. Passos, L.F.; Garcia, G.; Young, R. How Does Captivity Affect Skin Colour Reflectance of Golden Mantella Frogs? *Herpetol. J.* **2020**, *30*, 13–19. [CrossRef]
31. Bashaw, M.J.; Gibson, M.D.; Schowe, D.M.; Kucher, A.S. Does Enrichment Improve Reptile Welfare? *Leopard Geckos (Eublepharis macularius) Respond to Five Types of Environmental Enrichment. Appl. Anim. Behav. Sci.* **2016**, *184*, 150–160.
32. Burghardt, G.M. Environmental Enrichment and Cognitive Complexity in Reptiles and Amphibians: Concepts, Review, and Implications for Captive Populations. *Appl. Anim. Behav. Sci.* **2013**, *147*, 286–298. [CrossRef]
33. Rose, P.; Evans, C.; Coffin, R.; Miller, R.; Nash, S. Using Student-Centred Research to Evidence-Base Exhibition of Reptiles and Amphibians: Three Species-Specific Case Studies. *J. Zoo Aquar. Res.* **2014**, *2*, 25–32.
34. Hoehfurtner, T.; Wilkinson, A.; Nagabaskaran, G.; Burman, O.H.P. Does the Provision of Environmental Enrichment Affect the Behaviour and Welfare of Captive Snakes? *Appl. Anim. Behav. Sci.* **2021**, *239*, 105324. [CrossRef]
35. Bannister, C.C.; Thomson, A.J.C.; Cuculescu-Santana, M. Can Colored Object Enrichment Reduce the Escape Behavior of Captive Freshwater Turtles? *Zoo Biol.* **2021**, *40*, 160–168. [CrossRef]
36. Swaisgood, R.R.; Shepherdson, D.J. Scientific Approaches to Enrichment and Stereotypies in Zoo Animals: What's Been Done and Where Should We Go Next? *Zoo Biol.* **2005**, *24*, 499–518. [CrossRef]
37. Saudargas, R.A.; Drummer, L.C. *Single Subject (Small N) Research Designs and Zoo Research*; Wiley: Hoboken, NJ, USA, 1996.
38. Burghardt, G.M. A Brief Glimpse at the Long Evolutionary History of Play. *Anim. Behav. Cogn.* **2014**, *1*, 90–98. [CrossRef]
39. Januszczak, I.S.; Bryant, Z.; Tapley, B.; Gill, I.; Harding, L.; Michaels, C.J. s Behavioural Enrichment Always a Success? Comparing Food Presentation Strategies in an Insectivorous Lizard (Plica plica). *Appl. Anim. Behav. Sci.* **2016**, *183*, 95–103. [CrossRef]
40. Kanaan, V.T.; Hötzel, M.J. Does Environmental Enrichment Really Matter? A Case Study Using the Eastern Fence Lizard, Sceloporus Undulatus. *Appl. Anim. Behav. Sci.* **2011**, *135*, 169–170. [CrossRef]
41. Izzo, G.N.; Bashaw, M.J.; Campbell, J.B. Enrichment and Individual Differences Affect Welfare Indicators in Squirrel Monkeys (Saimiri sciureus). *J. Comp. Psychol.* **2011**, *125*, 347. [CrossRef]
42. Alligood, C.A.; Dorey, N.R.; Mehrkam, L.R.; Leighty, K.A. Applying Behavior-Analytic Methodology to the Science and Practice of Environmental Enrichment in Zoos and Aquariums. *Zoo Biol.* **2017**, *36*, 175–185. [CrossRef]
43. Learmonth, M.J.; Sherwen, S.; Hemsworth, P.H. Assessing Preferences of Two Zoo-Housed Aldabran Giant Tortoises (Aldabrachelys gigantea) for Three Stimuli Using a Novel Preference Test. *Zoo Biol.* **2021**, *40*, 98–106. [CrossRef] [PubMed]
44. Wilkinson, A.; Huber, L. Cold-Blooded Cognition: Reptilian Cognitive Abilities. In *The Oxford Handbook of Comparative Evolutionary Psychology*; Oxford University Press: Oxford, UK, 2012; pp. 129–141.
45. Reading, R.P.; Miller, B.; Shepherdson, D. The Value of Enrichment to Reintroduction Success. *Zoo Biol.* **2013**, *32*, 332–341. [CrossRef]
46. Westlund, K. Training Is Enrichment—And Beyond. *Appl. Anim. Behav. Sci.* **2014**, *152*, 1–6. [CrossRef]
47. Degregorio, B.; Sperry, J.; Tuberville, T.; Weatherhead, P. Translocating Ratsnakes: Does Enrichment Offset Negative Effects of Time in Captivity? *Wildl. Res.* **2017**, 44. [CrossRef]
48. Roe, J.H.; Frank, M.R.; Kingsbury, B.A. Experimental Evaluation of Captive-Rearing Practices to Improve Success of Snake Reintroductions. *Herpetol. Conserv. Biol.* **2015**, *10*, 711–722.
49. Roth, T.C.; Krochmal, A.R. The Role of Age-Specific Learning and Experience for Turtles Navigating a Changing Landscape. *Curr. Biol.* **2015**, *25*, 333–337. [CrossRef]
50. Meester, G.D.; Baeckens, S. Reinstating Reptiles: From Clueless Creatures to Esteemed Models of Cognitive Biology. *Behaviour* **2021**, *158*, 1057–1076. [CrossRef]
51. Mehrkam, L.R.; Dorey, N.R. Preference Assessments in the Zoo: Keeper and Staff Predictions of Enrichment Preferences across Species. *Zoo Biol.* **2015**, *34*, 418–430. [CrossRef]
52. Rosier, R.L.; Langkilde, T. Does Environmental Enrichment Really Matter? A Case Study Using the Eastern Fence Lizard, Sceloporus Undulatus. *Appl. Anim. Behav. Sci.* **2011**, *131*, 71–76. [CrossRef]
53. Cronin, K.; Ross, S. Technical Contribution: A Cautionary Note on the Use of Behavioural Diversity (H-Index) in Animal Welfare Science. *Anim. Welf.* **2019**, *28*, 157–164. [CrossRef]
54. Rosier, R.L.; Langkilde, T.L. In Response to the Letter to the Editor Regarding the Article: "Does Environmental Enrichment Really Matter? A Case Study Using the Eastern Fence Lizard, Sceloporus Undulatus". *Appl. Anim. Behav. Sci.* **2011**, *135*, 171–172. [CrossRef]
55. Tarou, L.R.; Bashaw, M.J. Maximizing the Effectiveness of Environmental Enrichment: Suggestions from the Experimental Analysis of Behavior. *Appl. Anim. Behav. Sci.* **2007**, *102*, 189–204. [CrossRef]
56. Bennett, D. *A Little Book of Monitor Lizards: A Guide to the Monitor Lizards of the World and Their Care in Captivity*; Viper Press: Aberdeen, UK, 1995.
57. Uetz, P.; Freed, P.; Hošek, J. The Reptile Database. Available online: https://reptile-database.reptarium.cz/ (accessed on 11 July 2021).
58. Ziegler, T.; Rauhaus, A.; Gill, I. A Preliminary Review of Monitor Lizards in Zoological Gardens. *Biawak* **2016**, *10*, 26–35.
59. Valdez, J.W. Using Google Trends to Determine Current, Past, and Future Trends in the Reptile Pet Trade. *Animals* **2021**, *11*, 676. [CrossRef]
60. CITES. *CITES Trade Database*; UNEP World Conservation Monitoring Centre: Cambridge, UK, 2021.
61. Pianka, E.; King, D. *Varanoid Lizards of the World*; Indiana University Press: Bloomington, IN, USA, 2004.

62. Welton, L.J.; Siler, C.D.; Bennett, D.; Diesmos, A.; Duya, M.R.; Dugay, R.; Rico, E.L.B.; Van Weerd, M.; Brown, R.M. A Spectacular New Philippine Monitor Lizard Reveals a Hidden Biogeographic Boundary and a Novel Flagship Species for Conservation. *Biol. Lett.* **2010**, *6*, 654–658. [CrossRef] [PubMed]
63. Law, S.J.; de Kort, S.R.; Bennett, D.; van Weerd, M. Diet and Habitat Requirements of the Philippine Endemic Frugivorous Monitor Lizard Varanus Bitatawa. *Biawak* **2018**, *12*, 12–22.
64. Dryden, G.; Green, B.; King, D.; Losos, J. Water and Energy Turnover in a Small Monitor Lizard, Varanus-Acanthurus. *Wildl. Res.* **1990**, *17*, 641–646. [CrossRef]
65. Frappell, P.B.; Schultz, T.J.; Christian, K.A. The Respiratory System in Varanid Lizards: Determinants of O_2 Transfer. *Comp. Biochem. Physiol. Part A Mol. Integr. Physiol.* **2002**, *133*, 239–258. [CrossRef]
66. Cross, S.L.; Craig, M.D.; Tomlinson, S.; Bateman, P.W. I Don't like Crickets, I Love Them: Invertebrates Are an Important Prey Source for Varanid Lizards. *J. Zool.* **2020**, *310*, 323–333. [CrossRef]
67. Cieri, R.L.; Farmer, C.G. Computational Fluid Dynamics Reveals a Unique Net Unidirectional Pattern of Pulmonary Airflow in the Savannah Monitor Lizard (Varanus exanthematicus). *Anat. Rec.* **2020**, *303*, 1768–1791. [CrossRef] [PubMed]
68. Schachner, E.R.; Cieri, R.L.; Butler, J.P.; Farmer, C.G. Unidirectional Pulmonary Airflow Patterns in the Savannah Monitor Lizard. *Nature* **2014**, *506*, 367–370. [CrossRef] [PubMed]
69. Clemente, C.J.; Withers, P.C.; Thompson, G.G. Metabolic Rate and Endurance Capacity in Australian Varanid Lizards (Squamata: Varanidae: Varanus). *Biol. J. Linn. Soc.* **2009**, *97*, 664–676. [CrossRef]
70. Northcutt, R.G. Variation in Reptilian Brains and Cognition. *Brain Behav. Evol.* **2013**, *82*, 45–54. [CrossRef] [PubMed]
71. Güntürkün, O.; Stacho, M.; Ströckens, F. The Brains of Reptiles and Birds. *Evol. Neurosci.* **2020**, 159–212.
72. Arena, P.; Warwick, C. Spatial and Thermal Considerations. In *Health and Welfare of Captive Reptiles*; Animal Welfare Series; Springer: Cham, Switzerland, 2022; Volume 22.
73. Shettleworth, S.J. Animal Cognition and Animal Behaviour. *Anim. Behav.* **2001**, *61*, 277–286. [CrossRef]
74. Mendyk, R.W.; Horn, H.G. Skilled Forelimb Movements and Extractive Foraging in the Arboreal Monitor Lizard Varanus Beccarii (Doria, 1874). *Herpetol. Rev.* **2011**, *42*, 343–349.
75. Cooper, T.; Liew, A.; Andrle, G.; Cafritz, E.; Dallas, H.; Niesen, T.; Slater, E.; Stockert, J.; Vold, T.; Young, M. Latency in Problem Solving as Evidence for Learning in Varanid and Helodermatid Lizards, with Comments on Foraging Techniques. *Copeia* **2019**, *107*, 78–84. [CrossRef]
76. Cooper, T.L.; Zabinski, C.L.; Adams, E.J.; Berry, S.M.; Pardo-Sanchez, J.; Reinhardt, E.M.; Roberts, K.M.; Watzek, J.; Brosnan, S.F.; Hill, R.L. Long-Term Memory of a Complex Foraging Task in Monitor Lizards (Reptilia: Squamata: Varanidae). *J. Herpetol.* **2020**, *54*, 378–383. [CrossRef]
77. Kane, D.; Davis, A.C.; Michaels, C.J. Play Behaviour by Captive Tree Monitors, Varanus Macraei and Varanus Prasinus. *Herpetol. Bull.* **2019**, *149*, 28–31. [CrossRef]
78. King, D.; Green, B. *Goannas: The Biology of Varanid Lizards*; UNSW Press: Kensington, Australia, 1999.
79. Korte, S.M.; Olivier, B.; Koolhaas, J.M. A New Animal Welfare Concept Based on Allostasis. *Physiol. Behav.* **2007**, *92*, 422–428. [CrossRef] [PubMed]
80. Meehan, C.L.; Mench, J.A. The Challenge of Challenge: Can Problem Solving Opportunities Enhance Animal Welfare? *Appl. Anim. Behav. Sci.* **2007**, *102*, 246–261. [CrossRef]
81. Myers, D.G.; Diener, E. Who Is Happy? *Psychol. Sci.* **1995**, *6*, 10–19. [CrossRef]
82. Nakamura, J.; Csikszentmihalyi, M. The Concept of Flow. In *Flow and the Foundations of Positive Psychology*; Springer: Berlin/Heidelberg, Germany, 2014; pp. 239–263.
83. Loop, M.S. Auto-Shaping: A Simple Technique for Teaching a Lizard to Perform a Visual Discrimination Task. *Copeia* **1976**, *1976*, 574–576. [CrossRef]
84. Firth, I.; Turner, M.; Robinson, M.; Meek, R. Response of Monitor Lizards (Varanus spp. *) to a Repeated Food Source: Evidence for Association Learning? Herpetol. Bull.* **2003**, *84*, 1–4.
85. Manrod, J.D.; Hartdegen, R.; Burghardt, G.M. Rapid Solving of a Problem Apparatus by Juvenile Black-Throated Monitor Lizards (Varanus albigularis albigularis). *Anim. Cognit.* **2008**, *11*, 267–273. [CrossRef]
86. Gaalema, D.E. Visual Discrimination and Reversal Learning in Rough-Necked Monitor Lizards (Varanus rudicollis). *J. Comp. Psychol.* **2011**, *125*, 246–249. [CrossRef] [PubMed]
87. Ward-Fear, G.; Thomas, J.; Webb, J.K.; Pearson, D.J.; Shine, R. Eliciting Conditioned Taste Aversion in Lizards: Live Toxic Prey Are More Effective than Scent and Taste Cues Alone. *Integr. Zool.* **2017**, *12*, 112–120. [CrossRef]
88. Andrews, R.M.; Pezaro, N.; Doody, J.S.; Guarino, F.; Green, B. Oviposition to Hatching: Development of Varanus Rosenbergi. *J. Herpetol.* **2017**, *51*, 396–401. [CrossRef]
89. Watterson, I.; Abbs, D.; Bhend, J.; Chiew, F.; Church, J.; Ekström, M.; Kirono, D.; Lenton, A.; Lucas, C.; McInnes, K. *Climate Change in Australia Projections for Australia's Natural Resource Management Regions*; CSIRO: Canberra, Australia, 2015.
90. Dayananda, B.; Webb, J.K. Incubation under Climate Warming Affects Learning Ability and Survival in Hatchling Lizards. *Biol. Lett.* **2017**, *13*, 20170002. [CrossRef] [PubMed]
91. Abayarathna, T.; Webb, J.K. Effects of Incubation Temperatures on Learning Abilities of Hatchling Velvet Geckos. *Anim. Cognit.* **2020**, *23*, 613–620. [CrossRef] [PubMed]

92. Kis, A.; Huber, L.; Wilkinson, A. Social Learning by Imitation in a Reptile (Pogona vitticeps). *Anim. Cognit.* **2015**, *18*, 325–331. [CrossRef] [PubMed]
93. Noble, D.W.; Byrne, R.W.; Whiting, M.J. Age-Dependent Social Learning in a Lizard. *Biol. Lett.* **2014**, *10*, 20140430. [CrossRef]
94. Iwaniuk, A.N.; Nelson, J.E.; Pellis, S.M. Do Big-Brained Animals Play More? Comparative Analyses of Play and Relative Brain Size in Mammals. *J. Comp. Psychol.* **2001**, *115*, 29–41. [CrossRef] [PubMed]
95. Stanner, M. Mammal-like Feeding Behavior of Varanus Salvator and Its Conservational Implications. *Biawak* **2010**, *4*, 128–131.
96. Burghardt, G.M. *The Genesis of Animal Play: Testing the Limits*; MIT Press: Cambridge, MA, USA, 2005.
97. Khandakar, N.; Jeny, K.N.; Islam, S.; Hakim, M.A.; Pony, I.A. Play Behavior by a Yellow Monitor, Varanus Flavescens (Hardwicke and Gray 1827). *Reptiles Amphib.* **2020**, *27*, 257–258. [CrossRef]
98. Bartholomew, G.A.; Tucker, V.A. Size, Body Temperature, Thermal Conductance, Oxygen Consumption, and Heart Rate in Australian Varanid Lizards. *Physiol. Zool.* **1964**, *37*, 341–354. [CrossRef]
99. Sweet, S.S. Spatial Ecology of Varanus Glauerti and V. Glebopalma in Northern Australia. *Mertensiella* **1999**, *11*, 317–366.
100. Eidenmüller, B. Bisher Nicht Beschriebene Verhaltensweisen von Varanus (Varanus) Flavirufus Mertens 1958, Varanus (Odatria) Acanthurus Boulenger 1885 Und Varanus (Odatria) Storri Mertens 1966 Im Terrarium. *Monitor* **1993**, *2*, 11–21.
101. D'Amore, D.C.; Clulow, S.; Doody, J.S.; Rhind, D.; McHenry, C.R. Claw Morphometrics in Monitor Lizards: Variable Substrate and Habitat Use Correlate to Shape Diversity within a Predator Guild. *Ecol. Evol.* **2018**, *8*, 6766–6778. [CrossRef]
102. Krebs, U.; Wünstel, U. Observations and Experiments on "Spinning Behavior" in Varanus Albigularis. *Biawak* **2019**, *13*, 54–61.
103. Mason, G.J.; Latham, N. Can't Stop, Won't Stop: Is Stereotypy a Reliable Animal Welfare Indicator? *Anim. Welf.* **2004**, *13*, S57–S69.
104. Tilly, S.-L.C.; Dallaire, J.; Mason, G.J. Middle-Aged Mice with Enrichment-Resistant Stereotypic Behaviour Show Reduced Motivation for Enrichment. *Anim. Behav.* **2010**, *80*, 363–373. [CrossRef]
105. Jackson, M.G.; Lightman, S.L.; Gilmour, G.; Marston, H.; Robinson, E.S. Evidence for Deficits in Behavioural and Physiological Responses in Aged Mice Relevant to the Psychiatric Symptom of Apathy. *Brain Neurosci. Adv.* **2021**, *5*, 23982128211015110. [CrossRef] [PubMed]
106. Meagher, R. Is Boredom an Animal Welfare Concern? *Anim. Welf.* **2018**, *28*, 21–32. [CrossRef]
107. Clark, F.E. Great Ape Cognition and Captive Care: Can Cognitive Challenges Enhance Well-Being? *Appl. Anim. Behav. Sci.* **2011**, *135*, 1–12. [CrossRef]
108. Walf, A.A.; Frye, C.A. The Use of the Elevated plus Maze as an Assay of Anxiety-Related Behavior in Rodents. *Nat. Protoc.* **2007**, *2*, 322–328. [CrossRef] [PubMed]

Journal of
Zoological and Botanical Gardens

MDPI

Article

Investigating the Effect of Disturbance on Prey Consumption in Captive Congo Caecilians *Herpele squalostoma*

Kimberley C. Carter [1,*], Léa Fieschi-Méric [2], Francesca Servini [1], Mark Wilkinson [3], David J. Gower [3], Benjamin Tapley [1] and Christopher J. Michaels [1]

1 Zoological Society of London Outer Circle, Regent's Park, London NW1 4RY, UK; francesca.servini@zsl.org (F.S.); ben.tapley@zsl.org (B.T.); christopher.michaels@zsl.org (C.J.M.)
2 Genetics and Ecology Research Group, Facility of Science, Engineering and Architecture, Laurentian University, Sudbury, ON P3E 2C6, Canada; leafieschimeric@gmail.com
3 Natural History Museum, London SW7 5HD, UK; apodauk@gmail.com (M.W.); d.gower@nhm.ac.uk (D.J.G.)
* Correspondence: kimberley.carter@zsl.org

Citation: Carter, K.C.; Fieschi-Méric, L.; Servini, F.; Wilkinson, M.; Gower, D.J.; Tapley, B.; Michaels, C.J. Investigating the Effect of Disturbance on Prey Consumption in Captive Congo Caecilians *Herpele squalostoma*. *J. Zool. Bot. Gard.* **2021**, 2, 705–715. https://doi.org/10.3390/jzbg2040050

Academic Editors: Kris Descovich, Caralyn Kemp and Jessica Rendle

Received: 28 October 2021
Accepted: 3 December 2021
Published: 14 December 2021

Publisher's Note: MDPI stays neutral with regard to jurisdictional claims in published maps and institutional affiliations.

Abstract: Maintaining Gymnophiona in captivity provides opportunities to study the behaviour and life-history of this poorly known Order, and to investigate and provide species-appropriate welfare guidelines, which are currently lacking. This study focuses on the terrestrial caecilian *Herpele squalostoma* to investigate its sensitivity to disturbances associated with routine husbandry needed for monitoring and maintaining adequate wellbeing in captivity. Fossorial caecilians gradually pollute their environment in captivity with waste products, and substrate must be replaced at intervals; doing so disturbs the animals directly and via destruction of burrow networks. As inappetence is frequently associated with stress in amphibians, the percentage consumption of offered food types, river shrimp (*Palaemon varians*) and brown crickets (*Gryllus assimilis*), was measured as an indicator of putative stress following three routine substrate changes up to 297 days post-substrate change. Mean daily variation in substrate temperatures were also recorded in order to account for environmental influences on food consumption, along with nitrogenous waste in tank substrate prior to a substrate change and fresh top soil in order to understand the trade-off between dealing with waste accumulation and disturbing animals. We found a significant negative effect of substrate disturbance on food intake, but no significant effect of prey type. Variations in daily soil temperatures did not have a significant effect on food intake, but mean substrate temperature did. Additionally, substrate nitrogenous waste testing indicated little difference between fresh and tank substrate. In conclusion, this study provides a basis from which to develop further welfare assessment for this and other rarely kept and rarely observed terrestrial caecilian species.

Keywords: amphibian; behaviour; diet; nitrogenous waste; welfare; zoo research

1. Introduction

Within zoos and other industries maintaining wild animals in captivity, there is a necessary balance to be struck between providing husbandry needs for captive animals and reducing negative effects that such provision may elicit [1]. Zoo licencing for the United Kingdom (UK), for example, outlines that animals should be checked twice daily while avoiding unnecessary stress or disturbance [2]. For some species, a frequency of twice daily checks is not feasible either due to the species' natural history making them difficult to visually monitor e.g., fossorial, or because such checks are intrusive and cause significant stress to the animals. Most disturbances such as enclosure changes, handling and restraining, and transportation are temporary and create a short-term change in behavioural responses and glucocorticoid hormone production [3,4]. However, frequent negative events can cause chronic stress, which in turn causes negative morphological and behavioural responses. Chronic stress may be visible in amphibians through reduced feeding, behavioural inhibition or decreased activity, hiding, fearfulness, frequency of startle responses, stereotypies, raised or changed posture and/or displacement behaviours [5–8].

However, behavioural responses can often be difficult to interpret and a good understanding of what is deemed an appropriate response and what are abnormal or deleterious responses for both the individual and the species is needed [3,9]. Changes in activity or arousal can equally be caused by positive or negative stimuli, and these should be interpreted in the context of what might be typical for that species and for the situation. The frequency of arousing events is also expected to impact activity and behavioural responses. A good knowledge of the species' natural history as well as individual animal history is needed to fully understand their husbandry needs [3], but for many rarely seen and understudied species this dearth of knowledge can create challenges for quantifying optimum requirements within captive settings.

One such group of little studied and poorly known animals are the elongated, limbless amphibians, caecilians (Order: Gymnophiona). There are about approximately 215 currently recognised species within the Order [10] with only six species currently being kept in zoos [11]. Many terrestrial caecilian species spend most of their lives in soil, making these animals difficult to monitor [12,13]. Alongside invertebrates, fossorial (burrowing) caecilian species may play an important part in engineering and maintaining ecosystems by influencing the structure of the soil and the distribution and cycling of organic matter [14–17]. However, caecilians are generally understudied, with most studies on fossorial vertebrate species focussing on burrowing mammals [15] in conjunction with a general overall bias away from studying amphibians [18]. Additionally, within amphibian research, caecilians, in general, are one of the least studied groups. As well as potentially providing direct benefits to species conservation, maintaining caecilians in captivity provides an opportunity to study various aspects of their biology, and develop and validate methods that can be used to understand and conserve them [13,19–25].

In this work, we studied the Congo caecilian (*Herpele squalostoma*), a fossorial caecilian from lowland forests and agricultural habitats across Nigeria, Cameroon, Central African Republic, mainland Equatorial Guinea and Bioko Island, Gabon, Congo, and the western Democratic Republic of Congo, with possible records in Angola [26]. *Herpele squalostoma* is reported to be locally abundant and sporadically traded (in large numbers on occasion) in the commercial international pet trade [12,20,27]. Despite this reported abundance, little is known about the ecology of this species [28]. Listed as Least Concern [26], *H. squalostoma* is not threatened, however this species can act as an analogue model to develop caecilian husbandry techniques to apply to more threatened taxa [29]. Currently this species is poorly represented in zoological collections that may aid in natural history research, with only 13 animals maintained between two institutions [11]. Due to the fossorial nature of this, and most, caecilian species it is difficult to monitor and assess behavioural responses that could inform welfare provisions within captive settings [13]. Prey consumption where food items are placed on the surface is one of the few visual and external measures of wellbeing for this study species.

This study evaluates the effects of disturbance from three substrate change events on the food consumption in *H. squalostoma*. We propose the proportion of prey consumption is a suitable, non-invasive, measure of putative stress in that species. Through this work, we aim to better understand the susceptibility of *H. squalostoma* to environmental disturbance and provide evidence to inform best husbandry practices that reduce negative welfare impacts and ensure that these needs are met.

2. Materials and Methods

2.1. Animal Models, Experimental Design, and Data-Collection

Four *H. squalostoma* of unknown sex and age were maintained at the Zoological Society of London (ZSL) London Zoo, on loan as part of a collaborative project with the Natural History Museum's Herpetology Research Group. Animal lengths ranged from 51.5–56.9 cm (as of February 2021) indicating that all animals were of adult age, though the exact ages are unknown. The enclosure was designed to mimic descriptions and photographs of wild habitat [9,28,29]. They were housed as a group in a 135 cm L × 71 cm H × 60.5 cm W glass enclosure (Custom Aquaria, Rushden, UK) with a substrate depth of 30 cm at the front of the tank, sloping upwards to a depth of 43 cm at the rear which was intended to improve the visitors view and the aesthetic of the enclosure. The enclosure had a small open column of water in the back right corner at a depth of 16–21 cm, permeable via a cork barrier to the substrate layer and planted with Radican Sword (*Echinodorus muricatus*). This allowed for a permanent layer of water in the base of the enclosure, ensuring the substrate layer retained moisture. Grasses (*Carex morrowii* and *C. m. variegata*) were planted in the terrestrial areas to provide surface cover, substrate structure and root structures for egg clutches to be laid around [30–32]. The substrate consisted of topsoil with buried masses of dried leaves of mixed tree species, cork tubes lined with clay and branches to provide potential nest sites.

Substrate minimal and maximal daily temperatures were recorded with a digital probe thermometer (ETI Ltd., Worthing, UK) at approximately 20 cm substrate depth from the surface, enabling us to approximate daily mean substrate temperatures as (Tmin + Tmax)/2 (thereafter referred to as "average temperature"), and the daily range of substrate temperature variation (Tmax–Tmin) on feeding days. Readings of maximum and minimum temperatures were taken only once per day. The room climate control provided ambient temperatures aligning with climate data for Yaounde, Cameroon based on field study sites [30]. Outdoor temperatures and sunlight influenced the substrate temperatures somewhat because the public-facing side of the tank is within 2 cm of the room show window.

The caecilians were fed a diet consisting of defrosted river shrimp (*Palaemon varians*) kept whole or halved if >2 cm total length; defrosted, gut-loaded, killed adult brown crickets (*Gryllus assimilis*); and small live worms (*Dendrobaena* sp.). Weights of whole shrimp were c. 0.3 g and brown crickets c. 0.5 g. This was designed to replicate the wild diet within the confines of what we can reasonably source [33,34]. Each prey type was given independently on a set schedule alternating between food types, and the animals were fed three times per week (Monday, Wednesday, and Friday) between 8:30 a.m.–5 p.m., most often before 12 p.m. Shrimp or crickets (Shrimp, Max = 54, Min = 12, Median = 27; Cricket, Max = 45, Min = 12, Median = 25) were offered on feed days, placed near burrow entrances to increase accessibility for the animals. The remaining number of prey items were counted the following day before being discarded. Live *Dendrobaena* worms were offered once per week on a set feed day but were excluded from this study as they could not be counted and removed without substantial disturbance to the animals. The number of prey items offered were relatively the same quantity irrespective of the number of tunnel entrances available, which greatly reduced post-substrate change. Because the quantity and size of the prey items offered varied between feed days, the proportionate consumption was calculated. Food intake was recorded after every non-worm feed, with a total of 147 observations over the course of our study, which lasted a total of 598 days (from 7 February 2020 to 27 September 2021).

Substrate changes for the study caecilians are typically performed every six to eight months to minimise frequent disturbance to animals and burrow structures while providing adequate environmental needs, for example by preventing the build-up of nitrogenous waste to detrimental levels. Due to an interruption from the 2020 COVID-19 pandemic, which created staff shortages and increased pressure on staff workload, a substrate change was postponed and occurred approximately ten months (297 days) after the previous change. During substrate changes the animals are caught, placed into separate plastic 9 L Really Useful Boxes (Really Useful Products Ltd., Castleford, UK) filled nearly full with tank substrate, to allow the animals to burrow, before being visually checked, weighed, and photographed for subsequent morphological measurements via ImageJ (https://imagej.nih.gov/ij, accessed on 21 September 2021) [35]. The enclosure was then stripped with all old substrate discarded and replaced with fresh 25 L bags of topsoil (B&Q, London, UK) that was pre-warmed to the same temperature as the enclosure substrate. Plants and furnishings were retained and replaced in the enclosure with the new substrate. Caecilians were then reintroduced to the enclosure by placing them on the surface of the new substrate. Substrate change duration were between four to five hours and animals were not fed while contained in the Really Useful Boxes. A subsequent change occurred within the usual time frame at around seven months (218 days) from the last change. Substrate changes were carried out on 26 November 2019, 21 April 2020 and 16 February 2021.

Substrate samples were taken from the tank during the removal of the substrate on the day of the most recent substrate change, and of fresh substrate taken directly from the bags of topsoil, provided by the supplier. Samples were taken from three locations within the tank, and from three different randomly selected bags of fresh substrate. Samples were only taken once, during this most recent change, and results did not include previous substrate changes. Samples of 10 g substrate were suspended in 100 mL reverse osmosis (RO) water before being filtered through coffee filter paper (Filtropa Unbleached Coffee Filter Papers, Size 4) overnight at a temperature of 0.3–0.5 °C. The sample water was removed with a pipette so as not to disturb the final sediment layer. This was tested with Salifert profitest (Duiven, Netherlands) nitrate water tests and Palintest (Gateshead, UK) ammonia and nitrite water tests using a Palintest Interface photometer 7500. The concentration of ammonia, nitrite and nitrate was recorded to assess waste build up in the substrate and the mean test results were calculated per condition. These variables were also measured against the RO water to control for any nitrogenous waste contamination.

2.2. Statistical Analysis

A full generalized linear mixed model (GLMM) was built with a binomial error distribution and a logit link function [36] for the proportion of food items eaten as the response variable, to test for the effect of temperatures on amount consumed. Days since the last substrate change, the food type (crickets or shrimps), the daily mean temperature, and the daily range of temperature variation were tested as covariates, and the substrate change number was implemented as a random effect to control for differences in intercepts due to repeated measures on the same group of individuals [37]. Using a frequentist hypothesis testing approach, the significance of each covariate was tested using Walt z-tests to determine the best structure for our final model. Our final model's assumptions were verified graphically (Appendix A, Figure A1) and its fit was assessed using Bolker's dispersion estimate and marginal and conditional R^2 metrics [38]. Parameter estimates were all calculated using Laplace approximation [39]. Analyses were conducted using the packages lme4 [40] and MuMIn [41] in the software R version 4.1.0 [42] and is available in open-access at https://github.com/LeaFieschiMeric/substrate_change_in_herpele (accessed on 28 October 2021).

3. Results

The daily variation in temperature and the type of food provided did not have a significant difference on the proportion of food items consumed (respectively, $z = 0.031$, p-value = 0.975; $z = 0.851$, p-value = 0.394). However, average temperature has a significantly negative effect on the food intake ($z = -3.424$, p-value < 0.05, Figure 1A). The proportion of food eaten significantly increases with time since last substrate change ($z = 7.624$, p-value < 0.05, Figure 1B).

Average substrate temperatures ranged from 23.3 °C to 30.4 °C with a daily substrate temperature variation of 0.1–5.8 °C on feeding days. The room climate control provided ambient temperatures between 25.7–33.3 °C day-time maximum and 20.9–30.5 °C night-time minimum. Outdoor temperatures and sunlight influenced the substrate temperatures somewhat as the public-facing side of the tank is within 2 cm of the room show window.

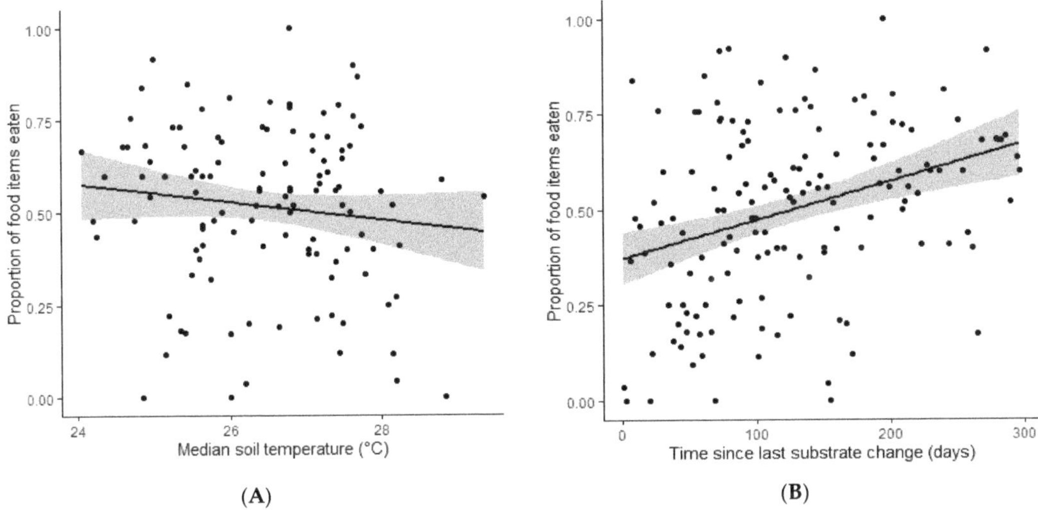

Figure 1. Scatterplots of the proportion of food items eaten by captive *Herpele squalostoma* depending on the significant regression parameters: (**A**) the average substrate temperature and (**B**) the time since the substrate was last changed.

Our final model includes the time since last substrate change and the average temperature as covariates, and the substrate change number as a random effect. The graphical assessment of the residuals (Appendix A, Figure A1) and the conditional R^2 (Table 1) in our final model suggest an acceptable fit.

Table 1. Regression parameters estimates on the log-odds scale, with their standard errors and z-values, for all covariates used in our final model, along with odds-ratios (OR) and confidence intervals given on the scale of the linear predictor. Model fit is acceptable according to the estimated measures of variance and dispersion.

Final Model	Estimate	Std. Error	z-Value	OR	2.5% CI	97.5% CI
Parameter estimates						
Intercept	2.580	0.848	3.041		2.506	69.789
Days since substrate change	0.004	0.0005	7.678	1.004	1.003	1.005
Average substrate temperature	−0.11	0.032	−3.524	0.896	0.838	0.951
Model fit						
R^2 marginal	0.453					
R^2 conditional	0.453					
Dispersion estimate (Chi^2)	5.49					

Every subsequent day after a substrate change, the proportion of food intake increases by 1.004, showing a cumulative increase of the proportion of food eaten over time after a substrate change. Conversely, for every 1 °C increase in the substrate temperature the caecilians consume 1.11 times less food.

Tests for ammonia, nitrite, and nitrate (mg/L) were carried out for tank substrate at 218 days after a substrate change and for fresh substrate from three random bags of commercially bought topsoil (Table 2). Water tests for the RO water used to suspend the substrate samples showed a mean ammonia of 0 mg/L (N = 1), nitrite of 0.01 mg/L (N = 1) and mean nitrate of 0 mg/L (N = 1).

Table 2. Concentrations of ammonia, nitrite and nitrate (mg/L) for tank substrate (at 218 days after the previous substrate change) and fresh substrate. Mean, median, range and N values are recorded.

	Tank Substrate			Fresh Substrate		
	Ammonia mg/L	**Nitrite mg/L**	**Nitrate mg/L**	**Ammonia mg/L**	**Nitrite mg/L**	**Nitrate mg/L**
Mean	0.03	0.03	0	0.04	0.06	2.00
Median	0.02	0.03	0	0	0.05	2.00
Range	0–0.07	0.02–0.04	0	0–0.12	0.05–0.07	2.00
N	3	3	3	3	3	3

Following the previous substrate change delayed by COVID mitigations and staff work demands, when removed from the enclosure all caecilians were considered healthy and increased in length and weight. Therefore, it is unlikely this prolonged period of an additional 2 months between changes had any visual detrimental effects.

4. Discussion

The complete destruction of the burrow systems of captive *Herpele squalostoma* in this study after a substrate change created an expected reduction in consumption due to the lack of accessibility to the surface and prey items left on the surface of the substrate (i.e., shrimp and crickets but not worms). However, this study shows that after the animals had re-built new burrow exits, consumption of the river shrimp and crickets remained reduced for an extended period. The exact timescale of the re-formation of burrow exits was not measured but anecdotally 1–2 exits were formed within a week and several exits ranging across at least half of the tank we made around 4–6 weeks after a substrate change occurred. This suggests that the disturbance from substrate changes did not only create short-term physical barriers to consumption but also longer-term psychological or behavioural barriers.

This species' natural history and the quantity of prey they typically consume in the wild is unknown. Therefore, low consumption is a relative term. In our study, we observe a large range of variation in the proportion of food eaten (we record values from 0 to 100% of offered items consumed) with a mean of 50% of offered items consumed. Indeed, these proportions are directly influenced by the total number of items offered, which varied greatly. There seems to be a plateau in the number of items eaten (maximum number of items eaten = 38, although maximum number of items offered = 54), corresponding to a mean of 9.4 items per individual. On average, 13.5 items were eaten in total, which corresponds to slightly more than 3 food items per individual. There was no trend in the number of items offered over time indicating that trends in consumption were not related to food increasing or decreasing over time. Future studies should use a fixed total number of food items proposed every day and try to determine typical food consumption per individual instead of in a group. Statistical tests confirm a significant effect of the length of time after a substrate change with percentage consumption with a cumulative increase over time post-disturbance. Daily variations in temperature did not significantly affect percentage of consumption. The average substrate temperatures did have a significant effect, with these individuals feeding less at higher temperatures.

The build-up of nitrogenous wastes in the substrate is one of the main concerns when providing adequate captive welfare and when determining the length between substrate changes in this species, because high levels of nitrogenous waste can have detrimental health effects for amphibians [43,44], such as increased mucous production, change in skin pigmentation as well as immunosuppression and increased vulnerability to disease [45]. Recommended nitrogenous waste concentrations for amphibians are <0.2 mg/L ammonia, <1.0 mg/L nitrites and <50 mg/L nitrates (both tank soil and fresh substrate in this study fall within these acceptable parameters) [46].

The nitrogenous substrate tests indicate that there is a small difference between values between substrate 218 days after the previous change and fresh substrate. Therefore, to reduce the disruptive effects of substrate changes, the frequency of substrate changes could be reduced in this captive group. Further experimentation needs to be carried out to determine the maximum time between changes before substrate quality becomes detrimental, but routine substrate tests could be used to inform substrate change frequency in the same way as is used as standard to inform aquarium water changes. In addition, the levels of nutrients in fresh, commercially bought topsoil is on average, slightly higher in all nitrogenous waste value concentrations. However, the concentration is also more variable between bags provided by suppliers, and the plants in this enclosure most likely reduced the build-up of nitrogenous waste and/or reduced the higher ammonia values from the fresh substrate [44]. Further experiments could be done to compare sparsely and heavily planted enclosures and the speed at which nitrogenous wastes build up over the same time periods. It is noted that some caecilians may favour and thrive in nutrient-rich substrates. For example, *Siphonops annulatus* is highly associated with organically rich, fertile, and humid soils in cabruca cacao plantations [17]. However, this preference for nutrient-rich microhabitats may be explained by humidity, temperature, or abundance of prey rather than nutrient richness, though cannot be confirmed. However, as the detrimental levels of nitrogenous waste for specifically caecilians are unknown and preference of soil richness varies between species, the natural history and wild habitat of each species should be considered when determining disturbance from substrate change frequency.

Some caecilian species might have (semi-) permanent burrow structures; therefore, the removal and disturbance of substrate could potentially be more detrimental to these species than those that do not have such permanent burrow structures [15,47,48]. *Boulengerula boulengeri*, for example, are more abundantly encountered during digging than during other sampling methods such as pitfall traps and visual surveys on the forest floor surface [49], possibly suggesting the use of permanent burrows in a particular soil depth range [50]. However, the movement of this species between burrows and the frequency and duration of use is unknown to confirm whether these are permanently used. Some species may show large home areas such as *Gegeneophis ramaswamii* which have been shown to have large movements in and out of a sampled study area of 100 m^2 [15]. Other species are epigeic for at least some part of the time, for example ichthyophiids or scolecomorphids [48,51]. Some caecilians may tolerate disturbed habitats; population densities of *B. taitanus* were greater in agricultural land than in forest [49]. The particular species' natural history is important to consider because disturbances in captivity to tunnels systems may have stronger welfare implications to some species over others.

Another variable that may also impact the determination of substrate change frequency is the preferred compression and hardness of the substrate [22]. Additionally, burrow permanence is likely dictated by soil type. The substrate is a basic factor in terrestrial caecilian husbandry; however, there is little data on preferences in the wild or in captivity [22]. For some species, it may be beneficial to have relatively frequent substrate changes if they prefer softer, less compacted substrate. Additionally, the composition of some softer substrates, such as wood pulp-based substrates like Megazorb Animal Bedding (Northern Crop Driers, Melbourne, York, UK) which has been used to house terrestrial caecilians [22,25], will decompose faster or allow for a faster build-up of nitrogenous wastes. Artificial paper-based substrates do not support live plant growth and accumulate bacterial growth much more rapidly [52]. The more rapid decomposition of some substrates again raises the dilemma of balancing the minimisation of disturbance and destruction of burrows against providing preferred substrate hardness or types and substrate chemical parameters. It has been noted that both *Geotrypetes seraphini* and *Microcaecilia unicolor* favoured Megazorb over coir in choice chambers [22,25].

There are several limitations to this study due to the lack of available knowledge on this species' natural history, the concealment of the usual behavioural indicators for assessing welfare due to their fossorial nature. Additionally, this study provides an indicator of group behaviour rather than individual behaviour; therefore, individual welfare cannot be assessed, and changes may not benefit all individuals equally [1]. Results may have skewed if one individual behaved largely different than its counterparts. However, as the study animals have been housed as a group for a long period and because no known recorded measure of welfare is currently available for this species, or any other caecilian in captivity, tracking food consumption changes in relation to disturbance does provide a basis to assess welfare on a group level. Furthermore, the overdispersion and the moderate conditional R^2 of our model indicate that it does not capture all the variation observed in the data. Other untested factors could explain some of the variability in the proportion of food eaten in *H. squalostoma*, and the feeding response cannot be used alone to predict putative welfare state. Indeed, the total amount of food items given (which ranged from 12–54) and their changing distribution within the enclosure could have introduced variation into the data. Food was positioned near tunnel entrances/exits, but consumption may have been affected by the proximity of the caecilians to these positions and their activity under the surface. Additionally, other covariates that may have impacted percentage consumption, such as ambient air humidity, were not tested here.

Zoo legislation in the UK calls for the daily check of all animals under a zoo's care, without causing unnecessary stress or disturbance [2]. Due to the fossorial nature of *H. squalostoma,* there are limitations on how frequently the animals can be checked and in how activity and stress can be monitored and assessed remotely to aid in welfare assessment tools. More research is needed to learn about this rarely maintained species, but this study demonstrates that simple captive experiments can provide opportunities for evidence-based husbandry and to improve the knowledge and welfare provision in captive caecilians.

Author Contributions: Conceptualization, K.C.C., C.J.M. and B.T.; methodology, K.C.C., C.J.M. and B.T.; formal analysis, L.F.-M.; investigation, K.C.C.; data curation, K.C.C. and F.S.; writing—original draft preparation, K.C.C. and L.F.-M.; writing—review and editing, D.J.G., K.C.C., F.S., B.T. and M.W.; supervision, B.T. and C.J.M.; project administration, K.C.C. All authors have read and agreed to the published version of the manuscript.

Funding: This research received no external funding.

Institutional Review Board Statement: Ethical review and approval were waived for this study, due to only naturally occurring behaviours being recorded during routine husbandry.

Informed Consent Statement: Not applicable.

Data Availability Statement: Analyses are available in open-access at https://github.com/LeaFieschiMeric/substrate_change_in_herpele (accessed on 1 December 2021).

Acknowledgments: We thank all members of the Herpetology team at ZSL London Zoo who assisted in the husbandry of the animals in this work and inputted into the data collection; in addition to the authors, Daniel Kane, Joe Capon, Unnar Ævarsson and Charlotte Ellis, and to Lewis Rowden for support in internal administration processes.

Conflicts of Interest: The authors declare no conflict of interest.

Appendix A

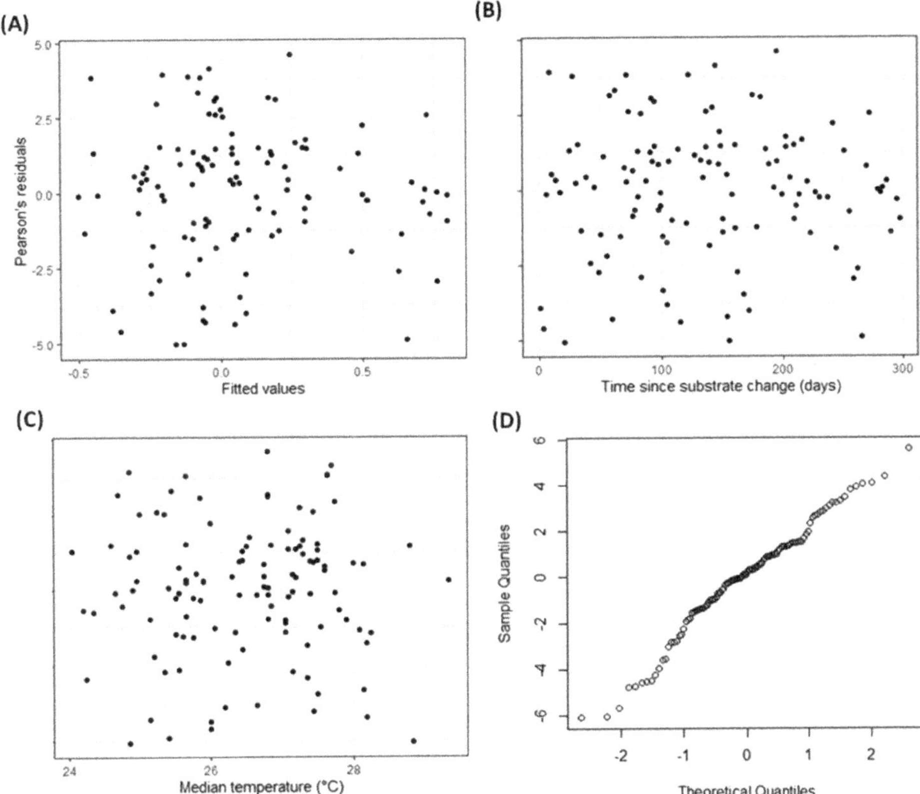

Figure A1. Graphical check of the final model' assumptions using Pearson's residuals against (**A**) fitted values and against the two covariates -the time since the last substrate change (**B**) and the median temperature (**C**) and QQ-plot of the residuals (**D**). The absence of pattern suggests that assumptions of homoscedasticity and normality of model residuals are verified.

References

1. Whitham, J.C.; Wielebnowski, N. New directions for zoo animal welfare science. *Appl. Anim. Behav. Sci.* **2013**, *147*, 247–260. [CrossRef]
2. Department of the Environment, Transport and the Regions. *Secretary of State's Standards of Modern Zoo Practice*; Department of the Environment, Transport and the Regions: London, UK, 2012.
3. Van Waeyenberge, J.; Aerts, J.; Hellebuyck, T.; Pasmans, F.; Martel, A. Stress in wild and captive snakes: Quantification, effects and the importance of management. *Vlaams Diergeneeskd. Tijdschr.* **2018**, *87*, 59–65. [CrossRef]
4. Narayan, E.J.; Forsburg, Z.R.; Davis, D.R.; Gabor, C.R. Non-invasive methods for measuring and monitoring stress physiology in imperiled amphibians. *Front. Ecol. Evol.* **2019**, *7*, 431. [CrossRef]
5. Morgan, K.N.; Tromborg, C.T. Sources of stress in captivity. *Appl. Anim. Behav. Sci.* **2007**, *102*, 262–302. [CrossRef]
6. Warwick, C.; Arena, P.; Lindley, S.; Jessop, M.; Steedman, C. Assessing reptile welfare using behavioural criteria. *Practice* **2013**, *35*, 123–131. [CrossRef]
7. Hossie, T.J.; Ferland-Raymond, B.; Burness, G.; Murray, D.L. Morphological and behavioural responses of frog tadpoles to perceived predation risk: A possible role for corticosterone mediation? *Ecoscience* **2010**, *17*, 100–108. [CrossRef]
8. Woody, S.M.; Santymire, R.M.; Cronin, K.A. Posture as a Non-Invasive Indicator of Arousal in American Toads (*Anaxyrus americanus*). *J. Zool. Bot. Gard.* **2021**, *2*, 1–9. [CrossRef]
9. Benn, A.L.; McLelland, D.J.; Whittaker, A.L. A review of welfare assessment methods in reptiles, and preliminary application of the welfare quality® protocol to the pygmy blue-tongue skink, *Tiliqua adelaidensis*, using animal-based measures. *Animals* **2019**, *9*, 27. [CrossRef]

10. Frost, D.R. *Amphibian Species of the World: An Online Reference*; Version 6.1; American Museum of Natural History: New York, NY, USA, 2021. Available online: https://amphibiansoftheworld.amnh.org/index.php (accessed on 17 August 2021). [CrossRef]
11. Species360. 2021. Available online: www.Species360.org (accessed on 15 August 2021).
12. Gower, D.J.; Wilkinson, M. The conservation biology of caecilians. *Conserv. Biol.* **2005**, *19*, 45–55. [CrossRef]
13. Tapley, B.; Michaels, C.J.; Gower, D.J.; Wilkinson, M. The use of visible implant elastomer to permanently identify caecilians (Amphibia: Gymnophiona). *Herpetol. Bull.* **2019**, *150*. [CrossRef]
14. Lavelle, P.; Bignell, D.; Lepage, M.; Wolters, V.; Roger, P.A.; Ineson, P.O.W.H.; Heal, O.W.; Dhillion, S. Soil function in a changing world: The role of invertebrate ecosystem engineers. *Eur. J. Soil Biol.* **1997**, *33*, 159–193.
15. Measey, G.J.; Gower, D.J.; Oommen, O.V.; Wilkinson, M. A mark–recapture study of the caecilian amphibian *Gegeneophis ramaswamii* (Amphibia: Gymnophiona: Caeciliidae) in southern India. *J. Zool.* **2003**, *261*, 129–133. [CrossRef]
16. Jones, D.T.; Loader, S.P.; Gower, D.J. Trophic ecology of East African caecilians (Amphibia: Gymnophiona), and their impact on forest soil invertebrates. *J. Zool.* **2006**, *269*, 117–126. [CrossRef]
17. Jared, C.; Antoniazzi, M.M.; Wilkinson, M.; Delabie, J.H. Conservation of the caecilian *Siphonops annulatus* (Amphibia, Gymnophiona) in Brazilian cacao plantations: A successful relationship between a fossorial animal and an agrosystem. *Agrotrópica* **2015**, *27*, 233–238. [CrossRef]
18. Bonnet, X.; Shine, R.; Lourdais, O. Taxonomic chauvinism. *Trends Ecol. Evol.* **2002**, *17*, 1–3. [CrossRef]
19. O'Reilly, J.C. Feeding in caecilians. In *Feeding: Form, Function and Evolution in Tetrapod Vertebrates*; Elsevier: Amsterdam, The Netherlands, 2000; pp. 149–166.
20. Wilkinson, M.; Müller, H.; Gower, D.J. On *Herpele multiplicata* (Amphibia: Gymnophiona: Caeciliidae). *Afr. J. Hepatol.* **2003**, *52*, 119–122.
21. Maddock, S.T.; Lewis, C.J.; Wilkinson, M.; Day, J.J.; Morel, C.; Kouete, M.; Gower, D.T. Non-lethal DNA sampling for caecilian amphibians. *Herpetol. J.* **2014**, *24*, 255–260.
22. Tapley, B.; Bryant, Z.; Grant, S.; Kother, G.; Feltrer, Y.; Masters, N.; Strike, T.; Gill, I.; Wilkinson, M.; Gower, D.J. Towards evidence-based husbandry for caecilian amphibians: Substrate preference in *Geotrypetes seraphini* (Amphibia: Gymnophiona: Dermophiidae). *Herpetol. Bull.* **2014**, *129*, 15–18.
23. Rendle, M.E.; Tapley, B.; Perkins, M.; Bittencourt-Silva, G.; Gower, D.J.; Wilkinson, M. Itraconazole treatment of *Batrachochytrium dendrobatidis* (Bd) infection in captive caecilians (Amphibia: Gymnophiona) and the first case of Bd in a wild neotropical caecilian. *J. Zoo Aquar. Res.* **2015**, *3*, 137–140.
24. Flach, E.J.; Feltrer, Y.; Gower, D.J.; Jayson, S.; Michaels, C.J.; Pocknell, A.; Rivers, S.; Perkins, M.; Rendle, M.E.; Stidworthy, M.F.; et al. Postmortem findings in eight species of captive caecilian (Amphibia: Gymnophiona) over a ten-year period. *J. Zoo Wildl. Med.* **2020**, *50*, 79–890. [CrossRef]
25. Whatley, C.; Tapley, B.; Michaels, C.J. Substrate preference in the fossorial caecilian *Microcaecila unicolor* (Amphibia: Gymnophiona, Siphonopidae). *Herpetol. Bull.* **2020**, *152*, 18–20. [CrossRef]
26. IUCN SSC Amphibian Specialist Group. *Herpele squalostoma*. The IUCN Red List of Threatened Species 2018: E.T59565A16958011. Available online: https://doi.org/10.2305/IUCN.UK.2018-1.RLTS.T59565A16958011.en (accessed on 1 August 2021).
27. Wilkinson, M.; Sherratt, E.; Starace, F.; Gower, D.J. A new species of skin-feeding caecilian and the first report of reproductive mode in *Microcaecilia* (Amphibia: Gymnophiona: Siphonopidae). *PLoS ONE* **2013**, *8*, e57756. [CrossRef] [PubMed]
28. Kouete, M.T.; Ndeme, E.S.; Gower, D.J. Further observations of reproduction and confirmation of oviparity in *Herpele squalostoma* (Stutchbury, 1836) (Amphibia: Gymnophiona: Herpelidae). *Herpetol. Notes* **2013**, *6*, 583–586.
29. Michaels, C.J.; Gini, B.F.; Preziosi, R.F. The importance of natural history and species-specific approaches in amphibian ex-situ conservation. *Herpetol. J.* **2014**, *24*, 135–145.
30. Channing, A.; Rödel, M.O. *Field Guide to the Frogs & Other Amphibians of Africa*; Penguin Random House: Cape Town, South Africa, 2019.
31. Kouete, M.T.; Wilkinson, M.; Gower, D.J. First reproductive observations for Herpele Peters, 1880 (Amphibia: Gymnophiona: Herpelidae): Evidence of extended parental care and maternal dermatophagy in H. squalostoma (Stutchbury, 1836). *Int. Sch. Res. Notices* **2012**, *2012*, 1–8. [CrossRef]
32. Kupfer, A.; Maxwell, E.; Reinhard, S.; Kuehnel, S. The evolution of parental investment in caecilian amphibians: A comparative approach. *Biol. J. Linn. Soc.* **2016**, *119*, 4–14. [CrossRef]
33. Measey, G.J.; Gower, D.J.; Oommen, O.V.; Wilkinson, M. A subterranean generalist predator: Diet of the soil-dwelling caecilian *Gegeneophis ramaswamii* (Amphibia; Gymnophiona; Caeciliidae) in southern India. *C. R. Biol.* **2004**, *327*, 65–76. [CrossRef] [PubMed]
34. Kouete, M.T.; Blackburn, D.C. Dietary partitioning in two co-occurring caecilian species (*Geotrypetes seraphini* and *Herpele squalostoma*) in Central Africa. *Integr. Org. Biol.* **2020**, *2*, obz035. [CrossRef]
35. ImageJ. Image Processing and Analysis in Java. Available online: https://imagej.nih.gov/ij/ (accessed on 16 August 2021).
36. Bolker, B.M.; Brooks, M.E.; Clark, C.J.; Geange, S.W.; Poulsen, J.R.; Stevens, M.H.H.; White, J.S.S. Generalized linear mixed models: A practical guide for ecology and evolution. *Trends Ecol. Evol.* **2009**, *24*, 127–135. [CrossRef]
37. Millar, R.B.; Anderson, M.J. Remedies for pseudoreplication. *Fish. Res.* **2004**, *70*, 397–407. [CrossRef]
38. Johnson, P.C. Extension of Nakagawa & Schielzeth's R^2GLMM to random slopes models. *Methods Ecol. Evol.* **2014**, *5*, 944–946. [PubMed]

39. Raudenbush, S.W.; Yang, M.L.; Yosef, M. Maximum likelihood for generalized linear models with nested random effects via high-order, multivariate Laplace approximation. *J. Comput. Graph. Stat.* **2000**, *9*, 141–157.
40. Bates, D.; Mächler, M.; Bolker, B.; Walker, S. Fitting Linear Mixed-Effects Models Using lme4. *J. Stat. Softw.* **2015**, *67*, 1–48. [CrossRef]
41. Barton, K. Mu-MIn: Multi-model inference. R Package Version 1.43.17. 2020. Available online: http://R-Forge.R-project.org/projects/mumin/ (accessed on 22 September 2021).
42. R Core Team. *R: A Language and Environment for Statistical Computing*; R Foundation for Statistical Computing: Vienna, Austria, 2021. Available online: http://www.R-project.org/ (accessed on 22 September 2021).
43. Rouse, J.D.; Bishop, C.A.; Struger, J. Nitrogen pollution: An assessment of its threat to amphibian survival. *Environ. Health Perspect.* **1999**, *107*, 799–803. [CrossRef]
44. Lentini, A.M. 34. Husbandry and care of amphibians. In *Zookeeping*; University of Chicago Press: Chicago, IL, USA, 2013; pp. 335–346.
45. Densmore, C.L.; Green, D.E. Diseases of amphibians. *ILAR J.* **2007**, *48*, 235–254. [CrossRef] [PubMed]
46. Odum, R.A.; Zippel, K.C. Amphibian water quality: Approaches to an essential environmental parameter. *Int. Zoo Yearb.* **2008**, *42*, 40–52. [CrossRef]
47. Voorhies, M.R. Vertebrate burrows. In *The Study of Trace Fossils*; Springer: Berlin/Heidelberg, Germany, 1975; pp. 325–350.
48. Gower, D.J.; Loader, S.P.; Moncrieff, C.B.; Wilkinson, M. Niche separation and comparative abundance of *Boulengerula boulengeri and Scolecomorphus vittatus* (Amphibia: Gymnophiona) in an East Usambara forest, Tanzania. *Afr. J. Herpetol.* **2004**, *53*, 183–190. [CrossRef]
49. Measey, G.J. Are caecilians rare? An east African perspective. *J. East Afr. Nat. Hist.* **2004**, *93*, 1–21. [CrossRef]
50. Kupfer, A.; Wilkinson, M.; Gower, D.J.; Müller, H.; Jehle, R. Care and parentage in a skin-feeding caecilian amphibian. *J. Exp. Zool. Part A Ecol. Genet. Physiol.* **2008**, *309*, 460–467. [CrossRef]
51. Kupfer, A.; Nabhitabhata, J.; Himstedt, W. Life history of amphibians in the seasonal tropics: Habitat, community and population ecology of a caecilian (genus *Ichthyophis*). *J. Zool.* **2005**, *266*, 237–247. [CrossRef]
52. Michaels, C.J.; Preziosi, R.F. Clinical and naturalistic substrates differ in bacterial communities and in their effects on skin microbiota in captive fire salamanders (*Salamandra salamandra*). *Herpetol. Bull.* **2020**, *151*, 10–16. [CrossRef]

Journal of
Zoological and Botanical Gardens

Article

Evaluating Environmental Enrichment Methods in Three Zoo-Housed *Varanidae* Lizard Species

James O. Waterman [1], Rachel McNally [1,2], Daniel Harrold [1], Matthew Cook [1], Gerardo Garcia [1], Andrea L. Fidgett [1,3] and Lisa Holmes [1,*]

[1] Chester Zoo, Upton-by-Chester, Chester CH2 1LH, UK; j.waterman@chesterzoo.org (J.O.W.); rachelmcnally@outlook.com (R.M.); dh16450.2016@my.bristol.ac.uk (D.H.); m.cook@chesterzoo.org (M.C.); g.garcia@chesterzoo.org (G.G.); AFIdgett@sdzwa.org (A.L.F.)
[2] Faculty of Biology, Medicine and Health, University of Manchester, Manchester M13 9PL, UK
[3] San Diego Zoo, Zoo Wildlife Alliance, San Diego, CA 92101, USA
* Correspondence: l.holmes@chesterzoo.org

Abstract: Environmental enrichment has been shown to enhance the behavioural repertoire and reduce the occurrence of abnormal behaviours, particularly in zoo-housed mammals. However, evidence of its effectiveness in reptiles is lacking. Previously, it was believed that reptiles lacked the cognitive sophistication to benefit from enrichment provision, but studies have demonstrated instances of improved longevity, physical condition and problem-solving behaviour as a result of enhancing husbandry routines. In this study, we evaluate the effectiveness of food- and scent-based enrichment for three varanid species (Komodo dragon, emerald tree monitor lizard and crocodile monitor). Scent piles, scent trails and hanging feeders resulted in a significant increase in exploratory behaviour, with engagement diminishing ≤330 min post provision. The provision of food- versus scent-based enrichment did not result in differences in enrichment engagement across the three species, suggesting that scent is just as effective in increasing natural behaviours. Enhancing the environment in which zoo animals reside is important for their health and wellbeing and also provides visitors with the opportunity to observe naturalistic behaviours. For little known and understudied species such as varanids, evidence of successful (and even unsuccessful) husbandry and management practice is vital for advancing best practice in the zoo industry.

Keywords: behavior; environmental enrichment; evidence-based; husbandry; reptile; lizard; welfare; zoo

Citation: Waterman, J.O.; McNally, R.; Harrold, D.; Cook, M.; Garcia, G.; Fidgett, A.L.; Holmes, L. Evaluating Environmental Enrichment Methods in Three Zoo-Housed *Varanidae* Lizard Species. *J. Zool. Bot. Gard.* **2021**, 2, 716–727. https://doi.org/10.3390/jzbg2040051

Academic Editors: Kris Descovich, Caralyn Kemp and Jessica Rendle

Received: 1 November 2021
Accepted: 26 November 2021
Published: 14 December 2021

1. Introduction

Environmental enrichment (referred to as enrichment hereafter) is used to improve the health and welfare of species managed ex situ, one desired outcome of which is the broadening of an individual's behavioural repertoire [1]. Reptile enrichment methods have historically been based on the anecdotal evidence of caregivers, often drawn from experience with a limited group of individuals [2,3]. As such, there is limited information describing the impacts (positive and negative) of these methods [4–7], including the extent to which different types of enrichment affect behaviour, and the longevity of these effects [8–10]. However responsible, modern collections require robust, quantitative evidence on which to base husbandry decisions; despite an increasing focus on herptile enrichment, this is still lacking for reptiles [1,3,6].

Two key points may account for this gap in our knowledge: (1) evaluating the welfare of reptiles is challenging [6] and/or (2) the cognitive sophistication of non-avian reptiles is often under-estimated (particularly compared to that of mammals and birds [8,11,12]). However, there is clear evidence that many captive squamates provided with enrichment display unexpected problem-solving skills, enhanced behavioural development and plasticity, and reduced stereotypies, as well as greater longevity, increased breeding success, and improved body condition [8,13–17].

Here, we examine the behavioural responses (exploratory behavior and engagement with enrichment objects) to enrichment of three Southeast Asian varanid species held at Chester Zoo, UK: Komodo dragons (*Varanus komodoensis*), emerald tree monitors (*V. prasinus*) and crocodile monitors (*V. salvadorii*). Varanid (monitor) lizards (*Varanus*; Merrem, 1820) are endemic to a variety of habitats in Afro-Eurasia [5] and common within zoological collections. Komodo dragons are largely terrestrial, inhabiting woodland and dry savannah habitats of the Eastern Indonesian islands. Emerald tree monitors and crocodile monitors are largely arboreal, inhabiting rainforests and mangroves on Papua [18]. All three species are predominantly carnivorous, intelligent, occupy large territories and have a high metabolic rate compared to other reptiles [16,18–20].

There are five main ways in which environmental enrichment may be provided: by (1) creating and managing a dynamic habitat, (2) encouraging social interactions between individuals, (3) encouraging foraging, (4) introducing novel objects and (5) training [21]. These methods aim to increase behavioural diversity, reduce abnormal behaviours, increase the range of natural behaviours demonstrated, increase the positive use of the environment, and increase the animal's ability to cope with challenges in a more natural way [22]. We focus here on whether, and to what extent, the provision of a variety of enrichment items encourages exploratory foraging behaviour and/or direct item engagement (see Table 1 for descriptions of enrichment items and Table 2 for behavioural definitions). The study individuals are routinely provided with (predominantly food-based) enrichment, but no long-term monitoring of its efficacy has been carried out. Additionally, because olfactory stimuli play an important role in foraging, mating and social interactions in these species [18], a mixture of food- and scent-based enrichment items was used for this investigation.

Table 1. Description of enrichment conditions and the species to which they were presented.

Enrichment Condition	Description	Species Sampled
Control	Keeper entered enclosure as per normal husbandry routine for two minutes.	*V. komodoensis* *V. prasinus* *V. salvadorii*
Furnishings	Bedding and enclosure furniture from four mammalian exhibits: Congo buffalo (*Syncerus caffer nanus*) bedding; red river hog (*Potamochoerus porcus*) browse logs; white-faced saki monkey (*Pithecia pithecia*) enclosure logs; mixed bedding from Bovidae species.	*V. komodoensis* *V. salvadorii*
Food (ground)	Hollowed log feeders filled with black crickets (*Gryllus* sp.) on enclosure floor.	*V. prasinus*
Food (suspended)	Hollowed log feeders filled with black crickets (*Gryllus* sp.) suspended on enclosure furniture.	*V. prasinus*
Scent (trail—food)	Blended food items (quail eggs, quail feathers, chicken eggs) spread throughout the enclosure. Food items (quail meat, day-old chickens) were also placed along the scent trail.	*V. salvadorii*
Scent (trail)	Liquids spread throughout the enclosure (blood–water solution, fish defrosting water, blended pinkie mice, blended-quail eggs and feathers).	*V. komodoensis* *V. prasinus* *V. salvadorii*
Scent (pile)	A blood–water solution spread throughout the exhibit (within leaf/litter, on logs, buried in substrate surface, as a frozen solution on ground).	*V. komodoensis*

Table 2. Varanid lizard ethogram.

Behaviour	Description
Bask	Individual stationary underneath a heat/UV lamp for a minimum of five seconds.
Rest	Individual stationary (not under a heat/UV lamp) for a minimum of five seconds.
Explore	Relaxed interest/awareness in proximate or novel objects, relaxed visual explorations. Calm chemical sampling of surrounding, e.g., smelling or tasting objects or air (tongue-flicking). Individual moves more than half a body length from its starting position.
Feed	Consumption of food items, including holding food in mouth, chewing and swallowing. Feeding was considered finished after swallowing had stopped.
Social	Touching, vocalising and/or signalling to a conspecific.
Enrichment engagement (interest)	Rapid chemical sampling of surrounding, e.g., smelling or tasting objects or air (tongue-flicking). Individual moves (more than half a body length from its starting position) directly towards and/or stares directly at enrichment item.
Enrichment engagement (use)	Direct manipulation of enrichment item, including attempts to reach the item and/or active following of scent trails. Where live food was presented, included chasing food items.

We predicted (P.1) that the provision of enrichment would increase exploratory behaviour, compared to control trials and (P.2) that the magnitude of any changes in exploratory behaviour would differ by enrichment type. We also predicted (P.3) that engagement with enrichment would differ by item type and (P.4) that engagement would diminish over time as the stimulating effect of novelty wore off. Finally, we predicted (P.5) that individuals would engage with food-based enrichment items for longer than with scent-based items. We made no predictions about species-specific responses to enrichment or enrichment type, because the data were too sparse (see Section 2.3) to support the inclusion of meaningful three-way interactions. However, we include raw data plots to highlight species-specific responses to enrichment provision and/or type.

2. Materials and Methods

2.1. Study Individuals and Housing

Five individuals from three species were studied: *V. komodoensis* (one male and one female, both 4 years old). During the enrichment trials, the male was housed in an on-show mixed-species exhibit (8.0 × 14.0 × 11.5 m) with Java sparrows (*Lonchura oryzivora*) with an average ambient temperature of 26–28 °C. The female was housed singly in an off-show exhibit (4.0 × 3.3 × 2.5 m) with an ambient temperature range of 22–28 °C and infra-red heaters providing basking spots with a temperature range of 27–44 °C; *V. prasinus* (one male and one female, 15 and 8 years old, respectively) was housed in a single-species enclosure (2.0 × 1.5 × 2.5m) with an ambient temperature range of 20–28 °C, a radiant panel heater and 160 w solar raptor spot lamps to provide basking spots with a temperature range of 30–35 °C; *V. salvadorii* (a single female, 11 years old) was housed in a single-species enclosure (5.5 × 2.0 × 4.0 m) with an ambient temperature range of 21–27 °C, a ceramic panel and infra-red heaters providing basking spots with a temperature range of 33–35 °C. All individuals were assessed to be in good clinical health prior to and throughout sampling.

2.2. Testing Protocol

Following a three-week pilot study in December 2014, the study individuals were presented with a randomized series of six enrichment conditions (Table 1), plus a control, over the course of 16 weeks (8 January 2015 through 6 May 2015). Different enrichment conditions were only presented once to each species, but not all species were presented with all possible enrichment types (see Table 1); the control was presented four times to each species. Once a week, each species was observed on one day, for a total of 80 min, split into four 20 min blocks: (1) pre-enrichment, (2) during-enrichment, (3) post-enrichment

1, and (4) post-enrichment 2. Blocks one and two were contiguous, spanning the 40 min prior to and following the introduction of the enrichment/control item (at approximately 1100 h). To examine the longevity of the enrichment effect, blocks 3 and 4 began at random times, 30–150 min and 151–330 min, respectively, after the introduction of the enrichment/control. Control trials were time-matched with enrichment trials to control for potential diel effects. Because captive animals often respond to known keepers and/or their distinctive uniformed appearance, the observer (N = 1) wore 'normal' clothing and carried out observations from the public viewing windows.

2.3. Data Collection

Following an adapted ethogram [23,24] (Table 2), behavioural data were collected (with pen and paper) using continuous all-occurrences focal animal sampling [25]. This yielded a total of 94 h and 24 min of data over 47 observation days and 71 observation sessions.

2.4. Data Preparation

To prepare our data for analysis, we removed any (20 min) observation blocks in which the study individuals were out of sight for the entire duration (N = 24). In order to compare the effect of food- and scent-based enrichment, we collapsed enrichment conditions into food-based (Food (ground), Food (suspended), Scent (trail–food)) and scent-based (Furnishings, Scent (trail), Scent (pile)). For our response variables, we calculated the proportion (expressed as a percentage throughout) of observation time individuals spent in exploratory and enrichment engagement behaviour (see Table 2). Beta regression is the most appropriate way to model proportion data (described in Section 2.5), but these models cannot handle values of exactly zero or one. Therefore, we compressed the range of the data according to the following equation: $p^{\circ} = (p(n-1) + 1/2)/n$, where p is the original proportion, and n is the sample size [26].

2.5. Statistical Analysis

All analyses were carried out using the software R, version 4.1.0 [27]. We used the package 'glmmTMB' [28] to fit three Beta GLMMs (Generalized Linear Mixed Models) with logit links. The Beta distribution is typically used to model continuous proportion data, and the logit link function ensures positive fitted values that range from 0 to 1 [29]. Proportion data are by definition limited to numerical values between, and including, 0 and 1, and their variance is rarely constant across the range of the predictor(s). As such, they typically violate two important assumptions of standard statistical techniques (normality of errors and constant variance) [30]. This makes analysis using familiar techniques (such as linear regression and ANOVA and their extensions) inappropriate. Transformations are often applied to proportion data, so that linear models can be used [31], but these result in biased estimates and difficulties in interpretation. However, after the appropriate adjustment (detailed above), beta regression provides a robust and easily interpretable approach to modelling proportion data.

We used a full model approach throughout, and model fit and assumptions were verified by plotting residuals versus fitted values with the package 'DHARMa' [32]. We determined the significance of the fixed effects using likelihood ratio tests. We fitted full and restricted models (models in which the parameter of interest, the fixed effect, are withheld, i.e., fixed to 0) and based test statistics on comparisons of the full model with the restricted models. The significance of the likelihood ratio test statistic is calculated using a chi-squared distribution with the appropriate degrees of freedom. Post-hoc tests were carried out using Tukey's HSD (honestly significant difference) tests, with the package 'emmeans' [33]. All statistical tests were two-tailed with α set to 0.05.

All models included the same control variables: sex (factor with two levels: female, male) and days since last feed (continuous numeric variable: range 0–31). Days since last feed was scaled and centred prior to analysis. To incorporate the dependency among observations of the same individuals, of the same species, across the four observation blocks, all models included the same random effects structure; trial nested in individual, nested in species was used as a random intercept.

2.5.1. Model 1: Effect of Enrichment on Exploratory Behaviour

To model the proportion of time that individuals spent in exploratory behaviour as a function of enrichment provision (P.1) and type (P.2), we included the interaction between the fixed-covariates observation block (factor with two levels: pre-enrichment, during enrichment) and enrichment type (factor with 7 levels: control, furnishings, food (ground), food (suspended), scent (trail–food), scent (trail), scent (pile)).

2.5.2. Model 2: Effect of Enrichment Type on Engagement Time and Longevity

To model the proportion of time that individuals spent engaging with enrichment items as a function of enrichment type (P.3) and to examine the longevity of this effect (P.5), we included the interaction between the fixed covariates observation block (factor with three levels: during enrichment, post-enrichment 1, post-enrichment 2) and enrichment type (this time excluding the control condition).

2.5.3. Model 3: Effect of Food- vs. Scent-Based Enrichment on Engagement Longevity

To model the effect of food- vs. scent-based enrichment over time (i.e., to compare the longevity of each) we included the interaction between the fixed-covariates observation block (factor with three levels: during enrichment, post-enrichment 1, post-enrichment 2) and enrichment type (factor with two levels: food, scent).

3. Results

3.1. *Effect of Enrichment on Exploratory Behaviour*

As predicted (P.1 and P.2), the provision of enrichment increased the exploratory behaviour, and the magnitude of this increase differed significantly by enrichment type (likelihood ratio test (LRT); $\chi^2(6) = 16.844$, $p = 0.001$: Figure 1a). Post-hoc testing revealed that the food (hanging) (Tukey HSD; P = 0.008), scent (pile) (Tukey HSD; $p < 0.001$), and scent (trail) (Tukey HSD; $p < 0.0001$) conditions were all associated with significant increases in exploratory behaviour (Figure 1a). However, no significant differences in exploratory time were observed between the pre- and during-enrichment blocks in response to the control (Tukey HSD; $p = 0.877$), furnish (Tukey HSD; $p = 0.110$), food (ground) (Tukey HSD; $p = 0.211$), or scent (trail–food) (Tukey HSD; $p = 0.666$) conditions. Furthermore, throughout the during-enrichment block, the food (hanging), scent (pile), and scent (trail) conditions were all associated with significant increases in exploratory behaviour compared to the control conditions (Tukey HSD; $p = 0.008$, $p = 0.004$, $p = 0.001$, respectively: Figure 1a). Species-specific responses are shown in Figure 1b.

3.2. *Effect of Enrichment Type on Engagement Time and Longevity*

As predicted (P.4), engagement with enrichment items diminished over time (LRT; $\chi^2(2) = 32.667$, $p < 0.0001$: Figure 2a). Engagement with enrichment was significantly lower in the post-enrichment 2 block than the during- and post-enrichment 1 blocks (Tukey HSD; $p < 0.001$, $p = 0.049$, respectively: Figure 2a). However, contrary to P.3, we found no significant effect of enrichment type on engagement longevity during or after enrichment provision (i.e., engagement with enrichment diminished more or less equally over time, regardless of enrichment type (LRT; $\chi^2(10) = 8.719$, $p = 0.559$: Figure 2). Species-specific responses are shown in Figure 2b.

3.3. *Effect of Food- vs. Scent-Based Enrichment on Engagement Longevity*

Contrary to P.5, we found no significant difference in the longevity of engagement associated with food- vs. scent-based enrichment items (LRT; $\chi^2(2) = 1.599$, $p = 0.452$: Figure 3a). Engagement with both enrichment types diminished, as described in Section 3.2 above. Species-specific responses are shown in Figure 3b.

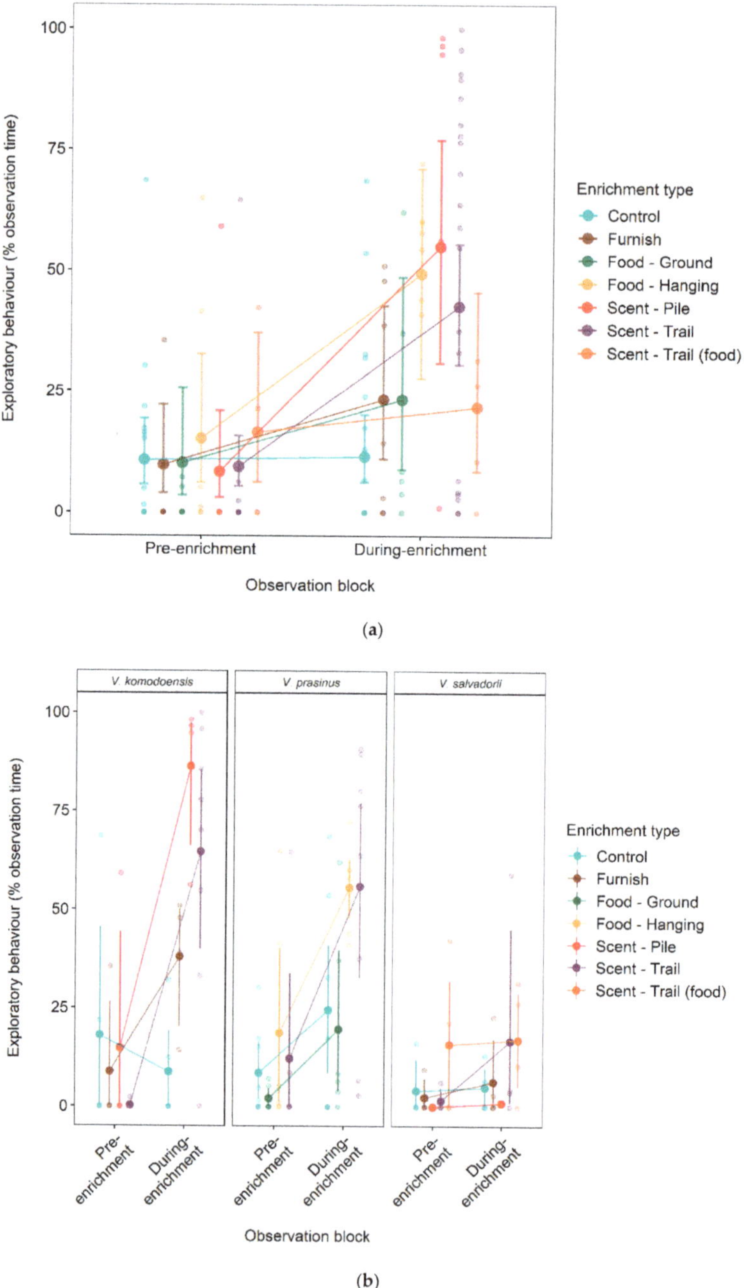

Figure 1. (**a**) Effect of enrichment provision on exploratory behaviour in three varanid lizard species. Large points and error bars represent predicted means ± standard error from a Beta GLMM. Small points represent individual trials (raw data). (**b**) Species-specific effect of enrichment provision on exploratory behaviour in three varanid lizard species. Large points and error bars represent raw data means and 95% non-parametric bootstrap (10,000 samples) confidence intervals. Small points represent individual trials (raw data).

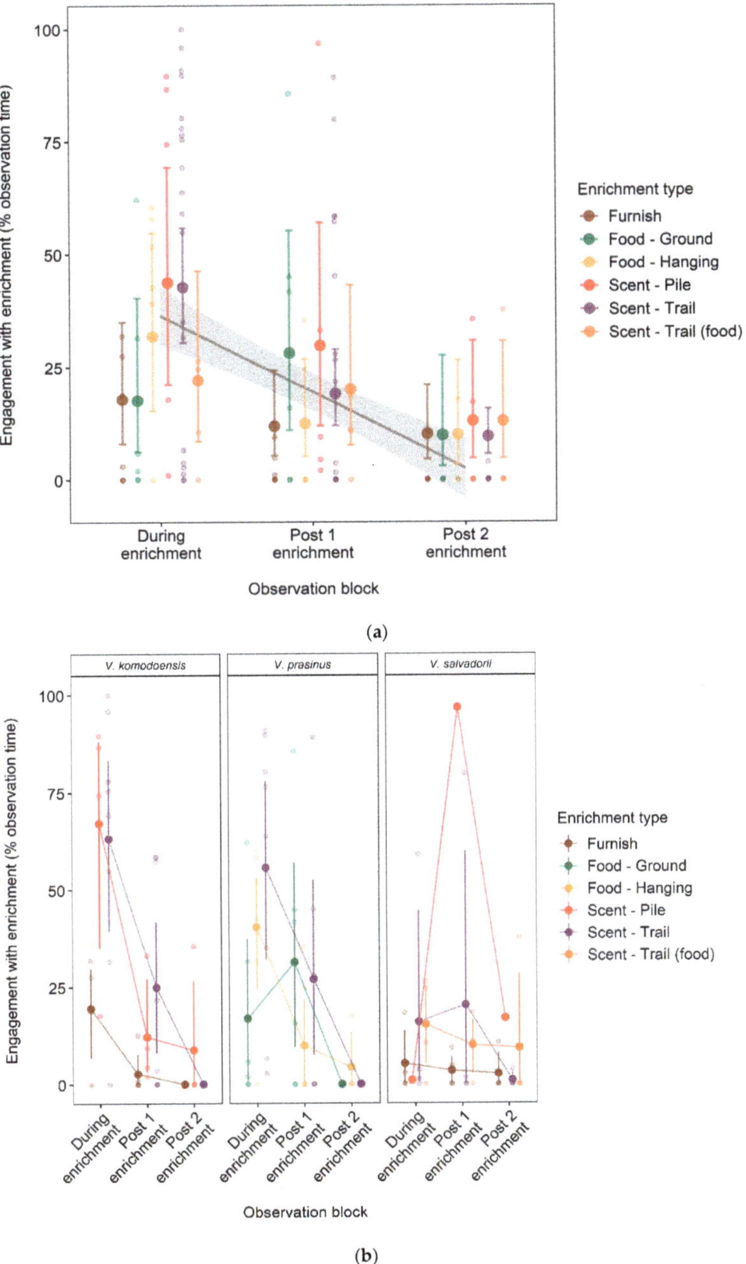

Figure 2. (**a**) Engagement with enrichment items over time in three varanid lizard species. Large points and error bars represent predicted means ± standard error from a Beta GLMM. Small points represent individual trials (raw data). Line of best fit through raw data to indicate trend. (**b**) Species-specific engagement with enrichment items in three varanid lizard species. Large points and error bars represent raw data means and 95% non-parametric bootstrap (10,000 samples) confidence intervals. Small points represent individual trials (raw data).

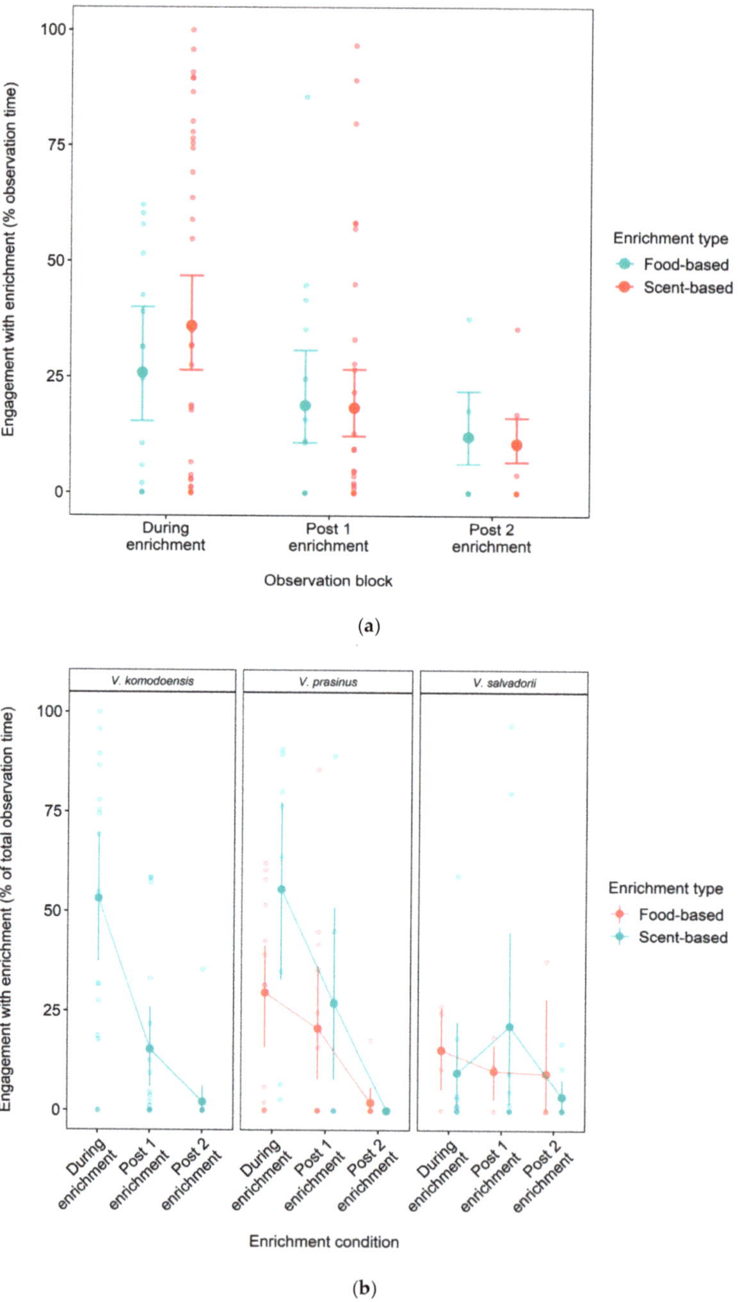

Figure 3. (**a**) Engagement with food- and scent-based enrichment items over time in three varanid lizard species. Error bars represent the standard error of the mean. Raw data plotted as small circles behind the main plot. (**b**) Species-specific engagement with food- and scent-based enrichment items in three varanid lizard species. Large points and error bars represent raw data means and 95% non-parametric bootstrap (10,000 samples) confidence intervals. Small points represent individual trials (raw data).

4. Discussion

This study examined the responses of three captive varanid lizard species to the provision of enrichment by comparing their exploratory behaviour and engagement with six different types of enrichment items and a control. Lizards exhibited significant increases in exploratory behaviour in response to hanging feeders, scent piles and scent trails. Contrary to our predictions, we found that engagement with these different enrichment types diminished more or less equally over time, returning to baseline levels by the post-enrichment 2 block (151–330 min after introduction). This finding held true when enrichment types were binned into food- and scent-based categories, i.e., no significant differences in the longevity of the enrichment effect were observed. Considered together, these results confirm that the provision of enrichment can be effective in promoting explorative behaviour and engagement in captive varanids. Specifically, our findings indicate that (a) not all enrichment types elicit similar behavioural changes, (b) that scent-based enrichment appears to provide the most effective cross-species stimulus and (c) that these effects can persist for up to 2.5 h.

With a few exceptions [13,14], there are still very few published studies about the effects of enrichment on non-avian reptiles. However, our findings contribute to, and are largely consistent with, those that exist: for example, food-based (problem tube) and scent-based (conspecific male scent) enrichment resulted in significant increases in tongue licking and exploratory activity in juvenile black-throated monitor lizards (*V. albigularis albigularis*) [16] and male brown wall lizards (*Podarcis liolepis*), [34] respectively. Similarly, a combination of sensory and physical enrichment increased the exploratory, focused swimming behaviour in sea turtles (*Caretta caretta* and *Chelonia mydas*), whilst reducing the occurrence of stereotypical behaviours [35]. The introduction of novel enrichment items was also shown to elicit play behaviour in Komodo dragons [36,37] and to facilitate training in crocodile monitors [38]; although we did not examine these behaviours in our study, our findings similarly support the provision of a complex captive environment to physically and cognitively stimulate reptiles [13].

More specifically, our results confirm that the most effective enrichment types may be those that mimic natural challenges routinely faced by lizards in the wild. Many reptiles rely primarily on chemical/olfactory senses to communicate and explore the environment [39,40], and this is particularly true of varanids [41,42]. Varanids vary enormously in body size (length, including tail: <300 mm to 3 m) and occupy a wide range of habitats and ecological niches including terrestrial predator/scavenger (*V. gigantius, komodoensis*), arboreal (*V. prasinus, gilleni, timorensis, tristis*), aquatic (*V. mertensi, salvator, niloticus*) and small insectivore (*V. brevicauda*) [43,44]. However, they are the only group of lizards that use the tongue exclusively for sensory function: unlike for all other lizards, it plays no part in food ingestion [42]. Indeed, although some debate exists, comparative studies suggest that the morphological specializations of varanid tongues (long, narrow, forked and deeply incised) relate to protrusability and sensory function [42]. Hence, the general agreement (yet to be rigorously tested) that the chemical/olfactory senses of varanids exceed those of other lizards. Komodo dragons can detect carrion from nearly 8 km away by virtue of airborne chemosensory signals and are reported to climb ridgelines to sniff/sample the wind for carrion odours over a large area [41]. Similarly, emerald tree monitors can detect the scent of prey hidden entirely inside tree branches [45].

Clearly, the varanid olfactory system is important in a wide range of feeding, social, territorial and courtship behaviours [39,41–43,46]. This may explain why the scent-based enrichment items in this study were consistently the most successful in promoting exploration and engagement, across the three varanid species: the provision of scent-based enrichment stimulates what is likely the most important, sensitive and evolutionarily conserved sensory system in this taxa [39,47]. With respect to food-based enrichment, it is harder to draw meaningful conclusions from this study. This is largely because *V. prasinus* was the only species to be presented with actual food-based enrichment: *V. salvadorii* was presented with a food-based scent trail, and *V. komodoensis* with only scent-based enrichment (Figure 2b). This highlights an important (solvable) shortcoming of this, and many other, captive enrichment studies, i.e., the use of unbalanced experimental designs. For example, it is important to note that our (between-species) raw data indicate a delayed response to scent-based enrichment in *V. salvadorii*, in contrast to the immediate responses of *V. komodoensis* and *V. prasinus* (Figures 2b and 3b). This (potential) discrepancy may be explained by the fact that we only sampled one individual of this species and that she was 11 years old, technically considered geriatric [7]. The delayed exploratory response may simply be a result of this individual struggling/declining to move at the same speed as other younger study individuals.

Small sample sizes, low replication and unbalanced designs are common problems in zoo-based studies, including enrichment work [8,48]. Our study clearly suffers from these issues, and while the results should therefore be interpreted with caution, we have compensated (as far as possible) by using the appropriate statistical methods. This has allowed us to analyse the pooled responses of three similar species to an unbalanced enrichment design, whilst still accounting for the similarities and differences between individuals and species (i.e., mixed models with random effect terms). The inclusion of more individuals is rarely a simple matter in zoo-based studies; however, future studies should (and can) insist on pre-determined balanced enrichment protocols, especially if we intend to extrapolate any findings to other collections and/or species.

Here, we have shown that enrichment in the form of hanging feeders, scent piles and scent trails effectively stimulate exploratory and engagement behaviours in three captive varanid species: *V. komodoensis*, *V. prasinus* and *V. salvadorii*. We also present preliminary evidence that scent-based enrichment may be particularly effective in promoting these behaviours, alongside an ecologically valid explanation of why this may be so. Varanids are clearly complex and intelligent animals, and enrichment should be designed to physically and cognitively stimulate them in ways that mimic the natural challenges they would otherwise face in the wild. This is particularly important for the efficacy of ex situ reintroduction of threatened species, especially as more lesser-known taxa are brought into captivity as assurance populations [3]. It is also important for the wellbeing of long-term captive individuals. Finally, effective enrichment can have positively influence on visitor perception, which may in turn promote in situ projects [21]. In sum, it is essential that further enrichment studies implement robust, well-balanced protocols that monitor and evaluate the impact of a range of randomly presented enrichment items over time. In doing so, we can gather a comparable, testable body of data that will allow us to improve the wellbeing of the wide variety of reptiles in captivity.

Author Contributions: Conceptualization, G.G., A.L.F. and L.H.; methodology, R.M., G.G., M.C., A.L.F. and L.H.; formal analysis, J.O.W.; investigation, R.M.; resources, G.G. and M.C.; data curation, R.M., L.H., J.O.W.; writing—original draft preparation, J.O.W., R.M. and D.H.; writing—review and editing, R.M., A.L.F. and L.H.; visualization, J.O.W.; supervision, A.L.F. and L.H.; project administration, R.M., A.L.F. and L.H. All authors have read and agreed to the published version of the manuscript.

Funding: This research received no external funding.

Institutional Review Board Statement: This study was approved by Chester Zoo's Scientific Committee (ref 2014.39 on 17 December 2014) and conducted in accordance with Chester Zoo's Animal Research Ethics Framework. The introduction of enrichment items mentioned here is part of normal husbandry routine, and individual animals were never food- or water-deprived.

Informed Consent Statement: Not applicable.

Data Availability Statement: Data are available upon request.

Acknowledgments: Thank you to Chester Zoo's Reptile Team for their assistance throughout planning and data collection.

Conflicts of Interest: The authors declare no conflict of interest.

References

1. Bashaw, M.J.; Gibson, M.D.; Schowe, D.M.; Kucher, A.S. Does enrichment improve reptile welfare? Leopard geckos (*Eublepharis macularius*) respond to five types of environmental enrichment. *Appl. Anim. Behav. Sci.* **2016**, *184*, 150–160. [CrossRef]
2. Arbuckle, K. Folklore husbandry and a philosophical model for the design of captive management regimes. *Herpetol. Rev.* **2013**, *44*, 448–452.
3. Januszczak, I.S.; Bryant, Z.; Tapley, B.; Gill, I.; Harding, L.; Michaels, C.J. Is behavioural enrichment always a success? Comparing food presentation strategies in an insectivorous lizard (*Plica plica*). *Appl. Anim. Behav. Sci.* **2016**, *183*, 95–103. [CrossRef]
4. Horn, H.-G.; Visser, G.J. Review of reproduction of monitor lizards *Varanus* spp. in captivity II. *Int. Zoo Yearb.* **1997**, *35*, 227–246. [CrossRef]
5. Horn, H.-G.; Visser, G.J. Review of reproduction of Monitor lizards *Varanm* spp. in captivity. *Int. Zoo Yearb.* **1989**, *28*, 140–150. [CrossRef]
6. Mendyk, R.W. Life expectancy and longevity of varanid lizards (Reptilia:Squamata:Varanidae) in North American zoos. *Zoo Biol.* **2015**, *34*, 139–152. [CrossRef]
7. Mendyk, R.W.; Newton, A.L.; Baumer, M. A Retrospective Study of Mortality in Varanid Lizards (Reptilia:Squamata:Varanidae) at the Bronx Zoo: Implications for Husbandry and Reproductive Management in Zoos. *Zoo Biol.* **2013**, *32*, 152–162. [CrossRef]
8. Burghardt, G.M. Environmental enrichment and cognitive complexity in reptiles and amphibians: Concepts, review, and implications for captive populations. *Appl. Anim. Behav. Sci.* **2013**, *147*, 286–298. [CrossRef]
9. Kuczaj, S.; Lacinak, T.; Fad, O.; Trone, M.; Solangi, M.; Ramos, J. Keeping environmental enrichment enriching. *Int. J. Comp. Psychol.* **2002**, *15*, 127–137.
10. Tarou, L.R.; Bashaw, M.J. Maximizing the effectiveness of environmental enrichment: Suggestions from the experimental analysis of behavior. *Appl. Anim. Behav. Sci.* **2007**, *102*, 189–204. [CrossRef]
11. Michaels, C.J.; Downie, J.R.; Campbell-Palmer, R. The importance of enrichment for advancing amphibian welfare and conservation goals: A review of a neglected topic. *Amphib. Reptile Conserv.* **2014**, *8*, 7–23.
12. Rosier, R.L.; Langkilde, T. Does environmental enrichment really matter? A case study using the eastern fence lizard, Sceloporus undulatus. *Appl. Anim. Behav. Sci.* **2011**, *131*, 71–76. [CrossRef]
13. Almli, L.M.; Burghardt, G.M. Environmental Enrichment Alters the Behavioral Profile of Ratsnakes (Elaphe). *J. Appl. Anim. Welf. Sci.* **2006**, *9*, 85–109. [CrossRef]
14. Burghardt, G.M.; Ward, B.; Rosscoe, R. Problem of reptile play: Environmental enrichment and play behavior in a captive Nile soft-shelled turtle, Trionyx triunguis. *Zoo Biol.* **1996**, *15*, 223–238. [CrossRef]
15. Doody, J.; Burghardt, G.; Dinets, V. Breaking the Social-Non-social Dichotomy: A Role for Reptiles in Vertebrate Social Behavior Research? *Ethology* **2013**, *119*, 95–103. [CrossRef]
16. Manrod, J.; Hartdegen, R.; Burghardt, G. Rapid solving of a problem apparatus by juvenile black-throated monitor lizards (*Varanus albigularis albigularis*). *Anim. Cogn.* **2008**, *11*, 267–273. [CrossRef]
17. Wilkinson, S.L. Reptile wellness management. *Vet. Clin. Exot. Anim. Pract.* **2015**, *18*, 281–304. [CrossRef]
18. Bennett, D. *Monitor Lizards: Natural History, Biology & Husbandry*; Chimaira Bucnhandelsgesellschaft: Frankfurt, Germany, 1998.
19. McBrayer, L.D.; McBrayer, L.B.; Miles, D.B. (Eds.) *Lizard Ecology*; Cambridge University Press: Cambridge, UK, 2007.
20. Pianka, E.R.; Pianka, E.R.; Vitt, L.J. *Lizards: Windows to the Evolution of Diversity*; University of California Press: Berkeley, CA, USA, 2003; Volume 5.
21. Hurme, K.; Gonzalez, K.; Halvorsen, M.; Foster, B.; Moore, D.; Chepko-Sade, B.D. Environmental Enrichment for Dendrobatid Frogs. *J. Appl. Anim. Welf. Sci.* **2003**, *6*, 285–299. [CrossRef] [PubMed]
22. Young, R.J. *Environmental Enrichment for Captive Animals*; John Wiley & Sons: Chichester, UK, 2013.
23. Warwick, C.; Arena, P.; Lindley, S.; Jessop, M.; Steedman, C. Assessing reptile welfare using behavioural criteria. *Practice* **2013**, *35*, 123–131. [CrossRef]
24. Whittaker, G.; Whittaker, M.; Coe, J. Prototyping Naturalistic Enrichment Features: A Case Study. In Proceedings of the Seventh International Conference on Environmental Enrichment, New York, NY, USA, 31 July–5 August 2005; p. 60.
25. Altmann, J. Observational study of behavior: Sampling methods. *Behaviour* **1974**, *49*, 227–266. [CrossRef] [PubMed]
26. Smithson, M.; Verkuilen, J. A Better Lemon Squeezer? Maximum-Likelihood Regression with BetaDistributed Dependent Variables. *Psychol. Methods* **2006**, *11*, 54. [CrossRef]
27. R Core Team. *R: A Language and Environment for Statistical Computing*; R Foundation for Statistical Computing: Vienna, Austria, 2021.
28. Brooks, M.E.; Kristensen, K.; van Benthem, K.J.; Magnusson, A.; Berg, C.W.; Nielsen, A.; Skaug, H.J.; Machler, M.; Bolker, B.M. glmmTMB Balances Speed and Flexibility Among Packages for Zero-inflated Generalized Linear Mixed Modeling. *R J.* **2017**, *9*, 378–400. [CrossRef]
29. Zuur, A.F.; Ieno, E.N.; Walker, N.; Saveliev, A.A.; Smith, G.M. *Mixed Effects Models and Extensions in Ecology With R*; Springer: New York, NY, USA, 2009; Volume 574.
30. Douma, J.C.; Weedon, J.T. Analysing continuous proportions in ecology and evolution: A practical introduction to beta and Dirichlet regression. *Methods Ecol. Evol.* **2019**, *10*, 1412–1430. [CrossRef]
31. Sokal, R.R.; Rohlf, F.J. *Introduction to Biostatistics*; Dover Publications: Mineola, NY, USA, 2009.

32. Hartig, F. DHARMa: Residual Diagnostics for Hierarchical (Multi-Level/Mixed) Regression Models. 2021. Available online: r-project.org (accessed on 31 October 2021).
33. Lenth, R.V.; Buerkner, P.; Herve, M.; Love, J.; Riebl, H.; Singmann, H. Emmeans: Estimated Marginal Means, Aka Least-Squares Means. 2021. Available online: uni-muenster.de (accessed on 31 October 2021).
34. Londoño, C.; Bartolomé, A.; Carazo, P.; Font, E. Chemosensory enrichment as a simple and effective way to improve the welfare of captive lizards. *Ethology* **2018**, *124*, 674–683. [CrossRef]
35. Therrien, C.L.; Gaster, L.; Cunningham-Smith, P.; Manire, C.A. Experimental evaluation of environmental enrichment of sea turtles. *Zoo Biol.* **2007**, *26*, 407–416. [CrossRef]
36. Burghardt, G.M. *The Genesis of Animal Play: Testing the Limits*; Bradford Books: Cambridge, MA, USA, 2005.
37. Murphy, J.B.; Ciofi, C.; de La Panouse, C.; Walsh, T. *Komodo Dragons: Biology and Conservation*; Smithsonian Institution: Washington, DC, USA, 2015.
38. Camina, Á.; Salinas, N.; Cuevas, J. Husbandry and breeding of the crocodile monitor Varanus salvadorii Peters & Doria, 1878 in captivity. *Biawak* **2013**, *7*, 56–62.
39. Mason, R.T. Reptilian pheromones. *Biol. Reptil.* **1992**, *18*, 114–228.
40. Ord, T.J.; Martins, E.P. Tracing the origins of signal diversity in anole lizards: Phylogenetic approaches to inferring the evolution of complex behaviour. *Anim. Behav.* **2006**, *71*, 1411–1429. [CrossRef]
41. Auffenberg, W. *The Behavioral Ecology of the Komodo Monitor*; University Press of Florida: Gainesville, FL, USA, 1981.
42. Smith, K.K. Morphology and function of the tongue and hyoid apparatus in Varanus (varanidae, lacertilia). *J. Morphol.* **1986**, *187*, 261–287. [CrossRef] [PubMed]
43. Cogger, H. *Reptiles and Amphibians of Australia*; AH and AW Reed: Dunedin, New Zealand, 1975.
44. Ziegler, T.; Schmitz, A.; Koch, A.; Böhme, W. A review of the subgenus Euprepiosaurus of Varanus (Squamata: Varanidae): Morphological and molecular phylogeny, distribution and zoogeography, with an identification key for the members of the *V. indicus* and the *V. prasinus* species groups. *Zootaxa* **2007**, *1472*, e28. [CrossRef]
45. Murphy, J.; Mendyk, R.; Miller, K.; Augustine, L. Tales of Monitor Lizard Tails and Other Perspectives. *Herpetol. Rev.* **2019**, *50*, 178–201.
46. Gaalema, D. Visual Discrimination and Reversal Learning in Rough-Necked Monitor Lizards (*Varanus rudicollis*). *J. Comp. Psychol.* **2011**, *125*, 246–249. [CrossRef] [PubMed]
47. Pepin, D.J. *Natural History of Monitor Lizards (Family Varanidae) with Evidence from Phylogeny, Ecology, Life History and Morphology*; Washington University: St. Louis, MO, USA, 2011.
48. Shyne, A. Meta-analytic review of the effects of enrichment on stereotypic behavior in zoo mammals. *Zoo Biol.* **2006**, *25*, 317–337. [CrossRef]

Journal of
Zoological and Botanical Gardens

Article

Investigating the Effect of Enrichment on the Behavior of Zoo-Housed Southern Ground Hornbills

James Edward Brereton [1,*], Mark Nigel Geoffrey Myhill [2] and James Ali Shora [3]

1 Zoo and Animal Studies, Higher Education, University Centre Sparsholt, Westley Lane, Sparsholt, Winchester SO21 2NF, UK
2 Beale Wildlife Park, Lower Basildon, Pangbourne, Reading RG8 9NW, UK; markmyhill@bealepark.org.uk
3 Battersea Park Children's Zoo, Battersea Park, Chelsea Bridge, London SW11 4NJ, UK; jameshora@gmail.com
* Correspondence: James.Brereton@sparsholt.ac.uk

Abstract: Enrichment is essential for the welfare of many zoo-housed animals, yet the value of enrichment is not well understood for all taxa. As an intelligent, long-lived species, the southern ground hornbill (*Bucorvus leadbeateri*) is a good model for enrichment research. A pair of southern ground hornbills, housed at Beale Wildlife Park and Gardens, were observed during study periods in 2014, 2018, and 2019. Three types of enrichment were provided for the birds; these enrichment types were developed based on information on the habits of the species as found in natural history papers. The enrichment types consisted of a pile of twigs, small animal carcasses, and plastic mirrors. Overall, the carcass feeds and the mirrors resulted in the greatest changes in behavior, with hornbills engaging in long periods of food manipulation with carcasses. For the mirror condition, hornbills spent time stalking around and pecking at mirrors, similar to the 'window smashing' behavior seen in wild hornbills. Overall, the research suggests that not only can enrichment modify the behavior of southern ground hornbills, but non-nutritional enrichment may be equally valuable to the animals. Natural history papers may have some value in inspiring novel enrichment items for zoo-housed animals.

Keywords: *Bucovus leadbeateri*; *Bucerotidae*; mirror; carcass feeding; spread of participation index

Citation: Brereton, J.E.; Myhill, M.N.G.; Shora, J.A. Investigating the Effect of Enrichment on the Behavior of Zoo-Housed Southern Ground Hornbills. *J. Zool. Bot. Gard.* **2021**, *2*, 600–609. https://doi.org/10.3390/jzbg2040043

Academic Editors: Kris Descovich, Caralyn Kemp and Jessica Rendle

Received: 6 September 2021
Accepted: 8 November 2021
Published: 13 November 2021

1. Introduction

Enrichment is fundamentally important for the welfare of many animals in zoos, yet there remain gaps in the knowledge of provision of enrichment for some taxa [1]. Many enrichment studies have been conducted for some taxonomic groups, such as the mammalian families, Felidae and Elephantidae [2]. The availability of studies allows researchers to evaluate and compare enrichment strategies, and therefore put in place the most effective plans. For some taxonomic groups, however, information on enrichment is more limited. This reduces the information available to practitioners to help improve the welfare of their animals.

Enrichment is particularly important for highly cognitive species that may otherwise become bored with a predictable or unstimulating environment. Many avian families are particularly susceptible, expressing unnatural behaviors, such as stereotypy, if enrichment is insufficient. Parrots (family Psittacidae) are a good example; birds in this family are typically intelligent, capable problem-solvers who live long lives. Where enrichment or social groups are not provided, feather-plucking and stereotyped behavior commonly occur [3]. Fortunately, enrichment has been well studied for the Psittacidae, and as a result, animal keepers have several effective strategies available to reduce the prevalence of stereotypy [1–3].

However, other intelligent, long-lived bird taxa are kept in zoos, some of which have received less focus in terms of enrichment research. One example is the hornbills (Order Bucerotiformes). Hornbills can be found throughout Africa and Asia, and are well-known for their unusual nesting habits, in which females often seal themselves into a tree cavity to incubate their chicks. Hornbills have been shown to be capable problem-solvers [4] and are also sensitive to both visitors and keepers when housed in zoos [5]. With a diverse range of hornbill species held in captivity, each representing different habitats, dietary niches and breeding strategies, there is a need to further investigate enrichment for this group of species.

The southern ground hornbill (*Bucorvus leadbeateri*) is one of the two largest extant hornbill species, reaching weights of up to 4 kg [6]. In recent years, the species has received some conservation attention in the wild, on account of decreasing population numbers [7–11]. In many parts of its historic range, the species has been persecuted because it is viewed as a bad omen [12]. The southern ground hornbill has a cooperative breeding strategy and a slow reproductive rate, leaving it vulnerable to extinction. As a result, zoo populations for this species are important, potentially providing a 'safety net'. A July 2021 search of Species360's [13] database revealed that at least 151 institutions globally keep this species, with an overall population size of over 390 birds. While these numbers are not excessively high in comparison to other zoo-housed birds [14–17], there is a need to further develop enrichment strategies for this species.

In the wild, southern ground hornbills typically forage on the ground, walking across the savannah [10]. The birds are entirely carnivorous, and feed on a range of foods including carrion, and invertebrates, reptiles, birds, and small mammals that are captured and killed [6]. The birds are particularly intelligent and social communication is advanced for this species, with small groups or 'mobs' developing, that work cooperatively to support female birds during incubation.

The purpose of this study was twofold. First, we aimed to develop an activity budget for zoo-housed southern ground hornbills. Second, we aimed to identify which enrichment types were most effective in encouraging exploratory behavior for the pair of birds.

2. Materials and Methods

2.1. Study Site and Subjects

The study was conducted according to the guidelines of the Association for the Study of Animal Behaviour and approved by the ethics committee of Beale Wildlife Park (A17, 19 January 2014). Following ethical approval from Beale Wildlife Park, the study commenced at Beale Wildlife Park and Gardens in Reading, United Kingdom. Three periods of data collection were undertaken: these were from 27 March 2014 to 30 June 2014, 15 January 2018 to 16 July 2018, and 1 November 2019 to 23 December 2019. Animals were observed during three observation periods: these were 08:00–10:00, 10:30–12:30, 13:00–15:00, and 15:30–17:30. Birds were observed based on the availability of the authors.

The study focused on two (1.1) parent-reared southern ground hornbills who were kept in a large single-species aviary in the 'Owl Walk' (Table 1, Figure 1). The exhibit consisted of several elevated perches, and one large barrel (for breeding purposes). The exhibit substrate was a mixture of leaf litter and soil. The birds were not flight restrained, but the exhibit was covered with nylon mesh to prevent escapes.

Table 1. Study subjects.

Sex	Date of Birth	Studbook Number	GAN	Movement into Collection
Male	20 May 2000	EAZA/73	MIG12-28772165	12 November 2010
Female	11 May 2001	EAZA/74	MIG12-28772164	12 November 2010

Figure 1. Hornbill enclosure, with male hornbill resting on log.

2.2. Enrichment Types

Enrichment was provided on a randomized schedule. Three enrichment types were provided for the animals; the enrichment types were inspired by the natural history documentation for the species. The first enrichment type consisted of a large pile of twigs and branches, into which several morio worms (*Zophobas morio*) were presented; this enrichment style was developed to encourage hornbills to forage using their talons and beak. The second enrichment type consisted of an entire rabbit carcass, as the species regularly feeds on large carcasses, and can hunt large prey in the wild [6,18,19]. The final enrichment style consisted of two large, non-shatter mirrors. Mirrors were used because they are frequently applied in bird husbandry to mimic the presence of conspecifics. There is also some evidence to suggest that wild hornbills interact with reflective surfaces such as on water, or in windows, in their native range.

The behavior of the birds was also observed during 'control' periods, when food was not present, and during normal feeding hours. Control periods were matched for time of day to the experimental treatments. The normal feed for the hornbills was provided at 16:00 and consisted of either day-old chicks or chunks of rabbit or quail, chopped into small pieces. When enrichment feeds were provided, they were deducted from the normal dietary provision.

2.3. Behavioral Recording

Behavior was recorded using instantaneous focal sampling of both birds simultaneously. One-hour observations were conducted, with the observer recording state behaviors at one-minute intervals. An ethogram was developed, containing behaviors which were adapted from Kemp & Kemp [6] (Table 2).

Table 2. Hornbill ethogram. Inspired by Kemp & Kemp [6].

Behavior	Description
Allopreening	The bird engages in preening of a conspecific.
Enrichment interaction	The bird uses its beak or feet to poke at or scratch an enrichment item.
Feeding	The bird takes food items from the exhibit and swallows them.
Flying	The bird lifts off the ground by raising and lowering its wings rapidly.
Object in beak	The bird is standing or walking with an item (e.g., food or nest material) in its beak.
Preening	The bird wipes its bill across its feathers in a repeated fashion.
Resting	The bird is motionless. The eyes may be either open or closed. Includes both standing and perching.
Sunbathing	The bird extends its wings in a fan and angles them toward the sun. The bird remains motionless.
Walking	The bird moves around the exhibit using its feet.

During observations, the observers partially concealed themselves behind a large oak tree at the front of the exhibit. Data were compiled onto paper observation sheets. In addition to behavioral data, observers also recorded the temperature and humidity (using BBC weather information for Pangbourne), weather conditions (e.g., rain, cloud), and the number of visitors that walked past the exhibit during the hour observation period.

2.4. Enclosure Use

In addition to hornbill behavior, the enclosure use of the two birds was recorded. The location of each bird was recorded using instantaneous focal sampling at one-minute intervals. The enclosure was separated into seven different zones, based on their biological value to the animals. The size of each useable surface per zone was measured using a tape measure (Table 3).

Table 3. Enclosure zones for the Southern ground hornbills.

Zone	Description	Size (m^2)
Elevated perches (left)	Large logs for perching, 2–2.5 m from ground.	12.5
Elevated perches (right)	Logs for perching, roughly 1.8–2.4 m from ground.	14.1
Central log	Long tree trunk, extending between left and right elevated perches, 0.8–1.5 m from ground.	15.6
Water	Small water pool and surrounding concrete.	3.5
Barrel	Barrel used by female during nesting.	2.1
Tree stump	Large tree stump turned upside down, with roots available for perching.	7.8
Ground	Substrate of enclosure, consisting of soil and leaf litter.	151.2
Total		206.8

Overall hornbill enclosure use was assessed for each bird using a modified spread of participation index (mSPI) [15]. The equation for mSPI is:

$$\frac{\Sigma|\text{fo} - \text{fe}|}{2(\text{N} - \text{femin})}$$

Here, N refers to the number of observations; fo and fe refer to the number of observed and expected observations in a given zone, respectively; femin refers to the expected observation in the smallest zone [15]. For mSPI, the maximum value of 1 indicates uneven enclosure use (the animal is using only the smallest zone and avoiding all other zones) whereas the minimum value of 0 indicates that animals are using all zones equally (in proportion to their size) [15].

2.5. Data Analysis

Data were compiled onto a Microsoft Excel™ 2010, Albuquerque, USA spreadsheet and then uploaded to Minitab version 21 for analysis. Analysis was conducted on the effect of the three enrichment types and the control condition on hornbill behavior. For analysis, behavioral data were tested for normality. Where data were normally distributed, one-way ANOVAs with Tukey post hoc tests were used to investigate the impact of enrichment. For non-normally distributed data, Friedman's ANOVAs with pairwise Wilcoxon tests were used [20].

3. Results

3.1. Behavior

A comparative activity budget was developed to demonstrate the effects of enrichment for the male and female hornbills (Figure 2). As data were non-parametric, Friedman's ANOVAs were run to determine the effect of enrichment on behavior (Table 4). Two behaviors, enrichment interaction and feeding, were significantly affected by enrichment type.

Table 4. Output of Friedman's ANOVAs on the effect of enrichment on hornbill behavior.

Behaviour	Test Statistic	p	Significant Post Hoc Tests
Allopreening	$X^2_{(3)} = 2.31$	0.412	
Enrichment interaction	$X^2_{(3)} = 99.62$	<0.001 *	Mirror-None, Mirror-Carcass, Mirror-Twigs
Feeding	$X^2_{(3)} = 75.16$	<0.001 *	Carcass-Mirror, Carcass-Twigs, Carcass-None
Flying	$X^2_{(3)} = 6.26$	0.096	
Object in beak	$X^2_{(3)} = 0.68$	0.718	
Preening	$X^2_{(3)} = 6.99$	0.099	
Resting	$X^2_{(3)} = 22.14$	0.127	
Sunbathing	$X^2_{(3)} = 4.16$	0.180	
Walking	$X^2_{(3)} = 11.41$	0.416	

* indicates significant values.

3.2. Enclosure Use

mSPI values were generated for all observations. A bar chart was developed to demonstrate the effect of enrichment on the mSPI values for the male and female hornbill (Figure 3). Whilst average mSPI scores differed slightly between enrichment types, the difference was not significant ($X^2_{(3)} = 6.06$, $p = 0.195$).

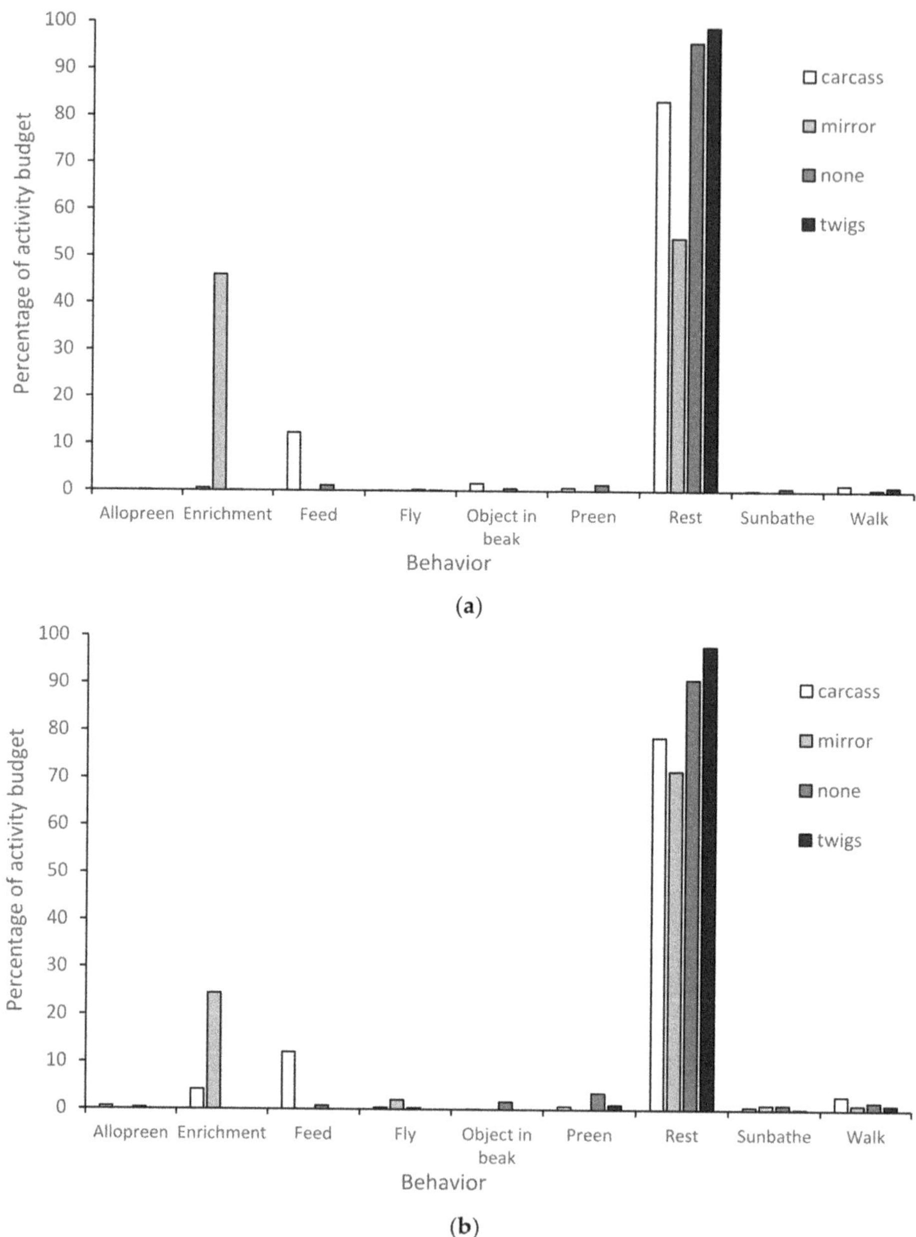

Figure 2. Activity budget for (**a**) female and (**b**) male hornbill.

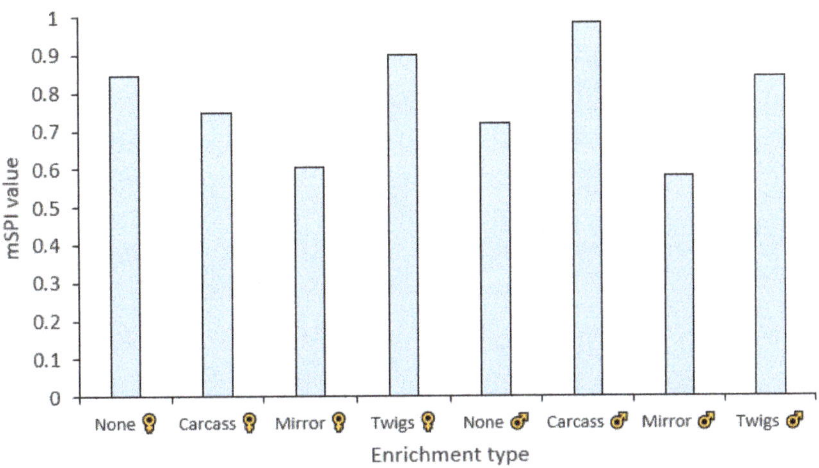

Figure 3. mSPI values for different enrichment types.

4. Discussion

Overall, the introduction of enrichment into the hornbill enclosures resulted in significant changes in feeding and enrichment interaction. Carcass provision resulted in hornbills spending much longer periods of time engaged in feeding and food manipulation, and mirrors were highly effective at engaging hornbills. No other significant changes in behavior occurred. While hornbills did appear to use their enclosure more evenly when enrichment was provided, this was not significant.

4.1. Carcass Enrichment

Significant differences in levels of food manipulation were noted when carcass enrichment was provided. Hornbills engaged in movements including stabbing and shaking of the carcass in order to remove pieces of meat. In the normal feed, typically consisting of chopped meat or day-old chicks, little food manipulation was observed. In the wild, southern ground hornbills may feed on large carcasses, and are also known to hunt animals such as hares and medium-sized snakes [6,9,20]. Occasional carcass feeds may therefore allow the birds to express a greater range of feeding-related behaviors. This could be used in tandem with small food items such as live invertebrates, which could be used to simulate hunting. Providing birds with the opportunity to express more natural behaviors is part of the five welfare domains [16].

Whilst some visitors may support this more natural feeding experience, there may also be a negative response from the public to these feeding techniques [17]. Carcass feeds may sometimes be met with disapproval from key visitor demographics, such as families with small children [17]. Whilst there is an educational value to the provision of carcass feeds, visitors may need to be made aware that carcass feeding is taking place.

Feeding whole foods could also reduce aggression between subjects [18] provided all animals are given access to food items simultaneously. In the current study, no aggression was observed between individuals, though this may pose a challenge if hornbills are kept in groups that simulate their wild social grouping [6,7,20]. Providing whole carcasses can save keepers food preparation time, but larger diet items may require significantly more storage, which could be more difficult for smaller collections. Consideration should also be given to exhibit cleaning once a large food item has been offered.

4.2. Mirror Enrichment

It is sometimes challenging to find non-food-related enrichment types for zoo-housed animals. Food-related enrichment may have drawbacks in that it must be deducted from the animal's normal rations [21]. Non-food-related enrichment, by contrast, can be used for long periods of time without reducing an animal's appetite or resulting in an imbalanced diet.

Mirrors are a common strategy employed by bird keepers for use as enrichment. The hornbill pair spent significantly longer interacting with mirrors than with any other enrichment type. This significant increase in interactions with mirrors could be considered beneficial, as levels of resting decreased while activity levels increased. The two birds in the study were typically inactive during visitor open hours, so an increase in activity may have a positive impact on physical fitness.

In the wild, hornbills have been noted to interact with reflective objects such as mirrors and windows, and even parts of cars [10]. The hornbill has been persecuted as a result of this behavior. The underlying purpose of the behavior is still not fully understood; the behavior may be related to curiosity or interaction with another hornbill [10]. Whilst the hornbills could in fact be aware that their reflection is harmless, this behavior could also be based around territorial displays, with the hornbills assuming they have another hornbill to defend against or compete with. This could indicate that the birds consider the mirror reflection to be a rival. While this condition could therefore be considered stressful to the birds, it does allow the birds to demonstrate natural behavior and potentially could improve pair bonding. Therefore, mirrors could play a similar role to the playback calls used in zoos for primate species, such as gibbons [21].

4.3. Enclosure Use

There was no significant difference in enclosure use for the southern ground hornbills as a result of enrichment type. Whilst the mSPI scores appeared lower for the enrichment and carcass feeds, this was not significant. This may be due to variability in mSPI scores as a result of other extraneous variables. For example, it is possible that visitor presence influenced the enclosure use of the birds. Other zoo-housed hornbills have been shown to respond to visitor presence [5]. Anecdotally, the southern ground hornbills appeared to favor elevated perches during time periods when visitor numbers were higher. Future studies could consider visitor presence and its influence on behavior.

Enrichment items encouraged the hornbills to use more of the ground substrate, walking around the exhibit. In the wild, southern ground hornbills spend much of their time walking, rather than flying, around grassland and savannah in search of prey [6,11,22,23]. The use of a greater range of zones, rather than primarily the elevated perches, could be beneficial in terms of physical movement for these birds.

4.4. Future Directions

Generally, birds seem to be a neglected taxa for enrichment, despite their prevalence in zoological collections [14]. Finding any objects that have significant impacts on activity is positive for keepers. Many enrichment items create animal interaction, but not for significant periods of time. Hornbills were observed interacting with mirrors for over 40 min, which is an extended period of activity for the animals. The public perception of birds in captivity can often be more negative than other taxa, as many captive birds lack the large amounts of space the public, with little knowledge of husbandry guidelines, believe they need for optimum health. Enrichment can improve public perception of welfare, especially considering birds are viewed much less emotively than other taxa, such as primates.

Future studies should consider use of tests that reduce issues with seudoreplication, such as G-tests. These studies could also utilize the extensive historical records of natural history, in order to identify novel enrichment practices. For example, early records of sightings of animals in their natural habitats, or interactions with other species, may help practitioners to identify novel husbandry practices to trial. In turn, the use of natural history documents may help zoos to provide more informed, evidence-based management for the animals that they keep.

5. Conclusions

Overall, provision of enrichment influenced some, but not all aspects of captive hornbill behavior. Interaction with enrichment varied between items, with the twigs pile receiving little attention and the carcasses resulting in considerable feeding activity. Mirrors were very well utilized by the birds, linking to the behavior of wild southern ground hornbills and their interest in reflective windows. Information on natural history may be useful in developing novel enrichment devices, especially enrichment types that do not involve food. Further inspiration for enrichment practices may be found in natural history books or papers that could, with controlled testing, be used to advance the state of current enrichment practice.

Author Contributions: Conceptualization, J.E.B.; methodology, J.E.B. and M.N.G.M.; software, J.E.B.; validation, J.A.S., J.E.B. and M.N.G.M.; formal analysis, J.E.B.; investigation, J.A.S. and J.E.B.; data curation, J.A.S.; writing—original draft preparation, J.A.S., J.E.B. and M.N.G.M.; writing—review and editing, J.A.S., J.E.B. and M.N.G.M.; visualization, J.A.S., J.E.B. and M.N.G.M.; supervision, J.E.B. All authors have read and agreed to the published version of the manuscript.

Funding: This research received no external funding.

Institutional Review Board Statement: The study was conducted according to the guidelines of the Declaration of Helsinki and approved by the ethics committee of Beale Wildlife Park (A17, 19 January 2014).

Informed Consent Statement: Not applicable.

Data Availability Statement: The data presented in this study are available on request from the corresponding author. The data are not publicly available due to privacy of zoo records.

Acknowledgments: The authors would like to thank staff at Beale Wildlife Park for their support during the project.

Conflicts of Interest: The authors declare no conflict of interest.

References

1. Swaisgood, R.R.; Shepherdson, D.J. Scientific approaches to enrichment and stereotypies in zoo animals: What's been done and where should we go next? *Zoo Biol.* **2005**, *24*, 499–518. [CrossRef]
2. Dos Santos, J.W.; Correia, R.A.; Malhado, A.C.; Campos-Silva, J.V.; Teles, D.; Jepson, P.; Ladle, R.J. Drivers of taxonomic bias in conservation research: A global analysis of terrestrial mammals. *Anim. Conserv.* **2020**, *23*, 679–688. [CrossRef]
3. Reimer, J.; Maia, C.M.; Santos, E.F. Environmental enrichments for a group of captive macaws: Low interaction does not mean low behavioral changes. *J. Appl. Anim. Welf. Sci.* **2016**, *19*, 385–395. [CrossRef] [PubMed]
4. Danel, S.; von Bayern, A.M.; Osiurak, F. Ground-hornbills (*Bucorvus*) show means-end understanding in a horizontal two-string discrimination task. *J. Ethol.* **2019**, *37*, 117–122. [CrossRef]
5. Rose, P.E.; Scales, J.S.; Brereton, J.E. Why the "visitor effect" is complicated. unraveling individual animal, visitor number, and climatic influences on behavior, space use and interactions with keepers—A case study on captive hornbills. *Front. Vet. Sci.* **2020**, *7*, 1–10. [CrossRef] [PubMed]
6. Kemp, A.C.; Kemp, M.I. The biology of the southern ground hornbill *Bucorvus leadbeateri* (Vigors)(Aves: Bucerotidae). *Ann. Transvaal Mus.* **1980**, *32*, 65–100. Available online: https://hdl.handle.net/10520/AJA00411752_1057 (accessed on 3 September 2021).
7. Combrink, L.; Combrink, H.J.; Botha, A.J.; Downs, C.T. Habitat preferences of Southern Ground-hornbills in the Kruger National Park: Implications for future conservation measures. *Sci. Rep.* **2020**, *10*, 1–9. [CrossRef] [PubMed]
8. Theron, N.; Grobler, P.; Kotze, A.; Jansen, R. The home range of a recently established group of Southern ground-hornbill (*Bucorvus leadbeateri*) in the Limpopo Valley, South Africa. *Koedoe Afr. Prot. Area Conserv. Sci.* **2013**, *55*, 1–8. [CrossRef]
9. Gula, J.; Phiri, C.G. Observations of Southern Ground-hornbill *Bucorvus leadbeateri* groups in the Kafue National Park, Zambia. *Ostrich* **2020**, *91*, 267–270. [CrossRef]
10. Msimanga, A. Breeding biology of Southern Ground Hornbill *Bucorvus leadbeateri* in Zimbabwe: Impacts of human activities. *Bird Conserv. Int.* **2007**, *14*, 63–68. [CrossRef]
11. Engelbrecht, D.; Theron, N.; Turner, A.; Van Wyk, J.; Pienaar, K. The status and conservation of Southern Ground Hornbills, *Bucorvus leadbeateri*, in the Limpopo Province, South Africa. *Act. Manag. Hornbills Habitats Conserv.* **2007**, *1*, 252–266.
12. Coetzee, H.; Nell, W.; van Rensburg, L. An exploration of cultural beliefs and practices across the Southern Ground-Hornbill's range in Africa. *J. Ethnobiol. Ethnomed.* **2014**, *10*, 28–36. [CrossRef] [PubMed]
13. Species360. Southern Ground Horn Bill. 2021. Available online: https://zims.species360.org/ (accessed on 26 July 2021).

14. Brereton, S.R.; Brereton, J.E. Sixty years of collection planning: What species do zoos and aquariums keep? *Int. Zoo Yearb.* **2020**, *54*, 131–145. [CrossRef]
15. Brereton, J.E. Directions in animal enclosure use studies. *J. Zoo Aquar. Res.* **2020**, *8*, 1–9.
16. Mellor, D.J. Operational details of the five domains model and its key applications to the assessment and management of animal welfare. *Animals* **2017**, *7*, 60. [CrossRef] [PubMed]
17. Gaengler, H.; Clum, N. Investigating the impact of large carcass feeding on the behavior of captive Andean condors (*Vultur gryphus*) and its perception by zoo visitors. *Zoo Biol.* **2015**, *34*, 118–129. [CrossRef]
18. Shora, J.A.; Myhill, M.G.N.; Brereton, J.E. Should zoo foods be coati chopped? *J. Zoo Aquar. Res.* **2018**, *6*, 22–25. [CrossRef]
19. Brereton, J.E. Challenges and directions in zoo and aquarium food presentation research: A review. *J. Zool. Bot. Gard.* **2020**, *1*, 13–23. [CrossRef]
20. Plowman, A.B. BIAZA statistics guidelines: Toward a common application of statistical tests for zoo research. *Zoo Biol.* **2008**, *27*, 226–233. [CrossRef]
21. Shepherdson, D.; Bemment, N.; Carman, M.; Reynolds, S. Auditory enrichment for Lar gibbons *Hylobates lar* at London Zoo. *Int. Zoo Yearb.* **1989**, *28*, 256–260. [CrossRef]
22. Kemp, L.V.; Kotze, A.; Jansen, R.; Dalton, D.L.; Grobler, P.; Little, R.M. Review of trial reintroductions of the long-lived, cooperative breeding Southern Ground-hornbill. *Bird Conserv. Int.* **2020**, *30*, 533–558. [CrossRef]
23. Galama, W.; King, C.; Brouwer, K. *EAZA Hornbill Management and Husbandry Guidelines*; EAZA: Amsterdam, The Netherlands, 2002.

MDPI
St. Alban-Anlage 66
4052 Basel
Switzerland
Tel. +41 61 683 77 34
Fax +41 61 302 89 18
www.mdpi.com

Journal of Zoological and Botanical Gardens Editorial Office
E-mail: jzbg@mdpi.com
www.mdpi.com/journal/jzbg

www.ingramcontent.com/pod-product-compliance
Lightning Source LLC
LaVergne TN
LVHW070959170726
843515LV00006B/1027

* 9 7 8 3 0 3 6 5 7 2 2 3 9 *